Handbook of
Disinfectants
and
Antiseptics

Handbook of Disinfectants *and* Antiseptics

edited by

Joseph M. Ascenzi

Director, Quality Assurance
Johnson & Johnson Medical, Inc.
Arlington, Texas

Marcel Dekker, Inc. **New York • Basel • Hong Kong**

Library of Congress Cataloging-in-Publication Data

Handbook of disinfectants and antiseptics / edited by Joseph M. Ascenzi.
p. cm.
Includes index.
ISBN 0-8247-9524-5 (hardcover : alk. paper)
1. Disinfection and disinfectants -- Handbooks, manuals, etc.
2. Antiseptic medication -- Handbooks, manuals, etc. I. Ascenzi, Joseph M.
RM400.H36 1996
614.4'8--dc20

95-35602
CIP

The publisher offers discounts on this book when ordered in bulk quantities. For more information, write to Special Sales/Professional Marketing at the address below.

This book is printed on acid-free paper.

Marcel Dekker, Inc.
270 Madison Avenue, New York, New York 10016

Current printing (last digit):
10 9 8 7 6 5 4 3 2

PRINTED IN THE UNITED STATES OF AMERICA

Preface

The field of antiseptics and disinfectants had its beginnings with the discovery that germs were associated with infection and that the use of certain chemicals prevented the spread of germs, resulting in fewer cases of postoperative infection. Lister's use of phenol around operating theaters and on surgical dressings and Semmelweis' use of handwashing were the beginning of the use of germicides as disinfectants and antiseptics. It is indeed a task to examine the use of chemicals as germicides in all fields (e.g., medicine, agriculture, industry, or others). It is the purpose of this book to look at those subjects related to infection control, the use of germicides in preventing the spread of infection, either by inanimate objects or from person to person. The use of disinfectants on inanimate objects (disinfectants) or on human tissue (antiseptics) has come under scrutiny in the past decade because of unsubstantiated claims made by manufacturers and because of the importance of effectiveness necessitated by diseases such as tuberculosis, acquired immunodeficiency syndrome (AIDS), and hepatitis.

The number of chemicals in use, like the potential number of applications, is too great to be covered in one book. I have attempted to include those chemicals that have the widest use for both disinfection of critical medical equipment and antisepsis. Some obvious omissions may be noted. Phenolics and quaternary ammonium compounds are not reviewed, primarily because they are not routinely used on critical medical equipment, and also because no new developments have been reported for some time. Also, several recent publications cover these chemicals in depth, and there is no need to repeat that information. Likewise, with iodine, although covered somewhat in Chapter 8, an exhaustive review at this time would be repetitious and unwarranted, as little new information has been generated.

This book covers the chemicals that have the most widespread use in current disinfection and antisepsis practices, both in the United States and abroad. The chemistry, microbiology, and toxicology of each agent is presented. In order to better understand the microbiology of these chemical agents, a section has been included to describe the procedures, both in-use and laboratory, that are used to generate microbiological claims for disinfectants and antiseptics. Because of the similarity in chemicals used in contact lens disinfectant solutions and antiseptics and disinfectants, an additional chapter has been included to discuss the methods used to evaluate these chemicals in this particular application.

It is my hope that this book will provide readers in the medical profession, academia, and industry with pertinent information in the field of disinfection and antisepsis.

Joseph M. Ascenzi

Contents

Contributors

Joseph M. Ascenzi, Ph.D. Johnson & Johnson Medical, Inc., Arlington, Texas

Sally F. Bloomfield, B.Ph., Ph.D., M.R.Pharm.S. Department of Pharmacy, King's College London, London, England

Mary K. Bruch Mary Bruch Micro-Reg, Inc., Leesburg, Virginia

Eugene C. Cole, Dr.Ph. Center for Environmental Technology, Research Triangle Institute, Research Triangle Park, North Carolina

James A. Cottone, M.S., D.M.D. The University of Texas Health Science Center at San Antonio, San Antonio, Texas

Andrea M. Lever, B.A. Bausch & Lomb, Inc., Rochester, New York

Michael J. Miller, Ph.D. Bausch & Lomb, Inc., Rochester, New York

John A. Molinari, Ph.D. Department of Biomedical Sciences, University of Detroit Mercy School of Dentistry, Detroit, Michigan

Daryl S. Paulson, Ph.D. BioScience Laboratories, Inc., Bozeman, Montana

N. S. Ranganathan, Ph.D. Technical Support/New Market Development, Amway Corporation, Ada, Michigan

Richard Robison, Ph.D. Brigham Young University, Provo, Utah

Manfred L. Rotter, M.D., Dip.Bact. Hygiene Institute, Allgemeines Krankenhaus, Vienna, Austria

Scott V. W. Sutton, Ph.D. Alcon Laboratories, Inc., Fort Worth, Texas

1

Test Methodology for Evaluation of Germicides

Eugene C. Cole
Research Triangle Institute
Research Triangle Park, North Carolina

Richard Robison
Brigham Young University
Provo, Utah

I. CHEMICAL GERMICIDE TESTING

The evaluation of chemical germicides predates the golden age of microbiology. The microbicidal effects of various chemicals have been investigated ever since microorganisms were first discovered. However, it was not until the late 1800s and early 1900s that any attempt was made to standardize these types of evaluations. The works of Koch [1], Rideal and Walker [2], Chick and Martin [3], and others formed the foundation of modern disinfectant test methodology. Current procedures can be classified into at least two main categories, based on their fundamental design: those that provide information on the antimicrobial activity of the product, and those that attempt to simulate real-life practical uses of the agent. Methods in this first category, sometimes referred to as in vitro tests, are usually suspension tests in which all test parameters are strictly controlled. These provide reliable information on the relative efficacy of a test agent. Methods in the second category, sometimes called practical tests, include carrier methods, capacity tests, and surface techniques that simulate actual disinfection protocols. These represent an evaluation of the specific disinfection procedure as much as they do the efficacy of a particular agent in use.

Although the evaluation of germicides is simple in principle, it is surprisingly difficult in practice. Several factors interact in complex ways to influence test results. The individual test organism used, the method of test organism preparation, the exposure conditions of the test, the method of disinfectant neutralization, and the particular subculture procedures employed, all are variables that can individually and cooperatively modulate test results.

A. Standard Laboratory Methods

Test methodology that has been accepted or sanctioned by the proper regulatory agency in a given country is considered a standard laboratory procedure. In the United States, the standard methods for the evaluation of environmental chemical germicides are those of the Association of Official Analytical Chemists (AOAC). The methods, along with their designated test organisms, are listed in Table 1. Data generated from these methods are used to determine the optimum use-dilution of a germicide to be used for a specific application, and they are also used to satisfy the U.S. Environmental Protection Agency (EPA) requirements for pesticide registration under the Federal Insecticide, Fungicide, and Rodenticide Act (FIFRA). With use of specified test organisms and, in some instances, representative environmental surfaces, the AOAC methods form the core of the EPA's Efficacy Data Requirements for chemical germicides. Some non-AOAC methods are specified by EPA to demonstrate efficacy against specific organisms. These include the EPA virucidal method, or a recently accepted alternative method for the quantitative assessment of tuberculocidal activity [4].

In Europe, a considerable effort has been made to arrive at a common standard for disinfectant evaluation. A suspension test based on the Dutch 5-5-5 test was formulated by a committee of experts from several countries. A collaborative trial of the method was performed in ten European countries, the outcome of which resulted in its adoption as an international standard [5]. The test has since been modified and is now commonly referred to as the European Suspension Test.

All of these methods, both qualitative and quantitative, are strictly laboratory methods to be used to ascertain the extent of antimicrobial activity of a particular chemical under controlled conditions. Results of such testing may not necessarily be indicative of how well the germicide performs under actual in-use conditions. One concern involves the use of "standard" laboratory-grown strains of test organisms and whether those organisms are as intrinsically resistant to

Table 1 EPA Recommended Methods for Testing Disinfectants Intended for Use on Hard Surfaces

Claim	EPA recommended method	Test organism
Sterilizer	AOAC Sporicidal Test	*Bacillus subtilis* *Clostridium sporogenes*
Tuberculocide	AOAC Tuberculocidal Activity Method Modified AOAC Tuberculocidal Activity Method Quantitative Tuberculocidal Activity Test Method	*Mycobacterium bovis*
Hospital disinfectant	AOAC Use-Dilution Method AOAC Germicidal Spray Products Test	*Salmonella chloeraesuis, Staphylococcus aureus*, and *Pseudomonas aeruginosa*
General disinfectant	AOAC Use-Dilution Method	*Salmonella choleraesuis*, or *Staphylococcus aureus*
Limited disinfectant	AOAC Use-Dilution Method Germicidal Spray Products Test	*Salmonella choleraesuis* and *Staphylococcus aureus*
Fungicide	AOAC Fungicidal Test AOAC Use-Dilution Method AOAC Germicidal Spray Products Test	*Trichophyton mentagrophytes*
Virucide	EPA Virucidal test parameters	Specific virus claimed
Sanitizing rinse (food contact surfaces)	AOAC Available Chlorine Germicidal Equivalent Concentration Method AOAC Germicidal and Detergent Sanitizers Method	*Escherichia coli* and *Salmonella typhi*; or *Staphylococcus aureus*
Sanitizer (inanimate, non–food-contact surfaces)	EPA sanitizer parameters	*Staphylococcus aureus* and *Klebsiella pneumoniae* aberrant, or *Enterobacter aerogenes*

Source: EPA (1984)

chemical inactivation as their naturally occurring counterparts [6]. Additional concerns about the variability of results generated from such standard methods have been recognized through recent research [7,8]. Because of these concerns and because neither EPA nor FDA verify disinfectant efficacy data that is submitted for registration, one's selection of a suitable germicide for a specific application should not be based solely on results from standard tests. Additional nonstandard methodology that simulates the actual use of the agent in question should also be performed.

B. Nonstandard Laboratory Methods

The investigation of microbial inactivation by chemical germicides is normally conducted through research that uses protocols designed for quantitative investigation of specific organisms, germicides, environmental conditions, or applications. Such methods include procedures for organism cultivation, challenge quantitation, material or disinfectant inoculation, replicate testing, germicide neutralization, organism recovery, data analysis, and interpretation. They may also focus on a variety of physicochemical factors, such as temperature, pH, and organic matter, and on hard water interference. When such investigations are peer-reviewed and subsequently published in a reputable scientific journal, they become an important part of a valuable and credible pool of technical, useful information. Such has been the case with several specific pathogens for which chemical inactivation data was needed to establish recommendations for disinfection in health care or animal facilities or for effective public water treatment. Specific germicidal research has thus been conducted with a host of human and veterinary pathogens, such as hepatitis A (HAV) and B (HBV) viruses [9,10], human immunodeficiency virus (HIV) [11,12], *Mycobacterium tuberculosis* [13,14], *Legionella pneumophila* [15], Norwalk virus [16], swine vesicular disease virus [17], and vesicular stomatitis virus [18].

Nonstandard methods are generally devised in an effort to add realism to disinfectant efficacy data. These methods often more closely simulate the intended use of an agent on a relevant surface. Good examples of these procedures include several quantitative surface disinfection protocols that use as test surfaces the actual surfaces intended for disinfection (i.e., formica, stainless steel, and such) [19,20]. It must be remembered that even if such nonstandard laboratory testing involves the use of naturally occurring strains of organisms, or if inoculation is carried out on a variety of materials or surfaces, it does not constitute in-use evaluation of a chemical germicide for a specific application.

C. In-Use Test Methods

In-use testing involves the antimicrobial evaluation of a chemical germicide under actual conditions of use on specified surfaces or materials in a designated environment. As with standard and nonstandard laboratory methods, prototype indicator organisms or actual organisms of concern may be used. If in-use testing is to be done frequently, then such routine monitoring should be conducted with appropriate organisms that are recognized as representative of a broad range of microorganisms, that are regarded as intrinsically resistant to the germicides being used, and that are nonpathogenic or are regarded as safe for monitoring and for laboratory personnel to use.

In-use tests of disinfectants can be performed by a variety of procedures. A convenient method involves sampling of a disinfectant solution following actual use on instruments or surfaces by membrane filtration [21]. The recovery of any viable non–spore-forming bacteria from these solutions after an appropriate recovery time indicates disinfectant failure. Another in-use test involves contact sampling of items after they have been disinfected [22]. As with the membrane filtration method, no vegetative organisms should be recovered. The major disadvantage of these methods is their failure to produce quantitative efficacy data. A hybrid-type test in which a disinfectant in actual use is periodically sampled, and the same quantitatively evaluated for antimicrobial activity, is a meaningful alternative [23].

II. TEST ORGANISMS

In the selection of suitable test organisms for the demonstration of germicide or chemical treatment system effectiveness, the concepts of indicator and surrogate pathogen organisms, as well as intrinsic germicide resistance become important factors.

A. Indicator Organisms

Specific organisms have long been used in microbial inactivation testing as indicators of broad-spectrum efficacy. Some indicator organisms are nonpathogenic, yet serve as effective indicators of antimicrobial activity because of their intrinsic resistance to a particular treatment. For example, *Bacillus stearothermophilus* is a nonpathogen the spores of which are most resistant to moist heat treatment. For this reason, it is used to validate the effectiveness of steam autoclaving of medical instruments. If a process was successful in inactivating a significant challenge of *B. stearothermophilus* spores, it is then recognized that all pathogenic bacterial spores, as well as viruses, fungi, and vegetative bacteria exposed to those conditions at that time were also inactivated.

Some indicator organisms function as surrogate pathogens. They are closely related to virulent human pathogens, yet are safer to work with because they are nonpathogens or are opportunistic pathogens. One may then assume that the inactivation profile of the indicator is reflective of the actual target pathogen. Such is the case with *Salmonella choleraesuis*, one of the three required test organisms specified in the AOAC use-dilution method for bactericidal activity. *Salmonella choleraesuis* is thus representative of all the *Salmonella* species, of which *S. typhi*, the causitive agent of typhoid fever, is the human pathogen of concern. The situation is the same with the AOAC and EPA tuberculocidal methods, wherein the surrogate pathogen indicator organism *Mycobacterium bovis* is representative of the virulent and more highly pathogenic *M. tuberculosis*.

B. Germicidal Resistance

1. Mechanisms of Resistance

The resistance of some microorganisms to chemical germicides is not well understood. Some organisms are intrinsically or naturally resistant to one or more classes of germicides owing to their physical characteristics. Thus, resistance of an organism, such as *Pseudomonas aeruginosa*, can be attributable to the impermeability of the cell wall to a specific agent. Likewise, bacterial spores have a natural structural resistance. It is well-documented that growth-related conditions, including media formulations, also can modulate microbial resistance to chemical germicides. Carson et al. demonstrated an extreme decrease in germicidal resistance when a naturally occurring *Pseudomonas aeruginosa* isolate was tested against four classes of disinfectants, passed one time on laboratory media, and then tested again [6]. Similarly, Harakeh et al. [25] showed that populations of *Yersinia enterocolitica* and *Klebsiella pneumoniae* were more resistant to chemical inactivation when grown under conditions most like natural aquatic environments, whereas Sobsey et al. [26] showed that MS2 coliphage was less resistant than HAV at low pH, but more resistant than HAV at high pH. For these reasons, it is important when selecting indicator organisms, that resistance under the designed test conditions be considered. One must also be cautious and properly separate intrinsically resistant microorganisms from those that survive germicidal exposure owing to inactivation of the test germicide by environmental conditions or because of a protective effect from protein or particulate materials.

2. Scale of Resistance

From many years of germicide research, as detailed in the published literature, a solid framework of resistance relative to the major groups of microorganisms has emerged. This general scale of resistance, from least to most resistant, is as follows: vegetative bacteria, vegetative fungi and fungal spores, viruses, mycobacteria, and bacterial spores. Relative to germicidal resistance, viruses are divided into two groups: the less resistant enveloped (lipophilic) viruses, and the more resistant nonenveloped (hydrophilic) viruses [27]. The scale can be useful in the selection of germicides for specific applications. Thus, an EPA-registered disinfectant with a fungicidal claim also predicts effectiveness against vegetative bacteria; a tuberculocidal claim predicts effectiveness against mycobacteria, bacteria, fungi, and viruses, with the exception of bacterial spores; and a sporicidal claim predicts the inactivation of all microbial forms. Although there are some exceptions in the nonenveloped viruses, this general scale holds true. Independent germicidal test research supports the concept. Grabow et al. [28], in studies of inactivation by free chlorine, found that HAV was more resistant than *Escherichia coli*, *Streptococcus faecalis*, coliphage MS2, and reovirus type 3, but less resistant than *Mycobacterium fortuitum*; Snead et al. [29] showed in their chlorine studies that poliovirus 1 and f2 bacteriophage were more resistant than *Shigella sonnei* and *Salmonella typhimurium*; Liu et al. [30] found that enteric viruses showed a resistance to chlorine about ten times greater than that of enteric bacteria; and Lensing et al. [31] investigated sporicidal and fungicidal activity of several germicides and found *Aspergillus fumigatus* and *A. niger* to be less resistant than spores of two *Bacillus* species. Prince, however, through standard efficacy testing, found that some tuberculocidal formulations were not effective against adenoviruses, enteroviruses, and rhinoviruses [32]. Although not normally placed on the scale of microbial resistance, the cyst forms of some parasites such as *Giardia* and *Cryptosporidium* species are of concern for infectious disease transmission. They are significantly resistant to inactivation [33,34] and should most likely be placed between the viruses and the mycobacteria.

C. Suggested Test Organisms

For each major group of microorganisms listed in the following, recommendations are made for indicator or surrogate pathogen organisms, based on peer-reviewed scientific data, as published in the literature. A listing of these recommended test organisms is shown in Table 2.

1. Bacteria

Prototype vegetative bacteria representative of both the gram-positive and the gram-negative species should be used. The gram-positive *Staphylococcus aureus* ATCC 6538 and gram-negative *Pseudomonas aeruginosa* ATCC 15442, currently required in the AOAC use-dilution method [35] and the European suspension test [8], are resistant to chemical inactivation, with *P. aeruginosa* significantly more resistant to a variety of classes of germicides than *S. aureus* [36].

2. Fungi

The fungi consist of both yeasts and molds. It is suggested that a representative of each group be used for testing. The yeast, *Candida albicans*, is used in the European standard methods, and a study of nine fungi exposed to seven germicides showed a strain of *C. albicans* (CBS 562) to be the most resistant [37]. Of the molds, the AOAC test organism, *Trichophyton mentagrophytes* ATCC 9533 may be used, although it appears preferable to use the more germicide-resistant *Mucor mucedo* or *Penicillium chrysogenum* as reported [37].

Table 2 Recommended[a] Test Organisms for Evaluation of Chemical Germicide Efficacy

Bacteria	*Staphylococcus aureus*	ATCC 6538
	Pseudomonas aeruginosa	ATCC 15442
Fungi	*Candida albicans*	ATCC 10231
	Mucor mucedo	
	Penicillium chrysogenum	
	Trichophyton mentagrophytes	ATCC 9533
Mycobacteria	*Mycobacterium terrae*	
	Mycobacterium bovis (BCG)	ATCC 35743
Parasites	*Giardia lamblia*	
	Cryptosporidium parvum	
Viruses	*Human*	
	Coxsackie B5	
	Coxsackie B3	
	Echovirus 1	
	Poliovirus 1	
	Human rotavirus	
	Norwalk virus	
	Animal	
	Porcine parvovirus	
	Theiler's murine encephalomyelitis virus	
	Reovirus 3	
	Feline panleukopeia virus	
	Feline calcivirus	
Bacterial spores	*Bacillus subtilis*	
	Bacillus stearothermophilus	

[a]Based on published research indicating significant resistance to chemical germicides or required in standard test methods.

3. Mycobacteria

For germicidal efficacy testing, *Mycobacterium bovis* (BCG) may be used for standard [35] or nonstandard [38] laboratory testing, or the less pathogenic *M. terrae*, a more rapid-growing organism with resistance equivalent to *M. tuberculosis*, can be used [39]. For safety during in-use testing, *M. terrae* is preferred. *Mycobacterium smegmatis* is no longer recommended as an indicator organism for tuberculocidal testing [39,40].

4. Parasites

Appropriate test organisms include *Giardia lamblia* cysts and *Cryptosporidium parvum* oocysts. Both organisms are considered significant environmental-source human pathogens. *Giardia* has been reported to show differing resistance to inactivation, depending on environmental factors such as temperature and pH [33]. *Cryptosporidium parvum* is 14 times more resistant to low-level chlorine dioxide exposure than are *Giardia* cysts [34].

5. Viruses

Virus susceptibility to chemical inactivation is primarily influenced by the presence or absence of a lipid protein envelope. Enveloped (lipophilic) viruses are generally more susceptible, hence less resistant, to chemical inactivation than the nonenveloped (hydrophilic) viruses [27]. Accordingly, such viral pathogens as coxsackievirus, echovirus, poliovirus, adenovirus, and rhinovirus, are significantly more resistant than the enveloped viruses, such as the AIDS virus (HIV), herpes simplex virus, Epstein–Barr virus, varicella–zoster virus, measles and mumps viruses, and respiratory syncytial and influenza viruses. However, viruses run a wide gamut of susceptibilities to chemical germicides. Although in a general sense, nonenveloped enteric viruses are most resistant to most agents, enough diversity exists to preclude a *complete* virucidal claim based upon data from a single representative organism. This has caused the EPA and FDA to limit virucidal claims to the specific virus(es) against which the germicide has been tested. Depending on a specific application and intended target organisms for a particular chemical germicide, a suitable prototype enveloped or nonenveloped virus can be selected. Guidance is provided by efficacy data generated through standard and nonstandard laboratory methods.

Human Disease Viruses

Harakeh and Butler [41] investigated the inactivation of enteric viruses against four chemical germicides and found the rank order of most to least resistance to chlorine dioxide to be human rotavirus (clinical isolate), coxsackievirus B5, echovirus type 1, poliovirus type 1, bacteriophage f2, and simian rotavirus SA11. Engelbrecht et al. [42] investigated the inactivation of enteric viruses by low levels of chlorine and found that coxsackievirus B5 was the most resistant across a range of pH values. Sattar et al. [43] investigated surface disinfection of four viruses with five disinfectants and found a field isolate of coxsackievirus B3 to be the most intrinsically resistant. Deswick et al. [16] found that Norwalk virus in water was more resistant to chlorine inactivation than poliovirus type 1, human rotavirus, simian rotavirus SA11, or f2 bacteriophage.

Animal Disease Viruses

Brown [44] evaluated the virucidal activity of 14 chemical germicides against the nonenveloped porcine parvovirus (PPV), the enveloped pseudorabies virus, and transmissible gastroenteritis virus of swine, and found that PPV was intrinsically resistant to 12 of the 14 test disinfectants. Watanabe et al. [45] investigated the effect of 14 disinfectant formulations and physical treatments (including heat) on the nonenveloped Theiler's murine encephalomyelitis virus (TMEV) and reovirus type 3 (RV), as well as the enveloped Sendai virus and canine distemper virus, and found that TMEV and RV showed distinct resistance to chemical inactivation and to heating at 45°, 56°, and 60°C. Scott [46] investigated the effects of a variety of chemical germicides on the enveloped feline viral rhinotracheitis virus (FVR) and the nonenveloped feline calicivirus (FCV) and feline panleukopenia virus (FPL) and found all disinfectants were virucidal for FVR, 11 of 35 for FCV, and only 3 of 27 for FPL.

6. Bacterial Spores

Bacterial spores are the most intrinsically resistant microorganisms to chemical germicide inactivation. However, differences in resistance within this group, do exist. Dye and Mead [47] investigated the sporicidal activity of chlorine on eight strains of *Clostridium. Clostridium welchii* was the most resistant and *C. bifermentans* the least resistant, whereas *Bacillus subtilis* var. *niger* was considerably more resistant than any of the clostridia. Similar data were generated when Kelsey et al. [48] tested two species of *Bacillus* and two species of *Clostridium* against mixtures of hypochlorite and methanol, and found the *Bacillus* species to be the most resistant. Brazis et al. [49] found that free available chlorine was much more sporicidal for the pathogenic *B. anthracis* than for the nonpathogenic *B. subtilis* var. *globigii*, except at pH levels higher than 9.5 Sykes [50] demonstrated that *B. stearothermophilus* spores were more resistant than *B. subtilis* spores when exposed to 5% phenol and 5% white coal tar disinfectant, and were essentially equivalent when exposed to 2% chlorhexidine digluconate and 1% hydrochloric acid in 70% alcohol. Thus, for evaluation of sporicidal efficacy as well as system monitoring, spores of *B. stearothermophilus* or *B. subtilis* may be used as suitable indicators. *Bacillus stearothermophilus* appears preferable for in-use testing, as it can be selectively isolated owing to its thermophilic property [50].

III. APPROACH

The evaluation of a medical instrument germicide system involves a two-phased approach: (a) germicide efficacy testing; and (b) in-use testing. Protocol outlines for each phase are described in the following.

A. Germicide Efficacy Testing

Data should be generated about the antimicrobial efficacy of the chemical (and its use-dilution and temperature) proposed for use in a system when using standard or nonstandard laboratory methods, from replicate testing with suitable indicator—surrogate pathogen organisms. The efficacy tests should yield quantitative data. The protocol should provide for quantitation of both the initial organism challenge before germicide exposure and of those organisms potentially recovered following germicide exposure.

1. Test Organisms

Efficacy should be demonstrated against vegetative bacteria, vegetative fungi and fungal spores, nonenveloped viruses, parasite cysts, and bacterial spores. Test organisms should be representative of all potential pathogens within each group and, thus, include a gram-positive and a gram-negative bacterium, a yeast and a mold, a mycobacterium, a nonenveloped human virus, a protozoan parasite, and a bacterial spore. For such a testing scheme, the following organisms from Table 2 are recommended: *S. aureus*, *P. aeruginosa*, *C. albicans*, *T. mentagrophytes*, *M. terrae*, poliovirus I, reovirus 3, *G. lamblia*, and *B. stearothermophilus*.

2. Hard Surface Testing

Since, in general, organisms are more resistant when dried on environmental surfaces than in suspension, testing should be carried out with test organisms inoculated and appropriately dried on a suitable environmental surface (such as metal, porcelain, or glass). The inclusion of organic matter (e.g., blood serum) will provide an additional significant challenge to the disinfectant.

3. Neutralization

An effective neutralizer for the test germicide should be identified and effective neutralization without residual toxic effects on surviving organisms should be demonstrated.

4. Quality Assurance

Quality assurance practices should be identified, carried out, and documented to include organism, media, reagent, and equipment controls.

B. In-Use or Simulated Use Testing of Germicide Efficacy

The current FDA guidelines for approval of disinfectants for use on medical instruments provides a good basis for the in-use or simulated in-use testing of disinfectants [51]. Although these are still guidelines at the time of this printing, they have been used for the 510(k) approval process of medical disinfectants. Additionally, the American Society for Testing and Materials (ASTM) has established a committee that has produced a draft document describing an in-use test method for the evaluation of disinfectants used on endoscopic equipment. Because there are a wide variety of materials and configurations that a disinfectant might come into contact with, it is not possible for all situations to be covered by the manufacturer of the disinfectant. Validation of disinfection or sterilization of a particular instrument by liquid chemical germicides should be part of the design of the instrument. Certain parameters should be addressed in the in-use evaluation of disinfectants.

1. Test Organism

Efficacy should be demonstrated against a wide variety of surrogate microorganisms, realizing that not all microorganisms can be addressed. As described in the foregoing section, test organisms should represent all the potential pathogens within each group and, therefore, include a gram-positive and gram-negative bacterium, a yeast and mold, a representative strain of mycobacteria, a nonenveloped human virus, a protozoan parasite, and if sterilization claim is made for the product, then bacterial spore-forming organisms should be included in the test regimen. These microorganisms should be the minimum for a complete evaluation of the disinfectant or sterilant.

It should be stressed that when circumstances warrant, a specific organism or organisms, other than the standard test organisms, should be included in the testing regimen. For instance, with a gastrointestinal endoscope, *Helicobacter pylori* or *Clostridium difficile* might be microorganisms that one would include in the test regimen, since they are pathogenic microorganisms specific to that area of the body and could be transmitted to other patients as a result of incomplete disinfection or sterilization of equipment.

One of the challenges with in-use or simulated testing is achieving a significant titer (i.e., $\geqslant 10^6$ colony-forming units; CFU) of organisms on or in the instruments. Achieving this titer may be difficult with some organisms that are difficult to grow under laboratory conditions or that are susceptible to drying conditions. The titer of the organism on the instruments, after drying, must be controlled such that the consistency within and between evaluations is achieved.

A second area of concern is the recovery efficiency of the procedure. Without the ability to recover a high percentage of the organisms applied, the method may be open for criticism, as a claim for efficacy would not be meaningful. A recovery efficiency greater than 90% is necessary to have meaningful results. Recovery may be achieved by various standard procedures, such as swabbing and exhaustive elution with saline solution. To enhance recovery, a surfactant may be added to the saline solution, with the proviso that it not have any adverse effect on the organisms being recovered.

The inoculum should always be added in combination with an organic soil load. This gives the method a degree of reality, since in real-use situations, instruments are usually contaminated with soil and microorganisms.

2. Inoculum Site

Both external surfaces and internal lumens, if they exist, should be inoculated and challenged with the disinfectant solution. The FDA guidelines state that the most difficult areas for the germicide to reach should be inoculated.

3. Replication

A sufficient number of replicates should be inoculated to obtain reliable results. The actual number of replicates performed should be statistically based, for the results to be meaningful [51].

4. Neutralization

As stated in Section III.A.3, appropriate controls for determining the efficacy of the neutralizer solution should be performed. This is to provide evidence to eliminate the potential for false-negative results caused by static or microbicidal activity of disinfectant carried over onto the recovery medium.

C. Future Work

The area of disinfectant testing has been under scrutiny for the past decade. Reports by the U. S. Government Accounting Office (GAO) indicate the lack of reliability of the then current methods [52]. In response, the U. S. Environmental Protection Agency (EPA) undertook a program to investigate the problems with the then current methods and awarded contracts for the study of each of the standard AOAC methods used for disinfectant testing. At the time of the writing of this chapter, none of the new methods have been published. Some incremental changes have been made to the methods over the past few years; however, it has been recognized that major changes to make the methods more quantitative, reproducible, and statistically sound are needed. These concerns still need to be examined, and it is hoped that in the new methods that come out the efforts at the EPA, these concerns will be addressed.

Both the efforts at FDA and ASTM have considered the issues of in-use efficacy of disinfectants. Translating the information received from in vitro data to in-use situations is not always possible or appropriate. The FDA guidelines do not outline specifics of a test method, but rather, give general guidance on what should be considered. The ASTM method, however, when finalized, should provide more specifics for the testing of disinfectants in conjunction with medical instruments. These in-use tests, when combined with in vitro test data, will provide a complete evaluation of germicidal potency and efficacy.

REFERENCES

1. R. Koch, Uber disinfektion. *Mitt. Kaiserlichen Gesundheitsam. 1*:234–282 (1881).
2. S. Ridel and J. T. A. Walker, The standardization of disinfectants. *J. R. Sanit. Inst. 24*:424–441 (1903).
3. H. Chick and G. J. Martin, The principles involved in the standardization of disinfectants and the influence of organic matter upon germicidal value. *J. Hyg.* (*Camb.*) *8*:654–697 (1908).

4. J. M. Ascenzi, R. J. Ezell, and T. M. Wendt, A more accurate method for measurement of tuberculocidal activity of disinfectants, *Appl. Environ. Microbiol. 53*:2189–2192 (1987).
5. B. Van Klingeren, A. B. Leussink, and W. Pullen, A European collaborative study on the repeatability and the reproducibility of the standard suspension test for the evaluation of disinfectants in food hygiene, *Report of the National Institute of Public Health and Environment Hygiene*, No. 35901001, Bilthoven, The Netherlands, 1981.
6. L. A. Carson, M. S. Favero, W. W. Bond, and N. J. Petersen, Factors affecting comparative resistance of naturally occurring and subcultured *Pseudomonas aeruginosa* to disinfectants, *Appl. Microbiol. 23*:863–869 (1972).
7. W. A. Rutala and E. C. Cole, Ineffectiveness of hospital disinfectants against bacteria: A collaborative study, *Infect. Control 8*:501–506 (1987).
8. S. F. Bloomfield and E. Looney, Evaluation of the repeatability and reproducibility of European suspension test methods for antimicrobial activity of disinfectants and antiseptics, *J. Appl. Bacteriol. 73*:87–93 (1992).
9. J. N. Mbithi, V. S. Springthorpe, and S. A. Satter, Chemical disinfection of hepatitis A virus on environmental surfaces, *Appl. Environ. Microbiol. 56*:3601–3604 (1990).
10. W. W. Bond, M. S. Favero, N. J. Petersen, and J. W. Ebert, Inactivation of hepatitis B virus by intermediate-to-high level disinfectant chemicals, *J. Clin Microbiol. 18*:535–538 (1983).
11. L. S. Martin, J. S. McDougal, and S. L. Loskoski, Disinfection and inactivation of the human T lymphotropic virus type III/lymphadenopathy-associated virus, *J. Infect. Dis. 152*:400–403 (1985).
12. S. F. Bloomfield, C. A. Smith-Burchnell, and A. G. Dalgleish, Evaluation of hypochlorite-releasing disinfectants against the human immunodeficiency virus (HIV), *J. Hosp Infect. 15*:237–278 (1990).
13. W. A. Rutala, E. C. Cole, N. S. Wannamaker, and D. J. Weber, Inactivation of *Mycobacterium tuberculosis* and *Mycobacterium bovis* by 14 hospital disinfectants, *Am. J. Med. 91*:(Suppl. 3B):267S–271S (1991).
14. A. Lind, M. Lundholm, G. Pedersen, V. Sundaeus, and P. Wahlen, A carrier method for the assessment of the effectiveness of disinfectants against *Mycobacterium tuberculosis, J. Hosp. Infect. 7*:60–67 (1986).
15. E. L. Domingue, R. L. Tyndall, W. R. Mayberry, and O. C. Pancorbo, Effects of three oxidizing biocides on *Legionella pneumophila* serogroup 1, *Appl. Environ. Microbiol. 54*:741–747 (1987).
16. B. H. Keswick, T. K. Satterwhite, P. C. Johnson, H. L. DuPont, S. L. Secor, J. A. Bitsura, G. W. Gary, and J. C. Hoff, Inactivation of Norwalk virus in drinking water by chlorine, *Appl. Environ. Microbiol. 50*:261–264 (1985).
17. J. H. Blackwell, J. H. Graves, and P. D. McKercher, Chemical inactivation of swine vesicular disease virus, *Br. Vet. J. 131*:317–323 (1975).
18. H. S. Wright, Test method for determining the virucidal activity of disinfectants against vesicular stomatitis virus, *Appl. Microbiol. 19*:92–95 (1970).
19. R. P. Christensen, R. A. Robison, D. F. Robinson, B. J. Ploeger, R. W. Leavitt, and H. L. Bodily, Antimicrobial activity of environmental surface disinfectants in the absence and presence of bioborden, *J. Am. Dent. Assoc. 119*:493–505 (1989).

20. S. F. Bloomfield, M. Arthur, B. Vanklingeren, W. Pulkn, J. T. Holah, and R. Elton, An evaluation of the repeatability and reproducibility of a surface test for the activity of disinfectants, *J. Appl. Bacteriol. 76*:86–94 (1994).
21. J. Prince and G. A. J. Ayliffe, In-use testing of disinfectants in hospitals, *J. Clin. Pathol. 25*:586–589 (1972).
22. E. Scott, S. F. Bloomfield, and C. G. Barlow, Evaluation of disinfectants in the domestic environment under "in use" conditions, *J. Hyg. (Camb.) 92*:193–203 (1984).
23. R. A. Robison, H. L. Bodily, D. F. Robinson, and R. P. Christensen, A suspension method to determine reuse life of chemical disinfectants during clinical use, *Appl. Environ. Microbiol. 54*:158–164 (1988).
24. J. F. Gardner, Principles of antimicrobial activity, *Disinfection, Sterilization and Preservation*, 2nd ed. (S. S. Block, ed.), Lea & Febiger, Philadelphia, 1977.
25. M. S. Harakeh, J. D. Berg, J. C. Hoff, and A. Matin, Susceptibility of chemostat-grown *Yersinia enterocolitica* and *Klebsiella pneumoniae* to chlorine dioxide, *Appl. Environ. Microbiol. 49*:69–72 (1985).
26. M. D. Sobsey, T. Fuji, and P. A. Shields, Inactivation of Hepatitis A virus and model viruses in water by free chlorine and monochloramine, *Water Sci. Technol. 20*:385–391 (1988).
27. M. Klein and A. DeForest, Principles of viral inactivation, *Disinfection, Sterilization and Preservation*, 3rd ed. (S. S. Block, ed.), Lea & Febiger, Philadelphia, 1983.
28. W. O. K. Grabow, V. Gauss-Muller, O. W. Prozesky, and F. Deinhardt, Inactivation of hepatitis A virus and indicator organisms in water by free chlorine residuals, *Appl. Environ. Microbiol. 46*:619–624 (1983).
29. M. C. Snead, V. P. Olivieri, K. Kawata, and C. W. Kruse, The effectiveness of chlorine residuals in inactivation of bacteria and viruses introduced by post-treatment contamination, *J. Clin. Microbiol. 14*:430–408 (1980).
30. O. C. Liu, H. R. Seraichekas, E. W. Akin, D. A. Brashear, E. L. Katz, and W. J. Hill, Jr., Relative resistance of twenty human enteric viruses to free chlorine in Potomac water, *Virus and Water Quality: Occurrence and Control* (V. Snoeyink, ed.), Proceedings of the 13th Water Quality Conference, University of Illinois, *69*:171–195 (1971).
31. H. H. Lensing and H. L. Oei, Investigations on the sporicidal and fungicidal activity of disinfectants, *Zentralbl. Bakteriol. Hyg. (B) 181*:487–495 (1985).
32. H. M. Prince, Disinfectant activity against bacteria and viruses: A hospital guide, *Particulate Microb. Control* Mar./Apr. (1983).
33. E. L. Jarroll, A. K. Bingham, and E. A. Meyer, Effect of chlorine on *Giardia lamblia* cyst viability, *Appl. Environ. Microbiol. 41*:483–487 (1981).
34. D. G. Korich, J. R. Mead, M. S. Madore, N. A. Sinclair, and C. R. Sterling, Effects of ozone, chlorine dioxide, chlorine, and monochloramine on *Cryptosporidium parvum* oocyst viability, *Appl. Environ. Microbiol. 56*:1423–1428 (1990).
35. AOAC, *Official Methods of Analysis*, 14th ed., Association of Official Analytical Chemists, Arlington, VA, 1984.
36. E. C. Cole, W. A. Rutala, and G. P. Samsa, Disinfectant testing using a modified use-dilution method: Collaborative study, *J. Assoc. Off. Anal. Chem. 71*:1187–1194 (1988).
37. H. Van de Voorde, G. Reybrouck, P. Van Dijck, and M. J. Vranckx, The choice of

fungi as test organisms in disinfectant testing, *Zentralbl. Bakteriol. Hyg. (B) 179*:125–129 (1984).
38. E. C. Cole, W. A. Rutala, L. Nessen, N. S. Wannamaker, and D. J. Weber, Effect of methodology, dilution, and exposure time on the tuberculocidal activity of glutaraldehyde-based disinfectants, *Appl. Environ. Microbiol. 56*:1813–1817 (1990).
39. B. van Klingeren and W. Pullen, Comparative testing of disinfectants against *Mycobacterium tuberculosis* and *Mycobacterium terrae* in a quantitative suspension test, *J. Hosp. Infect. 10*:292–298 (1987).
40. F. M. Collins, Bactericidal activity of alkaline glutaraldehyde solution against a number of atypical mycobacterial species, *J. Appl. Bacteriol. 61*:247–251 (1986).
41. M. Harakeh and M. Butler, Inactivation of human rotavirus, SA11 and other enteric viruses in effluent by disinfectants, *J. Hyg. (Camb.) 93*:157–163 (1984).
42. R. S. Engelbrecht, M. J. Weber, B. L. Salter, and C. A. Schmidt, Comparative inactivation of viruses by chlorine. *Appl. Environ. Microbiol. 40*:249–256 (1980).
43. S. A. Sattar, V. S. Springthorpe, Y. Karim, and P. Loro, Chemical disinfection of non-porous inanimate surfaces experimentally contaminated with four human pathogenic viruses, *Epidemiol. Infect. 102*:493–505 (1989).
44. T. T. Brown, Jr., Laboratory evaluation of selected disinfectants as virucidal agents against porcine parvovirus, pseudorabies virus, and transmissible gastroenteritis virus, *Am. J. Vet. Res. 42*:1033–36 (1981).
45. Y. Watanabe, H. Miyata, and H. Sato, Inactivation of laboratory animal RNA-viruses by physicochemical treatment, *Exp. Anim. 38*:305–311 (1989).
46. F. W. Scott, Virucidal disinfectants and feline viruses, *Am. J. Vet. Res. 41*:410–414 (1979).
47. M. Dye and G. C. Mead, The effect of chlorine on the viability of clostridial spores, *J. Food Technol. 7*:173–181 (1972).
48. J. C. Kelsey, I. H. MacKinnon, and I. M. Maurer, Sporicidal activity of hospital disinfectants, *J. Clin. Pathol. 27*:632–638 (1974).
49. A. R. Brazis, J. E. Leslie, P. W. Kabler, and R. L. Woodward, The inactivation of spores of *Bacillus globigii* and *Bacillus anthracis* by free available chlorine, *Appl. Microbiol. 6*:338–342 (1958).
50. G. Sykes, The sporicidal properties of chemical disinfectants, *J. Appl. Bacteriol. 33*:147–156 (1970).
51. United States Food and Drug Administration, Guidance on the content and format of pre-market notification [510(k)] submission for liquid chemical germicides. U.S. Food and Drug Administration, Washington, DC, 1992.
52. United States Government Accounting Office. Disinfectants. EPA lacks assurance they work. U.S. General Accounting Office, Washington, DC.

2

A Broad-Based Approach to Evaluating Topical Antimicrobial Products

Daryl S. Paulson
BioScience Laboratories, Inc.
Bozeman, Montana

I. INTRODUCTION

This chapter will present a broad, integrative perspective to evaluating topical antimicrobial products. Hence, it presents evaluating topical antimicrobial products from a biostatistical perspective, from a microbiological perspective, as well as from a marketing perspective. In many technical writings, the marketing aspects are omitted [1]. This is unfortunate, largely because the demands of the market constitute whether or not the antimicrobial product is accepted.

Let us now turn to the definition of an antimicrobial. "Antimicrobials," "antiseptics," and "anti-infectives" are terms commonly used to designate chemical compounds that have varying degrees of antimicrobial properties, in that they are either inhibitory or lethal to microorganisms [1].

The Food and Drug Administration (FDA), as published in the *Federal Register* of September 13, 1974, identified seven categories of topically applied products. They are defined as follows:

1. An *antimicrobial soap*: A soap, containing an antimicrobial ingredient or ingredients, which has demonstrated in vitro and in vivo antimicrobial activity against normal skin flora.
2. A *health care personnel hand wash*: A broad-spectrum antimicrobial formulation that is "fast-acting" and nonirritating, as it is designed for frequent use by humans.
3. A *patient preoperative skin preparation*: A broad-spectrum, fast-acting formulation containing an antimicrobial ingredient that results in a significant, immediate, and persistent reduction in the number of microorganisms residing on intact skin.
4. A *skin antiseptic*: A nonirritating preparation containing an antimicrobial ingredient designed to prevent clinical skin infections.
5. A *skin wound cleanser*: A nonirritating liquid preparation intended to remove foreign material from small superficial skin wounds. It may contain an antimicrobial ingredient and must demonstrate that it does not interfere with wound-healing time.
6. A *wound protectant*: A nonirritating preparation, containing an antimicrobial ingredient, that is intended to be applied to small, cleansed wounds for the provision of a protective physical chemical barrier that neither delays wound healing nor favors microbial growth.
7. A *surgical handscrub*: A broad-spectrum, fast-acting, persistent, and nonirritating antimicrobial preparation designed to significantly reduce the number of microorganisms residing on intact skin surfaces.

The primary focus in this chapter will be in three areas: (a) health care personnel handwash formulations, (b) preoperative skin preparation formulations, and (c) surgical scrub formulations. Before going on, let us briefly review how the performance of these formulations are evaluated.

II. GENERAL PERFORMANCE CHARACTERISTICS

Generally, topical antimicrobials are evaluated for their antimicrobial efficacy from three perspectives or parameters. These are (a) the immediate degerming efficacy of the formulation, (b) the persistent antimicrobial effectiveness of the formulation, and (c) the residual antimicrobial properties of the formulation when appropriate.

The product's immediate antimicrobial efficacy is a quantitative measurement of both the mechanical removal and immediate inactivation of microorganisms residing on the skin surface. The persistent antimicrobial effectiveness is a quantitative measurement of this topical product's ability to prevent microbial recolonization of the skin surfaces after application of the product either by microbial inhibition or lethality. The residual efficacy is a measurement of the product's cumulative antimicrobial properties after it has been used repeatedly over time. That is, as the antimicrobial product is used over time, it is absorbed into the stratum corneum of the skin and, as a result, prevents recolonization on the skin surfaces by microbes.

The ability to accurately measure these three parameters poses problems to the investigation [1, 2]. Since a product's antimicrobial activity is relative, it is of utmost importance that its efficacy measurements be well-defined and clearly stated before conducting an evaluation.

Let us turn our attention to some additional, important points to consider when evaluating topical antimicrobial products.

Traditionally, topical antimicrobial evaluations have been the domain of the microbiologist, which is logical and appropriate, since they are evaluated in terms of microbial parameters. However, although the focus is on microbiology, the efficacy of the product is measured mathematically, specifically through biostatistical methods. Because of the bipartite structure necessary in these evaluations, the investigator must have a background in clinical microbiology, as well as in biostatistics. This will enable the investigator to better design a study that will unambiguously answer the experimental questions of interest.

Since it is the experimental design and the statistical evaluation that give the investigator the most trouble, we will concentrate mainly in these areas. Our general strategy then will be to statistically design, conduct, and evaluate the study by using experimental design, microbiology, and statistics as appropriate.

A statistically designed study is one that systematically collects, organizes, analyzes, and draws valid conclusions about the antimicrobial products evaluated. When one designs a clinical trial, it is critical that the research objectives are explicitly stated and that the study is designed toward answering those objectives clearly, concisely, and unambiguously. This is a key to a successful evaluation program, and it has been our experience that this approach appreciably expedites FDA review. Now, let us look at the design of a study by looking a little closer at the study objectives.

Simply stated, the study objectives are the answers to the questions the study is intended to address. This will be perhaps the easiest of all the necessary documentation to draft relevant to clinical trials and one of the most important. Every other document or concern of the study will be in support of the objectives. Given the objectives, the general schema of this study will be to statistically design the study relative to those objectives, conduct the study employing microbiological methods within the framework of the experimental design and, finally, evaluate the results employing biostatistical methods to answer the study objectives. To accomplish this efficiently, several organizational mechanics will be required. They include a description of the study's purpose, a description of the experimental strategy used to achieve the purpose, a description of the microbial methods used to meet the study's purpose, and the statistical methods used to evaluate the study.

A. Description of the Study's Purpose

A concise description of the study's guiding purpose is critical. This point is so implicitly obvious that it is often explicitly ignored until the entire study has been completed, at which time the original objectives have become obscure and unclear. Then, the investigator must backtrack through the study to try to determine what these original objectives were. To avoid this, it is imperative to develop a statement of the study's purpose before implementation.

In determining the study's purpose, the three antimicrobial parameters (the immediate, persistent, and residual effects) must be taken into account. If one is evaluating a venipuncture antimicrobial product, the purpose will probably be concerned with the immediate antimicrobial effects only. However, if one is evaluating a surgical scrub product, the immediate, persistent, and residual antimicrobial effects will be of importance. Also, a reference (control) product needs to be used in the study as a gauge for comparison of the total antimicrobial effects.

B. Description of the Experimental Strategy

Once the study's purpose has been determined, one can begin to design the study to meet that purpose. The design of the study not only will answer the study's objectives and purpose, but also do it in a valid manner. Validity in experimental designs is concerned with two areas: internal and external validity.

1. Internal Validity

Usually, the internal validity of a study can be controlled using the appropriate statistical experimental design. Internal validity can then be built into the study by random sampling, by blocking, and by the use of other statistical and experimental controls [3]. There are various errors that cause a study to be internally invalid. Two of them are routinely encountered in clinical trials. They are as follows:

1. *History* as the specific events occurring between the first and subsequent measurements. Examples in clinical trials could be subjects washing their skin surfaces with antimicrobial compounds between the baseline and posttreatment measurements. The treatment measurement would be compounded with extraneous uncontrolled antimicrobial compounds. The evaluated antimicrobial product would rate higher in antimicrobial efficacy than it actually is.

2. *Instrumentation* in which changes in the observations or measuring instruments (agar plates) may produce changes in the study results. An example in clinical trials may be the use of different media lots that are significantly different in nutritional characteristics and, therefore, variable in their ability to support microbial growth. Another example is studies for which subjective evaluations must be provided, specifically in health care personnel skin irritation evaluations: here, two observers may score the same degree of chafing or redness differently.

2. External Validity

Unfortunately, no experimental designs have built-in controls for threats to external validity. External validity refers to the degree or extent to which the results of a specific study can be generalized to the population at large, or under other environmental conditions. External validity can be classified into two subgroups: population validity and ecological validity. Population validity deals with the generalization of the specific study to the general population, whereas ecological validity concerns the generalization of the specific study to other settings.

The easiest way to assure the external validity of the study is to have the same study conducted independently at different geographic locations. If the same results occur and the same conclusions are drawn at each site by different investigators, the external validity of the study can be considered, to a high degree of probability, satisfactory.

With these aspects in mind, let us return to the surgical scrub evaluation and some of the mechanics necessary to perform a successful evaluation. Recall that we are interested in the immediate, persistent, and residual antimicrobial effects of the test and control products. This may be considered a completely randomized design in relation to the products. That is, each subject in the study is equally likely to be assigned the test or control product to be used in the study.

In the process of comparing the two products for the immediate, persistent, and residual antimicrobial effectiveness, we need to compare the reduction in microorganism populations with some value. That value is a predetermined baseline value or the population numbers that normally reside on the surface of the hands.

In this study, the immediate antimicrobial effects will be measured just after the surgical scrub, and the persistent antimicrobial effects will be measured at both 3 and 6 h after the scrub. The residual antimicrobial effects will be measured over the course of 5 consecutive days of product use. This is represented using the following experimental design, which is a pretest–posttest design (Table 1).

By taking the time to sketch out basic experimental design procedures, one can clearly and simply address the design requirements necessary to achieve the study's objectives.

TABLE 1 Experimental Design for Surgical Scrub Evaluation

	Baseline measurement	Independent variable treatment	Dependent variables microbial counts over both hours and days		
(R)	O_1	A	O_2	O_3	O_4
(R)	O_5	B	O_6	O_7	O_8

where:

R = Completely randomized design
Each subject is equally likely to be selected into group A or B.
O_1, O_5 = Dependent variable (microbial counts during baseline)
Independent variables:
A = Treatment with test product (group A)
B = Treatment with control product (group B)
O_1 = Dependent variable (microbial counts after treatment with product A or B). $O_1 = O_2, O_3, O_4, O_6, O_7$, and O_8.

It is also important to know the data distribution of the results. Recall that the dependent variable is the variable one is interested in measuring after all other variables have been controlled. It is more often known, simply, as "the variable." Typically, in a topical antimicrobial evaluation, the response variable is simply the microbial colony counts. A problem that often occurs in the statistical analysis of microbial colony counts is that the data collected are nonlinear. They are exponential. Since the vast majority of statistical models are linear models, they cannot be used to evaluate nonlinear data. Therefore, the microbial counts must be transformed to a linear scale (most often, a $\log_{10}$ scale) to be statistically evaluated. This linearization of the data must be assured.

Also, in statistical designs, it is necessary to establish the levels of both the alpha (α) and the beta (β) error, so that the appropriate number of subjects to be tested and the number of replicate measurements to be taken in each sampling period, relative to the desired confidence level, can be determined [3]. Recall that α-error (type I error) is committed when one rejects a true-null hypothesis, and β-error (type II error) is committed by accepting a false-null hypothesis. In other words, α-error occurs when one states that there is a difference between products, when there really is not. β-Error occurs when one concludes that there is no difference between products, when there really is. The easiest way to control both α- and β-errors is to use more subjects (replicates) so that the possibility of both α- and β-errors is reduced. Otherwise, merely adjusting the α-error to a very small level will increase the probability of β-error.

Often the question arises in topical antimicrobial evaluations if the study should be single- or double-blinded. Recall that in single-blind situations, the participants in the study do not know if they are receiving the test product(s) or the control product, but the investigator does. In a double-blind study, neither the investigator nor the subjects know who is receiving which product. There is often pressure on an investigator to evaluate the sponsor's product in the best light, which builds a tacit biasing effect into the study. The use of a double-blind study can eliminate this biasing problem [4]. Once the study is completed and the data is evaluated, the true product identities may be safely revealed. For even greater protection from experimental bias, the various processes of the study need to be compartmentalized so that no one person has total control of all aspects of the study. This is also very important in cases where there is double-blinding, but the evaluated products are easy to distinguish. This would happen, for example, when the test product is a chlorhexidine gluconate product and the reference product is an iodophor. It is very easy to distinguish the two apart, no matter how they are labeled.

Commitment to designing an efficient statistically based experimental design will tend to provide a cost-effective model that can provide the investigator with valid results.

C. Description of The Microbial Methodology

The investigator must be familiar with the microorganism ecology of the skin to conduct these types of studies [2, 4]. The skin surfaces provide a unique habitat for microorganisms, and knowledge of histology, physical features, and nutrient factors of the skin are important.

The histological structure of the skin can be an aid in understanding both the physical and nutritional characteristics of the skin relative to microorganisms. The number of surface appendages to normal skin that must be taken into account relative to the microbial populations encountered at specific, anatomical skin sites includes eccrine sweat glands, sebaceous glands, apocrine glands, and hair. Other physical factors influencing microbial growth include skin pH, skin temperature, skin humidity, and the oxygen/carbon dioxide tension. Factors that need to be taken into account from a microbial nutritional perspective include age, diet, and anatomical site of the person. Finally, a knowledge of those microorganism types that normally inhabit and colonize the skin surfaces is valuable to the investigator. Those microorganisms include coryneform bacteria, the *Micrococcaceae*, including *Staphylococcus* strains; *Streptococcus*; gram-negative bacteria, including *Escherichia coli*; various fungi; virus particles; and *Mycoplasma* [4].

Within this context, the investigator can then determine what is the best culture media on which to grow the microorganisms that will be encountered at the anatomical sites sampled. It can also be determined what kind of dilution levels will be employed. What is the appropriate incubation temperature and incubation period? Again, these kinds of questions need to be addressed before beginning the study.

In the design of topical anti-infective evaluations, for whatever use, it is necessary to simulate in-use conditions with evaluations. For example, when one wants information concerning the antimicrobial properties of a new surgical scrub product, inoculating the hands with an extraneous indicator microorganism, such as *Serratia marcescens* may not be the most valid approach [5]. The use of normal flora residing on the subject's own skin surface is probably better, since it more closely approximates the actual conditions.

D. Statistical Methods Used to Evaluate the Study

It is critical to select statistical models (linear regression, analysis of variance, analysis of covariance, Student's *t*-test, or other) that comply with the experimental design [4]. If the investigator does not do this, it has the same effect as noncompliance to any protocol, which may invalidate the entire study.

Although the exact statistical model to be used within the framework of the experimental design often depends on the data distribution generated (normal, skewed, bimodal, exponential, binomial, or other), the use of exploratory data analysis (EDA) can help the investigator not only to select the appropriate statistical model, but to develop an intuitive "feel" for the data before the actual statistical analysis occurs. This helps the investigator because he or she can actually see the shape of the data distribution in terms of graphic displays and can identify any data trends that may not be obvious as well as determine the appropriate data distribution [6].

Also, it is important to identify the data-handling procedure. For example, if a prewritten statistical software package is not being used, the required statistical computer programs must be written as well as validated. If prewritten software packages, such as Statistical Package for the Social Sciences (SPSSx), Biomedical Data Program (BMDP), or Statistical Analysis System (SAS) are used, the software program used needs to be configured. Each program requires some careful thought in that the configuration depends on how the samples are to be taken, blocking procedures, and contrasts to be used. In addition, some thought must be given to the data input arrays to assure that the keyed input is compatible with the software's input capacity. Understandable graphic display formats will aid in interpreting even the most complex data analyses.

III. NEW PRODUCT DEVELOPMENT APPLICATIONS

Now, let us change our focus to the actual development of new products for the anti-infective market. We will begin with one of the most common problems in the industry, that of attempting to produce one antimicrobial product to meet the needs of the entire industry.

A common procedure used by the manufacturer in formulating topical antimicrobial products is to recommend their product as a surgical scrub, as a preoperative skin preparation formulation, and as a health care personnel handwash. The reason for this is simple. It takes less research and development work to offer one product as a surgical scrub, a preoperative skin preparation formulation, and a health care personnel handwash than it does to develop a specific surgical scrub product, a specific preoperative skin preparation formulation, or a specific health care personnel handwash product [6].

There is a very fundamental error with this rationale, however. Each application has its own unique needs and requirements. Therefore, the topical antimicrobial product needs to be designed for its specific, intended purpose as a health care personnel formulation, or a surgical scrub formulation, or a preoperative skin preparing formulation [2,6,7] For example, surgical scrub products, in general, are too harsh and potentially irritating to the skin for the large number of required, repeated washes necessary for health care personnel over the course of a day. A surgical scrub product must remove not only transient microorganisms, but also resident microorganisms [5]. The health care personnel handwash formulation needs only remove the transient microorganisms. The surgical scrub formulation is "stronger" than the health care personnel handwash formulation. Although a surgeon may scrub his or her hands two or three times a day in the surgical setting, it is not uncommon for health care personnel to wash their hands 25–30 times as they repeatedly interact with patients. Use of a formulation designed as a surgical scrub 25–30 times daily most often leads to skin irritation [8]. A product designed as a surgical scrub, but used as a health care personnel handwash formulation, will not be marketable, because it is too irritating to the hands.

To be marketable, the health care personnel product must be effective in reducing transient microorganisms, and it must be nonirritating after repeated and prolonged use. Health care personnel have strong preference for products that are mild and nonirritating to their skin as well as effective [8,9]. Generally, mildness can be built into the health care personnel handwash in three ways:

1. By proportionally reducing the amount of active ingredients contained in the product, such as chlorhexidine gluconate, iodophors, or alcohol. For example, instead of using the customary 4% level of chlorhexidine gluconate (CHG) found in many surgical scrub formulations, a 2% chlorhexidine gluconate or even less is used.
2. By adding skin conditioners or emollients to the formulation, thereby counteracting the irritating effects of the active compounds, making the product more gentle and mild to skin surfaces.
3. By using a combination of these two methods; that is, a reduction in the active ingredient levels and the addition of emollients and skin conditioners to the formulation. Since many of the health care personnel handwash products were originally designed as surgical scrubs, many

> of them are too harsh on the hands when used frequently and repeatedly. Hence, this market is vulnerable to a manufacturer who will develop a formulation specifically targeted for the health care personnel handwash market [10, 11].

Many suppliers are not the manufacturers of health care personnel products. As a result, they often feel that they are at a disadvantage in that they must "take" what products are offered to them by the manufacturers. However, the suppliers have more options than often realized, even when using preexisting formulations. It is suggested that they collect samples from several manufacturers for use in a pilot study to determine the product most suitable for their needs. Through this approach, the efficacy as well as the irritation potential of various different formulations can be evaluated. The optimum antimicrobial formulation, which is low in irritation potential as well as antimicrobially effective, can be readily identified.

Suppose a health care facility wants to use a handwash that has CHG as the active ingredient. Since there are several manufacturers currently producing CHG-based products, it would be wise to evaluate all of those products for their antimicrobial and irritation potential—evaluation is then based on your needs.

Suppose you want to evaluate both the 2% and the 4% CHG-based formulations of four different manufacturers. An efficient statistical evaluation can be

TABLE 2 Rating System for Evaluation of a Product's Skin Irritation Potential

Erythema	0 = No reaction
	1 = Mild or transient redness limited to sensitive areas
	2 = Moderate redness persisting over much of the product-exposed area
	3 = Severe redness extending over most or all of the product-exposed area
Edema	0 = No reaction
	1 = Mild (just perceptible) or transient
	2 = Moderate—definitely palpable
	3 = Severe
Rash	0 = No reaction
	1 = Mild—few, small eruptions
	2 = Moderate—scattered eruptions more than ten per hand
	3 = Severe
Dryness	0 = No reaction
	1 = Mild—transient, generally limited to cuticles, knuckles
	2 = Moderate—persistent, extending over much of the hand
	3 = Severe—persistent, extending over most of the hand, characterized by cracking and roughness

devised that will provide not only the information critical for evaluating the antimicrobial efficacy, but also the irritation factors of the products. A potentially useful basis for the evaluation of the irritation potential is the scoring of irritation factors in terms of edema, dryness, erythema, and skin eruptions, perhaps employing a 4-point rating system, such as is shown in Table 2 and utilizing a statistical model to compare the irritation potential of the products.

A particularly useful statistical model used for irritation evaluations is the χ^2 nonparametric statistic. This model allows the detection of significant differences between the products evaluated in terms of irritation potential. It is statistically robust (reliable), yet accurate and precise in detecting true significant differences between products.

Let us now turn our attention from the health care personnel handwash example to the more general statistical strategy of evaluating new product formulations for which no known antimicrobial characteristics are available to the investigator.

IV. STATISTICAL MODEL-BUILDING AND ANALYSIS: EXPLORATORY PHASE

In this example, a very small pilot study is employed. The study objectives, experimental design, and microbial methods can be determined from other study examples, since they are related to the independent or controlled variable. The dependent or response variable—the microbial counts—cannot be known; hence, one must explore the data collected before choosing the appropriate statistical models to evaluate the data.

This is done through exploratory data analysis (EDA). This phase is used to evaluate the data and to develop an intuitive "feel" for the data, to develop an appropriate statistical model for evaluating the product's performance.

Exploratory data analysis consists of four major facets:

1. *Data displays*, which allow the investigator to observe and intuitively comprehend the data distributions and patterns under study.
2. *Residual displays*, which enable the investigator to fit an appropriate statistical model to the data by interactive statistical model-building. From the residual values (the difference between the actual data values and the statistically expected numerical data values predicted by the statistical model) as an indicator of the statistical model's predictive ability, the investigator can generate a very effective, reliable statistical model.
3. *Reexpression* often simplifies the statistical analysis by rescaling the data through a variety of mathematical reexpressions, such as reciprocals, square roots, and $\log_{10}$ reexpressions. When working with non-

linear data, such as encountered in clinical trials, the reexpression procedure often linearizes the data so a simple linear statistical model can be used to evaluate the data instead of more complex nonlinear functions.

4. *Use of resistance and robust mathematical models*, which are affected little by the extreme data value outlines known to significantly affect parametric statistical models. Use of the robust models often produces more reliable evaluations, even when using very few subjects, as in pilot studies.

By employing EDA, one can often select a statistical model that is very representative of the real-world situation. This makes the statistical analysis of the product's performance more accurate and precise and, therefore, more realistic.

To select the most optimal statistical model—the one that realistically portrays the data—several EDA data-handling procedures and displays stand out. They include the following:

1. *Stem-and-leaf displays*, which provide a flexible, effective procedure for ordering raw data and, thereby, determining the distribution (e.g., the normal or a log linear distribution) the data demonstrate.
2. *Letter value displays*, which summarize and describe raw data in terms of their dispersion relative to the median value. These displays also show where the values lie, or do not lie, relative to the media. They also provide insight into the distributional shapes skewed left or skewed right of the data.
3. *Box plots* graphically display values in a format that resembles the standard Student *t*-confidence level diagram. The box plot display is strictly visual, with no numerical values provided. One can use them much like one uses a series of 95% confidence intervals (CI) in comparing groups of data

Once a good understanding of the data distribution is obtained, a statistical model to be used with an appropriate sample scheme is chosen.

V. STATISTICAL MODEL SELECTION

The statistical model, to be appropriate, must measure the data accurately and precisely. The test hypothesis should be stated as clearly and concisely as possible. If, for example, the statistical analysis is designed to test whether or not products A and B are equivalent over the course of multiple washings, the statistical model should accurately test that hypothesis and measure it.

Sampling techniques are also of great importance. No matter how objective the investigator is, experimental bias may creep into the study unless random-sampling methods are used. Often investigators create experimental bias when they try to randomize an experiment subjectively. The best way to reduce such bias is through a formal scheme of random sample selection, using a table of random digits. Such tables are available from various sources, including most basic statistical texts and random, number-generating computer programs.

Roger H. Green [12], in his book *Sampling Design and Statistical Methods for Environmental Biologist*, describes ten steps for effective statistical analysis. These steps are applicable to any topical antimicrobial product analysis:

1. State the test hypothesis concisely to be sure that what you are testing is what you want to test.
2. Always replicate the samples. Without replication, the variability measurements may not be reliable.
3. Keep the number of sample replicates equal throughout the study. This practice makes it much easier to analyze the study and produces more reliable results.
4. When testing whether a condition has a significant effect, be sure to take samples both where the test condition is present and where it is absent. (Example: If you find, through analysis, that a reduction in active ingredients neutralizes the effects, be sure to demonstrate that this problem can be corrected by increasing the level of active ingredients.)
5. Perform a small-scale study to provide a basis for sampling design and statistical model selection before analyzing the entire manufacturing system.
6. Verify that the sampling scheme actually is a representative measure of the population you want to measure. Guard against systematic and experimental bias by using techniques of random sampling.
7. Break a large-scale sampling process into smaller components.
8. Verify that the collected data meets the statistical distribution assumptions. In the days before computers were commonly used and programs readily available, assumptions had to be made about distributions. Now it is fairly easy to test these assumptions, at least in part, before accepting the statistical model as valid.
9. Test your model to make sure that it is useful for the process under study. If the model is satisfactory for one set of data, be certain that it is adequate for other sets of data from the same process.
10. Once these nine steps have been carried out and performed, accept the results with confidence. Much time, money and effort can be saved by following these ten steps to system analysis.

Once the investigator has a general understanding of the product's attributes, he or she must fit a statistical model to the data. At times, nonparametric models are better suited for this than parametric models. For example, if budgetary or time constraints force the investigator to use only a few subjects per product in the study, or if some requirements of the parametric model, such as a nonnormal distribution, cannot be achieved, then the nonparametric model is the model of choice.

Before the large-scale study, the investigator should reexamine (a) the test hypothesis, (b) the choice of variables, (c) the number of replicates required to protect against type I and type II errors, (d) the order of experimentation process, (e) the randomization process, (f) the statistical model used to describe the data, and (g) the data collection and data-processing procedures, to ensure that they continue to be relevant to the study.

Ultimately, the product will need to be marketed to be successful; therefore, a careful study of topical antimicrobial products in relation to the market place is in order.

VI. WORKING STUDY PROTOCOLS

Now that we have discussed various aspects of clinical trials, let us turn our attention to examples of standard protocols that can be used to evaluate topical antimicrobial products. Let us begin with the health care personnel handwash formulation.

A. Health Care Personnel Handwash Protocol

1. Purpose of Study

The study is intended to examine both the antimicrobial efficacy (immediate and persistent) and the skin irritation potential of two health care personnel handwash formulations in reducing transient microbial contamination of the hands over the course of 25 contamination and washing cycles.

2. Test Materials

The products to be evaluated are:

Test product

Name:
Lot number:
Expiration date:
Manufacturer:

Control product

Name:
Lot number:
Expiration date:
Manufacturer:

3. Test Methods

Subjects

A sufficient number of overtly healthy subjects, older than 18, but younger than 70, will be admitted into the study to ensure that 24 subjects complete the study, or 12 subjects per test group. Subjects will be of mixed sex and age; all will be free of clinically evident dermatoses or injuries to the hands and forearms. No immune-compromised patients will be admitted into the study, and all subjects will sign informed-consent forms before participating in the study.

Concurrent Treatment

No subject will be admitted into the study who is using oral contraceptives, topical or systemic antimicrobials, or any other medication known to affect the normal microbial flora of the skin.

Pretest Period

The 14 days before the test portion of the study begins will constitute the pretest period. During this time, subjects will avoid using medicated soaps, lotions, deodorants, and shampoos; and will avoid skin contact with solvents, detergents, acids, and bases. Bathing in chlorinated pools and hot tubs will be avoided. This regimen will allow stabilization of the normal microbial flora of the hands.

Experimental Period

The following 7 days will constitute the test week. Each subject will be tested only 1 day of that week for a 4- to 5-h period. On the designated test days, 5-ml aliquots of approximately 10^6 cells per milliliter of *Serratia marcescens* (red-pigmented strain) will be pipetted into each subject's cupped hands. The inoculum will then be distributed evenly over both hands, and to the area approximately 4 in above the wrist, by gentle massage. After a 1-min air dry, the glove juice sampling procedure will be performed.

The first inoculation cycle will constitute the baseline. It will be followed with a handwash with one of the two test products, according to the product's label directions. This inoculation and wash procedure will be repeated 25 times with a minimum of 5 min between washes. A transient microorganism count of the hands will be performed every five washes, using the glove juice-sampling procedure.

Glove Juice-Sampling Procedure

Following the prescribed wash and rinse, according to label instructions, non-powdered sterile surgical gloves will be donned. Seventy-five milliliters of sterile phosphate buffer (pH 7.8) aqueous solution, containing 0.1% Triton X-100, will be instilled into the glove. The glove will be secured at the wrist and the hand massaged through the glove for 60 s. Aliquots of the glove juice will be removed and directly plated or serially diluted in trypticase soy broth (TSB) containing the appropriate product neutralizer.

Triplicate, trypticase soy agar (TSA) spread plates containing the appropriate neutralizer will be prepared for each dilution. The plates will be incubated at 30–35°C until a distinguishable red color develops. Those plates providing between 25 and 250 red-pigmented colonies will be preferably utilized in this study. If no plates provide counts in the 25–250 range, the plates closest to that range will be used. The number of viable red-pigmented bacteria recovered will be determined using the formula: aliquot volume × dilution factor × mean plate count for the three plates.

Irritation Evaluation of the Product

The subjects' hands will be evaluated for product irritation potential. Scoring will be according to the schema represented in Table 2. In the event that a subject's hands develop significant irritation, the subject will be removed from the study.

Following the final wash and irritation examination, the subjects will be required to wash with 70% isopropyl alcohol for 2 min and rinse under tap water to degerm any remaining *S. marcescens* transiently on the hands.

4. Statistical Analysis

A pre–post experimental design will be used to evaluate the products' effectiveness and test for significant differences between the formulations over the course of the multiple washes. The 0.05 level of significance will be used in the study.

B. Surgical Scrub Evaluation Protocol

1. Purpose of Study

This study is designed to examine the immediate, persistent, and residual antimicrobial effectiveness of two test products and one reference product in reducing normal microbial flora residing on the hands.

2. Test Materials

Test Product

Test product 1

Name:
Lot number:
Expiration date:
Manufacturer:

Test product 2

Name:
Lot number:
Expiration date:
Manufacturer:

Reference Product

Name:
Lot number:
Expiration date:
Manufacturer:

3. Test Methods

Subjects

A sufficient number of overtly healthy subjects older than 18, but younger than 70 will be admitted into the study to ensure that 54 subjects complete the study. Each subject will be assigned to one of three test groups, comprising 18 subjects per test group. Insofar as possible, the subjects will be of mixed sex and age; all will be free of clinically evident dermatoses or injuries to the hands and forearms. All subjects will sign informed-consent forms before participating in the study.

Concurrent Treatment

No subject will be admitted into the study who is using topical or systemic antimicrobials, or any other medication known to affect the normal microbial flora of the skin.

Pretest Period

The two weeks before the test portion of the study begins will constitute the pretest period. During this time, subjects will avoid using medicated soaps, lotions, deodorants, and shampoos; and will avoid skin contact with solvents, detergents, acids, and bases. Bathing in chlorinated pools and hot tubs will be avoided. This regimen will allow stabilization of the normal microbial flora of the hands.

Baseline Period

The first week following the pretest period will constitute the baseline period. Baseline determinations will begin on day 1 of that week. This period will be used to determine eligibility for the study. Only those subjects with baseline counts of at least 1.0×10^5 organisms per hand will continue in the study. Baseline sampling will also be conducted on days 3 and 5.

Subjects will be required to remove all jewelry from their hands and arms.

The hands and forearms to two-thirds the distance from the wrist to the elbow will be washed for approximately 30 s by a trained attendant using a bland soap and tap water. Samples will be taken (within 5 min) according to the glove juice-sampling procedure.

Both hands will be sampled in the baseline determination. Subjects will not wash for at least 2 h before baseline sampling on days 1, 3, and 5 to assure that the baseline microbial populations are stable. The baseline determination portion of the study will employ the same sampling and recovery techniques, including the same media that will be used in the test portion of the evaluation.

Test Period

The test products will be used once before sampling on days 1, 2, and 5; two additional times after sampling on test day 2; and three times on test days 3 and 4. Products will be used accordingly to the directions supplied by the sponsor.

Sampling of the hands will be conducted at the times indicated in the sampling schedule (Table 3) according to glove juice-sampling procedures.

Glove Juice-Sampling Procedure

Following the prescribed handwash, sterile latex gloves will be applied. At the appropriate sampling times, 75 ml of sterile 0.1% Triton X-100 is instilled into the sterile latex glove. The glove is secured at the wrist, and the hand is massaged through the glove in a standardized manner for approximately 60 s. Aliquots of the "glove juice" (10^0) will be removed and serially diluted in trypticase soy broth

TABLE 3 Sampling Schedule

	Number of hands sampled at indicated time after treatment			
	Day	0 h	3 h	6 h
Baseline	1, 3, 5	54	0	0
Test product 1	1	18	9	9
Test product 1	2	18	9	9
Test product 1	5	18	9	9
Test product 2	1	18	9	9
Test product 2	2	18	9	9
Test product 2	5	18	9	9
Reference product	1	18	9	9
Reference product	2	18	9	9
Reference product	5	18	9	9

(TSB) containing the appropriate neutralizer, to provide appropriate dilutions. Duplicate pour plates will be prepared from each of the dilutions, using trypticase soy agar (TSA) with appropriate neutralizers. The trypticase soy agar (TSA) plates will be incubated at 30–35°C for 24–48 h. The plates are then counted, using only those plates giving counts between 25 and 250 colonies. The number of viable microorganisms recovered will be estimated using the formula: 75 × dilution factor × mean plate count.

4. Statistical Analysis

The $\log_{10}$ plate counts for the three product groups will be compared in terms of (a) immediate antimicrobial effectiveness (as measured for the first scrub), (b) persistent antimicrobial effectiveness (as measured by the 0- to 6-h period), and (c) residual antimicrobial effectiveness (as measured by the 5-day use regimen; Table 4).

An analysis of variance (ANOVA) statistical model will be used to evaluate the three areas of concern. The level of significance is set at $\alpha = 0.05$ for type I error.

C. Preoperative Skin Preparation Protocol

1. Purpose of Study

Before surgery, the proposed operative site is topically prepared with an antimicrobial product to reduce the number of microorganisms residing on the skin and, thereby, the potential for surgically associated infections.

TABLE 4 Experimental Design

	Baseline sampling	Day 1 wash	Day 1 sampling	Day 2 wash	Day 2 sampling	Day 5 wash	Day 5 sampling
(R)	O_1	A_1	O_1	A_1	O_1	A_1	O_1
(R)	O_2	A_2	O_2	A_2	O_2	A_2	O_2
(R)	O_3	A_3	O_3	A_3	O_3	A_3	O_3

O_1 = Dependent variable ($\log_{10}$ average microbial counts per hand)
A_1 = Independent variable
1 = Test product with 1 wash regimen
2 = Test product with 2 wash regimen
3 = Reference product wash regimen
(R) = Completely randomized design. Each subject has equal probability of being assigned to one of the three products.

This study is designed to evaluate:

1. The immediate antimicrobial effectiveness of the test and control products in reducing the normal microorganism populations residing on the skin.
2. The persistent microbial effects of the test and reference products over a 4-h postpreparation period.

2. Test Materials

Test product 1:
Name:
Number:
Expiration date:
Manufacturer:
Test product 2
Name:
Lot number:
Expiration date:
Manufacturer:
Control product
Name:
Lot number:
Expiration date:
Manufacturer:

3. Test Methods

Subjects

A sufficient number of overtly healthy subjects older than 18, but younger than 70 will be admitted into the study to ensure 24 subjects complete the evaluation which, using a bilateral comparison, will be prepared with the reference product. Twelve subjects will be prepared with each of the two test products. Insofar as possible, the subjects will be of mixed age, sex, and race; all will be free of clinically evident dermatitis or injuries to the hands and forearms. All subjects will sign informed-consent forms before participating in the study.

Concurrent Treatment

No subject will be admitted into the study who is currently using topical or systemic antimicrobials, or any other medication known to affect the normal microbial flora of the skin.

Pretest Period

The 1-week period before the product-use portion of the study will be designated the pretest period. During this time, subjects will avoid using medicated soaps, lotions, deodorants, and shampoos; and will avoid skin contact with solvents, detergents, acids, and bases. Bathing in chlorinated pools and hot tubs will be avoided. This regimen will allow stabilization of the normal microbial flora of the skin at the test sites. The study description and informed-consent forms, containing a list of restricted and permitted products, are to be supplied to all subjects. Subjects will not be allowed to bathe or shower within 24 h of the beginning of the study. Additionally, subjects must not shave the treated areas within 48 h of the beginning of the study.

Experimental Period

As subjects are admitted to the study, they will be questioned concerning adherence to protocol product restrictions to assure protocol adherence. Subjects will be physically examined to ensure no evidence of injury or dermatosis is present at the sampling sites.

A screening baseline sample of the abdomen will be taken when the subjects volunteer for the study. Additionally, a baseline sampling of (a) the upper, inner aspect of the thigh, and (b) the abdomen in the vicinity of the umbilicus, will be performed, using the cylinder-sampling technique just before preparing the subject. These two sites will then be prepared bilaterally with the test products following the instructions provided by the sponsor.

For a particular subject to be used in the study, their baseline colony counts at the upper, inner aspect of the thigh must be at least 1.0×10^5 microbial organisms per square centimeter. Subjects must demonstrate baseline colony counts at the abdomen of at least 1.0×10^3 to be eligible for this study.

A sample of the prepared area will then be taken by the cylinder-sampling technique 10 min after preparation. After sampling, a sterile bandage will be placed over the prepared area to prevent microbial contamination. Additional samples will be taken at 30 min and 4 h after preparation, using the cylinder-sampling technique.

Cylinder-Sampling Technique

The cylinder-sample technique will be performed as follows:

At the appropriate sampling times, a sterile cylinder will be held firmly to the test site to be sampled. Five milliliters of sterile stripping fluid (SSF) will be instilled into the cylinder and the skin area inside the cylinder will be circumferentially massaged for 2 min with a sterile rubber policeman. One milliliter aliquots of the fluid–microorganism suspension (10^0 dilution) will be removed, plated, or serially diluted, or both, (as appropriate) with recovery broth containing 1.0% Tween 80 and 0.3% lecithin as the product neutralizer.

Equivalent dilutions will be prepared for test and control products. Triplicate 1-ml pour plates will be prepared from each of these dilutions using recovery agar containing 1.0% Tween 80 and 0.3% lecithin.

The recovery agar plates will then be incubated at 30–35°C for 48–72 h.

Those dilutions yielding 25–250 colonies per plate will be counted. The number of viable organisms on the sampling site will be estimated using the formula: dilution factor × average plate count. If no plates provide counts in the 25–250 range, those plates with the highest number of countable colonies closest to that range will be used.

The adequacy of neutralization will be confirmed before performing the evaluation. The results will be presented in the final report.

Media

Sterile stripping fluid is 0.1% Triton X-100 in 0.1 *M* phosphate-buffered (pH 7.8) water.

Recovery broth/agar consists of trypticase soy broth/agar containing 1.0% Tween 80 and 0.3% lecithin.

4. Statistical Analysis (Table 5)

This experimental design allows the assessment and comparison of both the immediate and persistent antimicrobial effectiveness among the three products. The three products will be compared relative to their immediate antimicrobial effectiveness, and their persistent antimicrobial effectiveness over the course of 4 h, after preparation. A multivariate parametric statistical model will be used to evaluate the products' effectiveness. The specific model used will depend on the "data fit." The level of significance is set at $\alpha = 0.05$ for type I error.

The contrasts used to detect significant differences among the three products will be the Tukey method of least significant difference (LSD)

TABLE 5 Pre–Post Experimental Design

Baseline		Post-prep samples		
time 0	Prep	10 min	30 min	4 hour
O_1	A_1	O_1	O_1	O_1
O_2	A_2	O_2	O_2	O_2
O_3	A_3	O_3	O_3	O_3

O_1 = Dependent variable. Log_{10} microorganisms count per cubic centimeter at the *i*th sample time.
A_1 = Independent variable.
1 = Test product 1 treatment
2 = Test product 2 treatment
3 = Reference product treatment

VII. MARKETING CONCERNS

Although this is not often considered in the research and development effort, it is critical. The most effective topical antimicrobial product is not of much value if there is no market. So the question becomes, "What are the market needs?"

The topical anti-infective market, which mainly comprises health care personnel handwashes, surgical handscrubs, and preoperative skin preparation formulations, is experiencing "new" product developments on an almost monthly basis. A casual overview of the current advertising literature demonstrates this [1,13]. But after critical review, are these developments really focused on "new" products, or are they the standard products delivered in a new manner? The answer is "yes," to both these questions. There are new products being developed for the anti-infective market as well as new systems of delivery and new users for standard products.

Let us now take a closer look at the research and development activity in the surgical scrub and preoperative skin preparation markets. Since we have previously discussed the health care personnel handwash formulations in some detail, it will be omitted from this section.

A. Surgical Scrub

Recall that surgical scrub formulations are designed to remove both the transient microorganism population as well as a large proportion of the normal endogenous microorganism population residing on the hands. To be considered effective, surgical scrub formulations must demonstrate both immediate and persistent antimicrobial effectiveness (up to 6 h postscrub) and optimally demonstrate a "residual effect" by becoming more effective antimicrobials with repeated use over time because the active ingredient(s) are absorbed directly into the skin [1].

Over the years, the povidone–iodine market share has been significantly eroded by chlorhexidine gluconate formulations because of their residual properties and, recently, there has been interest in developing low-level (2 or 3%) chlorhexidine gluconate formulations for the surgical scrub segment. Although at the time of this writing, there were no low-level chlorhexidine gluconate surgical scrub formulations approved for use by the Food and Drug Administration, they have shown impressive antimicrobial performance in clinical trials.

The bulk of the surgical scrubs marketed in the early 1990s will probably be 4% chlorhexidine gluconate products. The market will also likely see the introduction of several low-level chlorhexidine gluconate solutions.

B. Preoperative Preparative Solutions

Recall that the preoperative skin preparative solution is designed to both degerm an intended anatomical surgical site as well as provide a high level of persistent antimicrobial activity (up to 4 h after preparation). In the past, these solutions have been the domain of the iodine products, but these are being aggressively challenged by several chlorhexidine gluconate formulations. In addition, several formulations providing long-term persistent antimicrobial activity (up to 96 h after preparation) are likely to be introduced.

There is also interest in developing new product delivery systems. For example, one company recently launched a highly successful iodine–alcoholic film barrier system. The active iodine ingredient is only 0.5% and yet is very effective in preventing microbial recontamination of the prepared site by its patented "film" barrier system.

Additional developments in new delivery systems, which effectively dispense preoperative preparative formulations to the intended surgical site, but with a "no-mess application," will most likely be introduced.

Although there is some market activity in developing and introducing new anti-infective formulations for the health care personnel handwash, surgical scrub, and preoperative skin preparative products, most of the efforts will be focused on using the common antimicrobials (e.g., iodophors and chlorhexidine gluconates) as the active ingredients, but applied with new and novel delivery systems.

Health care personnel formulations will increasingly be developed as health care personnel products, not relabeled surgical scrubs. Surgical scrub formulations introduced in the early 1990s will, again, tend to be modifications of existing products. The preoperative skin preparative solution area will experience the most activity, with new product delivery systems.

Also of note, several new applications will be developed and introduced for existing products. One will be a "full-body shower wash" to be used in conjunction with preoperative preparing solution regimens. The full-body shower wash will be employed to reduce the microbial baseline counts on the patient's body before being prepared for surgery. The preoperative skin preparation formulation will then have a reduced microorganism population to contend with, making it more effective in its intended use.

We are often asked, "How does one market topical anti-infective products?" There is no esoteric "trick" to successful marketing of topical anti-infective products such as surgical handscrubs, health care personnel handwash formulations, and preoperative skin preparative formulations. Thorough planning and implementation of a creative marketing program are required [14].

An analysis of new product development programs has led to the conclusion that programs are well thought out and ultimately very successful; however, the majority of programs are based on good intentions, but do not meet the actual market requirements [15].

Other companies have problems when they enter the topical antimicrobial market with a product that is essentially the same as a competitor's product and, in doing so, have an uphill battle trying to market a product that is not unique. There are many suppliers of surgical scrubs, health care personnel, and preoperative skin preparative formulations, and most of their products are basically identical with one another. For example, the vast number of CHG products used for surgical scrub formulations are essential identical 4% CHG solutions. Since there is no real difference between these products, the major selling point ultimately becomes the sales price.

This situation is particularly unfortunate when the industry has such tremendous potential for new, innovative products as well as new applications for existing ones. For example, there are several alcohol-based products that do not need a water rinse. There is also a need for an adjunct procedure to be used in conjunction with the preoperative skin preparative procedure to enhance the total antimicrobial effect. The adjunct procedure is known as the full-body shower wash and may be used by patients at home before elective surgical procedures. Neither is there a long-acting, broad-spectrum topical antimicrobial product for use with patients requiring long-term, venous catheterization. These and many more creative new applications are waiting to be developed.

And finally, if new product evaluation methods are designed for a new product, these methods will very likely be used as the procedures for evaluation of future "me too" products. If a full-body shower product application significantly reduces the microbial flora residing on the skin, the following preoperative skin preparation procedure may be even more effective, since fewer microorganisms are left with which the preoperative formula must contend. Hence, a new standard surgical procedure may be implemented to employ both the full-body shower wash and the preoperative skin preparative procedures in elective surgery.

VIII. CONCLUSION

We have concentrated on a broad, integrative perspective to evaluating topical antimicrobial products in this chapter. We have discussed these evaluations from an experimental design perspective, a microbiological perspective, a biostatistical perspective, and a market perspective.

It is my belief and experience that committing to such a program significantly aids in evaluating products as well as the FDA approval process because of the straight-forward, concise, and unambiguous manner of conducting valid topical clinical trials.

REFERENCES

1. D. S. Paulson, The anti-infective market: Where is it going? *Soap Cosmet. Chem. Spec.* Mar. (1990).
2. D. S. Paulson, Evaluation of three microorganism recovery procedure used to determine handwash efficacy, *Sanitation 13*:520–523 (1993).

3. D. J. Campbell and J. C. Stanley, *Experimental and Quasi-experimental Designs for Research*, Houghton-Mifflin, Boston, 1963.
4. D. S. Paulson and J. R. Gillis, Glove juice studies: A statistical approach, *Soap Cosmet. Chem. Spec.* Aug. (1986).
5. ASTM, *Standard Test Method for Evaluation of Healthcare Personnel Handwash Formulation*, American Society for Testing Materials, 1987.
6. D. S. Paulson, Successfully marketing topical anti-infective products, *Soap Cosmet. Chem. Spec.* Feb. (1991).
7. D. S. Paulson, Efficacy evaluation of a 4% chlorhexidine gluconate as a full body shower wash, *Am. J. Infect. Control 21*:205–209 (1993).
8. D. S. Paulson, Designing a healthcare personnel handwash, *Soap Cosmet. Chem. Spec.* Aug. (1988).
9. D. S. Paulson, Evaluation of three handwash modalities commonly employed in the food processing industry, *Sanitation 12*:615–918 (1992).
10. D. S. Paulson, In-house market survey, Skyland Scientific Services, Bozeman, MT, 1988.
11. A. F. Peterson, Personal communication, Xttrium Laboratories, Chicago, IL, 1988.
12. R. Green, *Sampling Design and Statistical Methods for the Environmental Biologist*, Wiley, New York, 1979.
13. D. S. Paulson, Lessons from my entrepreneurial experience, *World Bus. Acad. Perspect. 7*(2): (1993).
14. R. McKenna, *The Regis Touch*, Addison–Wesley, Reading, MA, 1986.
15. T. Peters, *Thriving on Chaos*, Knopf, New York, 1988.
16. M. K. Bruch, Methods of testing antiseptics: Antimicrobials used topically in humans. *Disinfection, Sterilization, and Preservation* (S. S. Block, ed.), Lea & Febiger, Philadelphia, 1983.

3

Neutralizer Evaluations as Control Experiments for Antimicrobial Efficacy Tests

Scott V. W. Sutton
Alcon Laboratories, Inc.
Fort Worth, Texas

I. INTRODUCTION

The study of biocidal efficacy requires a distinction between the potential forms of antimicrobial effects. An antimicrobial agent can be either biocidal or biostatic. A disinfectant or preservative is said to be *biocidal* when exposure of the index microorganism to the biocide results in cell death. A *biostatic* agent prevents the growth of the organism under conditions that would normally allow growth. A biocidal test measures the efficacy of an antimicrobial agent by an apparent decrease in the number of viable microorganisms. The primary goal of a biocidal efficacy assay is the accurate determination of cell survival with time. This requires effective neutralization of the biocidal agent at the specific sampling time points [1–4].

A carefully designed biocidal test measures microbial kill, not biostasis.

This is done by measuring the numbers of organisms able to grow after exposure to the antimicrobial agent. Carryover of residual disinfectant from the test could inhibit growth in the recovery medium, leading to an overestimation of kill. It is necessary to demonstrate the adequacy of neutralization to establish the reliability of the biocidal data [1,4].

Three methods have recently been published detailing neutralizer evaluation protocols. These methods have different goals, and strengths. Dey and Engley describe a procedure, with *Staphylococcus aureus* as the index organism, that measures survival with time. The challenge organism is inoculated directly into the disinfecting solution, then sampled with time [5]. The efficacy of the neutralizer was measured by increased recovery of the challenge organism among different treatments. This protocol is useful in identifying neutralizers. However, it lacks the ability to distinguish between improved neutralization of the disinfectant and improved recovery of the index organism.

Terleckyj and Axler describe a second method that uses *Candida albicans* as the index organism [6]. This protocol was designed as a control procedure to demonstrate neutralization for a fungicidal experiment. The initial step of this method was a neutralization period. The biocide was exposed to the neutralizer before addition of the challenge microorganism. The challenge organism was then added at a concentration of approximately 10^6 colony-forming units (CFU)/ml to the suspension after this initial incubation. Survival of the challenge organism was assayed after an additional 15-min incubation. The basic design was first described in 1972 by Bergan and Lystad [7]. An assumption of these methods is that all index organisms will behave identically, and so only one organism needs to be tested. This is not a valid assumption, as different organisms will differ in their sensitivity to biocides, and it is this sensitivity that is of concern in a neutralizer evaluation study. In addition, the method of Terleckyj and Axler involves a centrifugation step after exposure to the biocide. The cells are resuspended before plating, diluting the biocidal agent. Further dilution occurs because of the high number of cells inoculated in the solution. This high concentration (10^6 CFU/ml) requires dilution for the determination of viable colony counts. These dilutions compromise the stringency of the procedure.

The final method was described by myself and colleagues [8]. This method is similar in overall design to that described by Bergan and Lystad. However, it employs a smaller inoculum size, statistical analysis, and evaluation of the neutralizer with all index organisms. This procedure lends itself to specific modifications of the general procedure that are included to mimic the different biocidal assays. Details of these methods are described in Section V.

One specific function of a neutralizer evaluation is to serve as a control experiment to the biocidal evaluation. Therefore, replication of all critical parameters of the biocidal experiment is critical to the neutralizer evaluation. The importance of this point cannot be overstated.

This review will examine the neutralizer evaluation solely as a control experiment of the biocidal study. However, it must be noted, in passing, that many of the same concerns apply to sterility testing of biocides, disinfectants, and preserved solutions [9].

II. METHODS OF NEUTRALIZING BIOCIDES

A. Neutralization by Chemical Inhibition

Many biocides can be chemically inactivated. Table 1 provides a listing of known neutralizers and the biocides affected. Several neutralizing and dilution broth media have been formulated to take advantage of these neutralizers. Among the more popular of these broths are Dey-Engley (D/E), Letheen, and thioglycolate. A full listing of the formulations for currently used neutralizing broths is provided in Table 2. It is important to remember that the efficacy of the compounds listed in Table 1 and of the broths listed in Table 2, were established at specific concentrations of neutralizer and biocide. Additionally, particular challenge organisms were employed to demonstrate efficacy. The efficacy of any neutralizer must be demonstrated for the specific test system in use.

The D/E broth was formulated to neutralize a wide range of biocidal agents (compare ingredients as listed in Table 2 with Table 1) [3,5]. The neutralizer efficacy of the formulation was originally demonstrated with *Staphylococcus aureus* [5], then later with a variety of microorganisms [6,8]. Letheen broth is effective in recovering bacteria exposed to quaternary ammonium compounds and biguanides [1,43]. The thioglycolate medium is effective against mercurials [1]. Other recommended neutralization (or dilution) broths include TAT [44], and AOAC Disinfectant Neutralization Solution [45].

One concern with chemical inactivation of biocides is that the neutralizer itself may be harmful to some of the bacteria of interest [1,2,46]. Table 3 lists bacteria sensitive to specific neutralizers.

Despite the concerns over neutralizer toxicity, the convenience and quick action of chemical neutralizers encourage their use. Chemical inactivation of biocides is the preferred method for most applications. Performing the proper neutralizer evaluation before the biocidal assay is critical to any demonstration of disinfection efficacy.

B. Neutralization by Dilution

The concentration of a disinfectant exerts a large effect on its potency. The relation between concentration and antimicrobial effect differs among biocidal agents, but is relatively constant for a particular biocide. This relation is exponential in nature, with the general formula:

$$C^{\eta}t = k$$

Table 1 Biocide Neutralizers

Biocide	Neutralizers/Inactivators	Ref.
Alcohols		
Isopropanol, phenoxyethanol	Polysorbate 80	10
	Dilution	11
Aldehydes		
2-Bromo-2-nitropropane-1,3-diol (bronopol)	Serum, cysteine, thiosulfate, thioglycolate, metabisulfite	12
Formaldehyde	Sodium sulfite, ammonia	1
	Histamine	13
Glutaraldehyde	Dilution	14
	Sodium bisulfite	15
	Sodium sulfite, glycine	2
	Cystine or cyteine	16
	Glycine	17
Chlorallytriazaazoniadamantane (Dowicil 200)	Dilution	18
Dimethylol dimethyl hydatoin (Glydant)	Dilution	19
Biguanides and bis-biguanides		
Chlorhexidine	Lecithin/polysorbate 80, 0.5% polysorbate 80	20
		21,22
Polyhexamethylene biguanide HCl (Cosmocil CQ)	Polysorbate 80/lecithin	4
Phenolics		
Phenylphenol, chloroxylenol, cresols, chlorocresols, phenol	Nonionic surfactants	1
	Polysorbate 80	7
	Dilution	23

Quaternary ammonium compounds		
Cetrimide, benzalkonium and benzethonium chloride	Lecithin/polysorbate	24
	Suramin sodium	25
	Organic material	26
	0.5% polysorbate 80	27
	Cyclodextrins	28
Mercurials	Sulfhydryl compounds	1
	Thioglycolic acid	26
	Thiosulfate, bisulfite	29
	Ammonium sulfite	30
Organic acids		
Benzoic, propionic sorbic	Nonionic surfactants	10
	Dilution	1
	pH 7 or above	24
Halogens		
Hypochlorite	Thiosulfate	31
	Dilution	32
Iodine	Thiosulfate	33
	Polysorbate 80	32
	Skim milk	34
EDTA	Mg^{+2} or Ca^{+2} ions	35
Imadazolidinyl urea	Dilution	19
Diazolidinyl urea (Germall 115 or II)		
Methyl-, and methylcholoroisothiazolinone (Kathon)	Amines, sulfites, mercaptans, sodium bisulfite	36
	Heparin	37
Parabens		
methyl-, ethyl-, propyl-, butyl-parahydroxybenzoic	Lecithin, filtration, dilution	1
	Polysorbate surfactants	38
	1% polysorbate 80 or 20	21,39–41
	Dilution	42

TABLE 2 Composition of Available Neutralizer Broths

Ingredient	D/E Broth	Letheen	TAT + Tween 20	NIH thioglycolate	AOAC disinfectant neutralizer solution
Cystine				0.5 g	
Polysorbate 80	5.0 g	5.0 g			28.0 ml
Polysorbate 20			40.0 ml		
Lecithin	7.0 g	0.7 g	5.0 g		4.0 g
Sodium thioglycolate	1.0 g			0.5 g	
Sodium thiosulfate	6.0 g				
Sodium bisulfite	2.5 g				
Tryptone	5.0 g		20.0 g		
Yeast extract	2.5 g			5.0 g	
Dextrose	10.0 g			5.5 g	
Peptamin		10.0 g			
Beef extract		5.0 g			
Sodium chloride		5.0 g		2.5 g	
Soytone					
Casitone				15.0 g	
KH_2PO_4					42.5 mg

TABLE 3 Potential Toxicity of Neutralizers

Inactivating agent	Disinfectant	Potential toxicity	Ref.
Sodium bisulphite	Glutaraldehyde	Nonsporing bacteria	14
Sodium thioglycolate	Mercurials	Bacteria and spores	47,48
Sodium thiosulfate	Iodine and chlorine	Staphylococci	49
Lecithin + Lubrol W	QACs	Bacteria	50
Glycine	Glutaraldehyde	Growing cells	51
Lubrol W	QACs	*Pseudomonas* spp.	4

where

C is the concentration
t is the time required to kill a standard inoculum
k is a constant
η is gradient of the plot of log t against log C

The letter η is described as the *concentration exponent*. Excellent reviews on this subject have been prepared by Cowles [52], Tilley [53], and Hugo [54]. These articles are recommended to the interested reader for more discussion.

Concentration exponents are determined experimentally. Two concentrations of biocide are tested in the same formulation, C_1 and C_2. Identical inocula are added to each, and the minimum "kill time" determine for each (t_1 and t_2): η is then determined by the equation

$$\eta = \frac{\log(t_2) - \log(t_1)}{\log(C_1) - \log(C_2)}$$

The concentration exponents for several biocides are provided in Table 4. Note that biocides with high η values are rapidly neutralized by dilution, whereas those with low η values are less dramatically effected.

It is important to demonstrate that dilution is sufficient for neutralization in the test system even with biocides of high η values. A high η value is no assurance of adequate neutralization without experimental data.

C. Neutralization by Filtration

Dilution of the biocide by filtration is another means to neutralize a disinfecting solution. This procedure relies on filtration to separate the microorganisms in

Table 4 Concentration Exponent of Common Biocides

Compound	η Value	Increased time factor when concentration reduced to 1/2	1/3
Phenolics	6	2^6	3^6
Alcohol	10	2^{10}	3^{10}
Parabens	2.5	$2^{2.5}$	$3^{2.5}$
Chlorhexidine	2	2^2	3^2
Mercury compounds	1	2	3
QACs	1	2	3
Formaldehyde	1	2	3

Source: Ref. 54.

suspension from the disinfecting solution. The filter is then removed, and placed on the surface of an agar plate for incubation. Nutrients leach up through the membrane, and discrete colonies arise on the surface of the filter, allowing quantification of survivors.

Filtration alone may not remove sufficient quantities of the biocidal agent to allow growth of surviving microorganisms. Growth inhibition may occur owing to adherence of residual preservative to the filter membrane [55–60]. Filtration through a low-binding filter material, such as polyvinylidene difluoride, helps lessen this adherence [9]. Additionally, the preservative may be diluted or flushed from the filter by rinsing with a benign fluid, such as 0.1% peptone [61]. Chemical neutralizers included in a rinse of the filter can be useful in assuring complete neutralization. Filtration alone cannot be assumed to be an effective means of neutralization. Effective recovery of survivors by this procedure requires demonstration of neutralization efficacy in the test system.

III. FACTORS AFFECTING NEUTRALIZATION OF ANTIMICROBIALS

There are four criteria to be met in designing a study of potential neutralizers [2]:

1. The neutralizer must effectively inhibit the action of the biocidal solution.
2. The neutralizer must not itself be unduly toxic to the challenge organisms.
3. The neutralizer and active agent must not combine to form a toxic compound.
4. These first three criteria must be demonstrated under conditions that mimic the actual conditions of the assay for disinfecting efficacy.

These criteria can be met by consideration of three factors. The first is *neutralizer efficacy*, or the ability of the neutralizer to inhibit the action of the biocidal agent against a specific microorganism. The second is *neutralizer toxicity*, or the inherent toxicity of the neutralizer for the organisms under study. These two criteria can be analyzed by the comparison of three populations [2,8, 9,62,63] (Fig. 1). The final factor addresses the adequate recovery of organisms sublethally injured by exposure to the biocidal agent.

A. Neutralizer Efficacy

The inherent efficacy of neutralizers used in biocidal experiments will vary with both the organism under study and the biocide. In addition, the sensitivity of the organism to that biocide and the concentration of the biocide to be neutralized can have an enormous effect on the efficacy of the neutralizing treatment. The efficacy of neutralization can be demonstrated by comparing recovery of organisms from

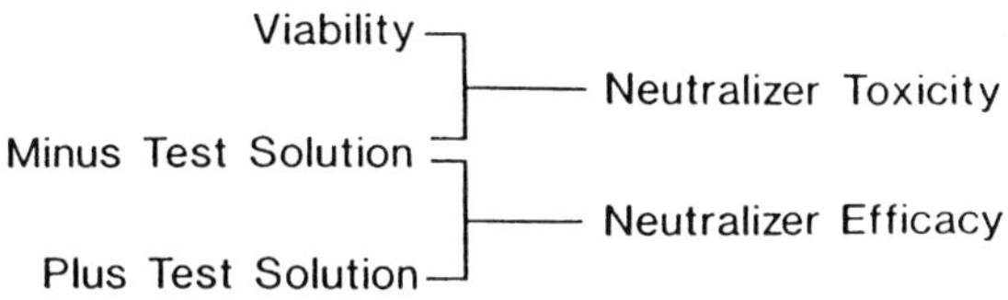

FIGURE 1 Population comparisons in neutralizer evaluation studies designed to control assays of antimicrobial efficacy. Neutralizer toxicity and neutralizer efficacy are defined as the similarity in the recovery between the two population pairs described.

two treatment populations. The first treatment consists of the neutralizer with the appropriate volume of biocide. This is the "plus-disinfectant" group. The second treatment consists of the neutralizer diluted with an equivalent volume of phosphate-buffered saline ("minus-disinfectant" group). Both populations are then inoculated with low numbers of microorganisms and assayed for survivors.

It is important to perform this experiment several times to allow statistical treatment of the recovery data. It is also important to compare the recovery of microorganisms from the neutralizer in the presence and the absence of the biocide. This comparison avoids confusing toxic effects of the neutralizer with inadequate inhibition of the biocidal agent (see following section).

B. Neutralizer Toxicity

Several chemical inhibitors of antimicrobials are themselves toxic (see Table 3). Care must be taken to avoid enhancing the apparent kill by an artifact of the recovery system. This factor can be estimated as a comparison of recovery between two populations, as was the factor of neutralizer efficacy. The two treatment groups are somewhat different. The first treatment consists of the minus-disinfectant group from the foregoing. The second treatment is the viability control and consists of an equivalent volume of phosphate-buffered saline (PBS) or other benign diluent. Both populations are then inoculated with low numbers of microorganisms and incubated for survivors. As in the previous section, it is important to perform this experiment several times to allow statistical treatment of the recovery data.

If the neutralizer is both effective and nontoxic, then all treatment groups will behave in identical manners. However, this is rarely so, and usually, some measure of neutralizer toxicity must be accepted to allow adequate neutralization.

C. Recovery of Injured Organisms

The recovery of injured organisms is not addressed by neutralizer evaluation studies. This limitation of a neutralizer evaluation study must be addressed at some point in preparation for the biocidal experiment. Although a sensitive assay for the neutralizing efficacy of a medium, a neutralizer evaluation study uses healthy cells that have not been exposed to disinfectants. This is not the situation in the actual experiment, where many viable cells will be damaged to varying degrees by exposure to the biocidal agent. The recovery of damaged microorganisms is extremely difficult to quantify [64,65]. One method used to estimate the number of crippled organisms in a biocidal study compares recovery in a rich nutrient medium with recovery in a stressful medium [66,67]. Trypticase soy agar medium is an example of a nutrient medium suitable for a variety of bacteria. Examples of stressful media might be medium supplemented with 5.5% KCl, or a medium with a low pH value. The fraction of the population unable to grow on the stressful medium is defined as the fraction of crippled organisms.

One way to demonstrate recovery of crippled organisms in a biocidal assay would be to compare the "kill curves" generated by plating on several different nutrient media. The optimal conditions for growth are those that give the least apparent kill [69]. Obviously, the neutralizer evaluation described earlier is a necessary prerequisite to this experiment.

IV. FACTORS AFFECTING NEUTRALIZER EVALUATIONS

The purpose of a neutralizer evaluation is to serve as a control experiment for the biocidal efficacy assay. Therefore, it is important to test the neutralizer under the conditions of the biocidal test. In general, any factor that affects the apparent disinfecting efficacy of a treatment must be a concern for the neutralizer evaluation. These factors include the nature of the challenge organism, pretreatment factors, treatment factors, and method of recovery [70,71].

A. Nature of the Challenge Organism

The nature of the challenge organism exerts a strong effect on the response to antimicrobial challenge. The *United States Pharmacopeia* (*USP*) [72] and American Society for Testing and Materials (ASTM) [73] preservative challenge tests use five different microorganisms. Represented among these index organisms for each test are gram-positive bacteria, gram-negative bacteria, yeast, and molds. These particular species serve as index organisms for the major prokaryotic and eukaryotic groups. Similarly, the manual of the Association of Official Analytic Chemists (AOAC) uses different tests for specific biocidal claims [45]. In other words, antistaphylococcal activity is a specific claim and is not taken as proof of tuberculocidal activity. These differences must be taken into account in the design of neutralizer evaluation studies.

Differences among the index organisms may well influence the efficacy and toxicity of the neutralizer, as discussed earlier. All microorganisms to be used as index organisms in the biocidal assay must be tested in the neutralizer evaluation. There are two reasons for this concern. First, many commonly used neutralizers are toxic to certain species (see Table 3). Second, the differing efficacy of the biocide among challenge organisms will require differing levels of neutralization. These differences among organisms can be clearly seen in differing responses to neutralization.

B. Pretreatment Factors

The growth and preparation of the challenge organism determines the physiological state of the cell. This state has a direct influence on the results of any assay of disinfecting efficacy. The conditions of organism preparation and storage must be standardized for the neutralizer evaluation and reflect the conditions of the antimicrobial assay.

Biocidal tests do not use individual cells. Rather, populations of cells are harvested for study. The data generated from these studies is less variable if the cell populations are homogeneous. Liquid cultures or confluent growths on solid medium are best suited for the reproducible preparation cultures [74].

C. Treatment Factors

All factors of the biocidal test must be duplicated in the neutralizer evaluations. The constituents of the suspension can have a dramatic effect on the efficacy of the test solution; organic load decreases the efficacy of oxidative antimicrobials [75]. Additionally, the manner and storage of organic load can affect the efficacy of antimicrobials. Even factors such as the water used to make up the solutions need to be standardized.

The concentration of the biocidal agent exerts a logarithmic effect on the efficacy of the formulation. This effect, known as the concentration exponent, was discussed earlier. A biocidal evaluation may involve plating 1 ml or 10^{-1}–10^{-5} dilutions onto an agar recovery medium. The neutralizer evaluation should test this recovery under the most stringent conditions to be seen. This is reflected by the 10^{-1} dilution, requiring growth in the highest concentration of biocide.

D. Posttreatment Factors

Two posttreatment factors influence neutralizer evaluations. The first is the recovery medium used to support the growth of survivors. This concern was discussed in the foregoing. The second consideration is the incubation conditions. Optimal conditions for growth must be employed to ensure complete growth and reproducible results.

V. NEUTRALIZER EVALUATION DESIGNS

Three methods to estimate the survivors of exposure to a disinfectant will be discussed. These are differentiated by recovery media; recovery on agar plates, recovery in liquid medium, and the use of membrane filtration as a means to concentrate survivors for enumeration on agar plates.

A. Recovery on Solid Medium

Suspension tests of antimicrobial efficacy usually quantify surviving microorganisms on solid (agar) medium. Challenge organisms are physically suspended in the test solution, and aliquots are removed with time to determine the number of surviving microorganisms by their recovery on agar plates.

The general procedure for these tests involves preparing a suspension of challenge organism in the test solution, usually at a concentration of approximately 10^6 CFU/ml. The samples are removed with time and diluted tenfold in a suitable neutralizing and dilution broth, typically Dey-Engley broth. This step should accomplish the neutralization of the biocide. The broth dilution is continued in series to allow appropriate dilution of the sample for plating. Several dilutions are plated in a recovery agar to quantify the number of viable organisms in the original sample.

Figure 2 describes one method for the evaluation of neutralizers in this system. The plus-disinfectant population for each test organism is constructed by adding 1 ml of biocide to 9 ml of neutralizing medium. These mixtures are incubated on the benchtop for approximately 5 min, then inoculated with challenge organism. The inoculated suspensions are incubated for approximately 10

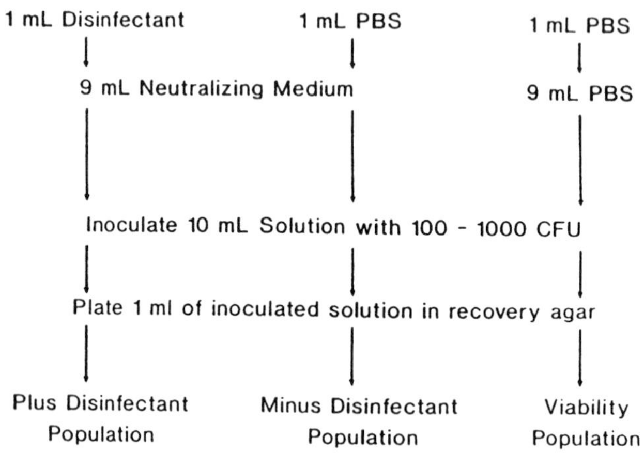

FIGURE 2 Diagram of a design for a neutralizer evaluation study to control a quantitative assay of antimicrobial efficacy. The important considerations in this design are the ratio of disinfectant/neutralizer-broth, recovery medium, and index organisms. This design can be used to validate a protocol for a test of preservative efficacy.

min, and plated in replicate. The minus-disinfectant population is treated in a similar fashion, substituting PBS for the disinfectant. Several different recovery media may be employed to ensure maximal recovery.

This technique has the advantage of allowing the separation of various critical operations in recovery. These operations include biocide neutralization, dilution of survivors to countable levels, and plating survivors for recovery. This separation is important, as optimal recovery of survivors may require neutralization in one medium, and plating for recovery on a second. Comparisons among the recoveries in each population can be statistically analyzed as described in Section V.D.

This procedure uses low numbers (30–100 CFU) of index organisms. The use of a low inoculum provides two advantages. First, the use of low numbers increases the effect of low levels of biocides, as a small reduction in measured CFU takes on increased significance. Second, these low numbers allow direct plating of solutions, avoiding dilution of the disinfectant during plating.

B. Recovery in Liquid Medium

Two common disinfection efficacy tests recover surviving organisms in liquid medium. The multi-item microbial challenge test is a carrier assay of contact lens disinfection regimens [76]. The use-dilution test is a carrier assay of surface disinfectants [45,77]. Both tests involve inoculating a carrier, then assaying the carrier for viable microorganisms after disinfection in liquid medium. The assumption made in these tests is that the recovery medium permits growth of all surviving microorganisms at the end of the disinfection period. The final determination of efficacy is based on the absence of growth in liquid culture. Therefore, this broth must serve both to neutralize the disinfectant and to support the growth of all index organisms. The quality and properties of the recovery medium are of critical importance to the accuracy of the test.

A method to examine neutralizer toxicity and efficacy for this type of test is shown in Figure 3. A small inoculum (10–100 CFU) in each of the three different media configurations (populations) is incubated under the proper conditions. Comparisons of growth between the different media configurations provide an appropriate measure of both neutralizer efficacy and neutralizer toxicity.

The viability population consists of an inoculum in a rich broth medium, establishing the viability and growth characteristics of the particular challenge organism. The minus-disinfectant population consists of organisms in the neutralizing–recovery broth diluted with phosphate buffered saline at the same dilution ratio as in the plus-disinfectant population. This must also be no less than the concentration of biocide seen in the final use-dilution or multi-item microbial challenge test. Comparison between growth in the rich broth and growth in the neutralizing–recovery broth medium without the disinfectant allows the deter-

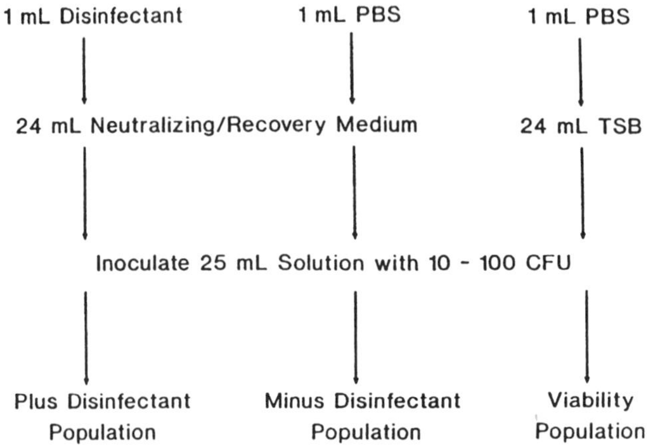

FIGURE 3 Diagram of a design for a neutralizer evaluation study to control a qualitative assay of antimicrobial efficacy. The important considerations in this design are the broth nature of the recovery medium, requiring a broth suitable for both neutralization and recovery, the ratio of disinfectant/neutralizer broth, and the index organisms.

mination of the neutralizer toxicity. The comparison of growth with, and without, the disinfecting solution determines neutralizer efficacy.

C. Recovery by Membrane Filtration

A neutralizer evaluation for a membrane filtration protocol requires the same comparisons and involves concerns similar to those previously described. Considerations in the neutralization evaluation include the test system, the membrane type, the filtration apparatus, the medium, the diluting (neutralization) fluid, and the solution being tested. Additionally, an estimate of the loss of cells through the mechanics of filtration should be determined [9,78–80].

One protocol of neutralizer evaluation, as patterned on the *USP* sterility test, is described in Figure 4 [9]. This procedure requires filtration of the determined amount of biocidal solution, followed by two, 100-ml volumes of diluting–neutralizing fluid. A third 100-ml volume, inoculated with 10–100 CFU of the index organism is then passed through the filter. Finally, the inoculated filter is placed on the surface of a freshly poured agar plate and incubated for growth.

Neutralizer toxicity is evaluated by comparison between the recovery of the membrane without the test solution and those filtered with *USP* diluting fluid A (DFA–0.1% meat peptone) [61]. Neutralizer efficacy is estimated by comparison between the recovery on the membrane both with and without the test solution. Filtration may lead to reduced recovery of organisms through death or adherence

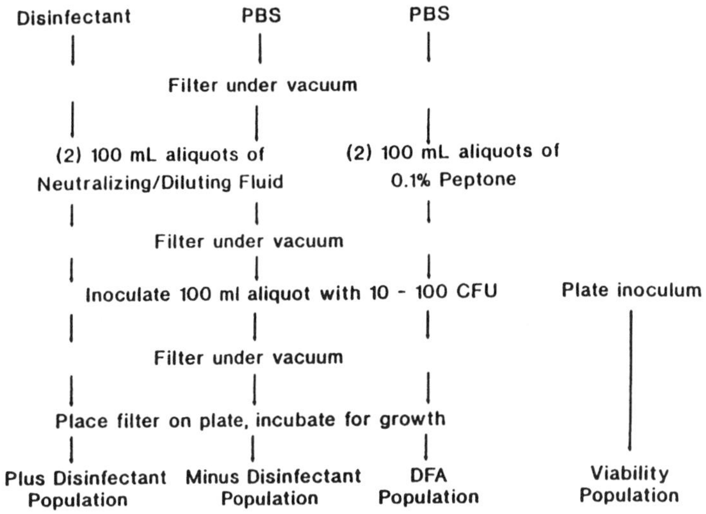

FIGURE 4 Diagram of a design for a neutralizer evaluation study to control an assay involving membrane filtration of an antimicrobial. The important considerations in this design are the amount of antimicrobial to be filtered, the amount of neutralizing/diluting fluid used to wash the membrane, the nature of the membrane filter material, the recovery medium, and the index organisms. This design can be used to validate a protocol for sterility testing, or an assay of antimicrobial efficacy.

to the filtration vessel walls. This technique-specific loss can be estimated by comparing recovery in the DFA population with the viable count. Membrane filtration allows for two methods to enhance neutralizer efficacy. The first is to incorporate a specific neutralizer in the diluting fluid to overcome the effect of the antimicrobial agent. The second is to test several different types of filters to find a filter type that is effective in the specific system. Several filter types are now available with different surface properties. These different properties will affect the relative binding of the active agent to the membrane.

D. Statistical Analysis

Assays such as the use-dilution test and the multi-item microbial challenge test require recovery in liquid media. These data are nominal (growth or no growth) and so can be analyzed by a χ^2 test [81], the sign test [82], or McNemar's test [83; reviewed in Ref. 84].

A second type of data is provided by tests measuring recovery as colony-forming units on agar plates. These natural data (CFU recovered) follow a Poisson distribution. They can be transformed to approximate a normal distribution either by taking the $\log_{10}$ value of each datum, or by the modified square root transformation of Anscombe [85]. These transformed data can be analyzed in several ways. Student's *t*-test can be used for simple pair-wise comparisons in this analysis. However, if several different neutralizers and recovery media are being evaluated, it would be more appropriate to analyze the data initially by analysis of variance (ANOVA). If significant differences were indicated by this analysis, further information on specific differences could be determined by Dunnett's test using the minus-disinfectant population as the control [81].

ACKNOWLEDGMENTS

The author thanks David W. Proud and Dr. Daniel Brannan for their contributions of time and ideas to this manuscript.

REFERENCES

1. A. D. Russell, I. Ahonkhai, and D. T. Rogers, Microbiological applications of the inactivation of antibiotics and other antimicrobial agents, *J. Appl. Bacteriol. 46*:207–245 (1979).
2. A. D. Russell, Neutralization procedures in the evaluation of bactericidal activity, *Disinfectants: Their Use and Evaluation of Effectiveness* (C. H. Collins, M. C. Allwood, S. F. Bloomfield, and A. Fox, eds.), Academic Press, London, 1981, pp. 45–59.
3. F. B. Engley and B. P. Dey, A universal neutralizing medium for antimicrobial chemicals, Proceedings of the 56th Mid-year Meeting of the Chemical Specialties Manufacturers Association, New York, 1970, pp. 100–106.
4. I. H. MacKinnon, The use of inactivators in the evaluation of disinfectants, *J. Hyg. (Camb.) 73*:189–195 (1974).
5. B. P. Dey and F. B. Engley, Methodology for recovery of chemically treated *Staphylococcus aureus* with neutralizing medium, *Appl. Environ. Microbiol. 45*:1533–1537 (1983).
6. B. Terleckyj and D. A. Axler, Quantitative neutralization assay of fungicidal activity of disinfectants, *Antimicrob. Agents Chemother. 31*:794–798 (1987).
7. T. Bergan and A. Lystad, Evaluation of disinfectant inactivators, *Acta Pathol. Microbiol. Scand. (B) 80*:507–510 (1972).
8. S. V. W. Sutton, T. Wrzosek, and D. W. Proud, Neutralization efficacy of Dey-Engley medium in testing of contact lens disinfecting solutions, *J. Appl. Bacteriol. 70*:351–354 (1991).
9. D. W. Proud and S. V. W. Sutton, Development of a universal diluting fluid for membrane filtration sterility testing, *Appl. Environ. Microbiol. 58*:1035–1038 (1992).
10. H. B. Kostenbauder, Physical factors influencing the activity of antimicrobial agents, *Disinfection, Sterilization and Preservation*, 2nd ed. (S. S. Block, ed.), Lea & Febiger, Philadelphia, 1977, pp. 912–932.
11. H. E. Morton, The relationship of concentration and germicidal efficiency of ethyl alcohol, *Ann. N.Y. Acad. Sci. 53*:191–196 (1950).

12. B. Croshaw, M. J. Groves, and B. Lessel, Some properties of bronopol, a new antimicrobial agent active against *Pseudomonas aeruginosa*, *J. Pharm. Pharmacol. 16*(Suppl.):127T–130T (1964).
13. T. Nash and A. Hirch, The revival of formaldehyde-treated bacteria, *J. Appl. Chem. 4*:58–461 (1954).
14. T. J. Munton and A. D. Russell, Aspects of the action of glutaraldehyde on *Escherichia coli*, *J. Appl. Bacteriol. 33*:410–419 (1970).
15. A. D. Russell and D. Hopwood, The biological uses and importance of glutaraldehyde, *Progress in Medicinal Chemistry*, Vol. 13 (G. P. Ellis and G. B. West, eds.), North Holland Publishers, Amsterdam, 1976, pp. 271–301.
16. S. P. Gorman and E. M. Scott, Evaluation of potential inactivators of glutaraldehyde in disinfection studies with *Escherichia coli*, *Microbiol. Lett. 1*:197–204 (1976).
17. S. P. Gorman and E. M. Scott, A quantitative evaluation of the antifungal properties of glutaraldehyde, *J. Appl. Bacteriol. 43*:83–89 (1977).
18. S. C. Marouchoc, Dowicil 200 preservative, *Cosmetic and Drug Preservation* (J. J. Kabara, ed.), Marcel Dekker, New York, 1984, pp. 143–164.
19. M. Rosen, Glydant and MDMH as cosmetic preservatives, *Cosmetic and Drug Preservation* (J. J. Kabara, ed.), Marcel Dekker, New York, 1984, pp. 165–190.
20. E. D. Weinberg, The antimicrobial activity of N^1,N^5-substituted biguanides, *Antibiot. Chemother. 11*:572–582 (1961).
21. M. R. W. Brown, Turbidimetric method for the rapid evaluation of antimicrobial agents. Inactivation of preservatives by non-ionic agents, *J. Soc. Cosmet. Chem. 17*:185–195 (1966).
22. M. R. W. Brown and R. M. E. Richards, The effects of Tween 80 on the resistance of *Pseudomonas aeruginosa* to chemical inactivation, *J. Pharm. Pharmacol. 16*:51–55 (1964).
23. H. S. Bean, Types and characteristics of disinfectants, *J. Appl. Bacteriol. 30*:6–16 (1967).
24. A. D. Russell, Factors influencing the efficacy of antimicrobial agents, *Principles and Practices of Disinfection, Preservation and Sterilization* (A. D. Russell, W. B. Hugo, and G. A. J. Ayliffe, eds.), Blackwell Scientific Publications, Oxford, 1982.
25. C. A. Lawrence, Inactivation of the germicidal action of quaternary ammonium compounds, *J. Am. Pharm. Assoc. (Sci. Ed.) 37*:57–61 (1948).
26. B. Croshaw, Evaluation of non-antibiotic antimicrobial agents, *Pharmaceutical Microbiology* (W. B. Hugo and A. D. Russell, eds.), Blackwell Scientific Publications, Oxford, 1983, pp. 185–201.
27. M. R. W. Brown and R. M. E. Richards, Effect of polysorbate (Tween) 80 on the resistance of *Pseudomonas aeruginosa* to chemical inactivation, *J. Pharmacol. 16*:511–551 (1964).
28. W. J. Simpson, Neutralization of the antibacterial action of quaternary ammonium compounds with cyclodextrins, *FEMS Microbiol. Lett. 90*:197–200 (1992).
29. C. B. Cox, J. C. Feeley, and M. Pittman, Sterility testing: Detection of fungi and yeasts in the presence of preservatives, *J. Biol. Stand. 1*:11–19 (1973).
30. J. Geppert, Zur lehre von den Antiseptics, *Berl. Klin. Wochenschr. 26*:7789 (1989).
31. G. R. Weber and M. Levine, Factors effecting germicidal efficiency of chlorine and chloramine, *Am. J. Public Health 34*:719–728 (1944).

32. C. L. Kelsey, I. H. MacKinnon, and I. M. Mauer, Sporicidal activity of hospital disinfectant, *J. Clin. Pathol. 27*:632–638 (1974).
33. W. B. Hugo and J. M. Newton, The adsorption of iodine from solution by microorganisms and by serum, *J. Pharm. Pharmacol. 16*:49–55 (1964).
34. C. A. Lawrence, C. M. Carpenter, and A. W. C. Naylor-Foote, Iodophors as disinfectants, *J. Am. Pharm. Assoc. (Sci. Ed.) 46*:500–505 (1957).
35. J. Dankert and I. K. Schut, The antibacterial activity of chlorzylonol in combination with EDTA, *J. Hyg. (Camb.) 76*:11–22 (1976).
36. A. B. Law, J. N. Moss, and E. S. Lashen, Kathon CG: A new single-component, broad spectrum preservative system for cosmetics and toiletries, *Cosmetic and Drug Preservation* (J. J. Kabara, ed.), Marcel Dekker, New York, 1984, pp. 129–142.
37. J. Favet, W. Griffiths, P. A. Amacker, and E. Schorer, Utilisation de la cinetique de l'effet bactericide du Kathon CG sur de germes connus, afin d'en determiner l'efficacite dans un milieu aquex et dans une creme hydrophile a l'heparine, *Pharm. Acta Helv. 60*:83–91 (1985).
38. R. L. Abshire and B. A. Schelech, A method for evaluating the effectiveness of preservative systems in water-immisicible ointments, *J. Parenter. Sci. Technol. 36*:216–221 (1982).
39. G. W. Miyawaki, N. K. Patel, and H. B. Kostenbauder, Interaction of preservatives with macromolecules. II, *J. Am. Pharm. Assoc. (Sci. Ed.) 48*:315–318 (1959).
40. N. K. Patel and H. B. Kostenbauder, Interaction of preservatives with macromolecules. I, *J. Am. Pharm. Assoc. (Sci. Ed.) 47*:289–293 (1959).
41. F. D. Pisano and H. B. Kostenbauder, Interaction of preservatives with macromolecules. II, *J. Am. Pharm. Assoc. (Sci. Ed.) 48*:310–314 (1959).
42. M. Lang and R. M. Rye, A correlation between the antibacterial activity of some *p*-hydroxybenzoate esters and their cellular uptake, *Microbios 7*:199–207 (1973).
43. R. A. Quisno, I. W. Gibby, and M. J. Foter, A neutralizing medium for evaluating the germicidal potency of the quaternary ammonium salts, *Am. J. Pharm. 118*:320–323 (1946).
44. U. S. Food and Drug Administration Center for Food Safety and Applied Nutrition, *Bacteriological Analytical Manual*, 5th ed., Association of Analytical Chemists, Washington, DC, 1978, pp. XXIII-6–XXIII-7.
45. A. Beloian, Disinfectants, *Official Methods of Analysis AOAC*, 15th ed., Association of Official Analytic Chemists, Arlington, VA, 1990, pp. 133–146.
46. G. Reybrouck, Bactericidal activity of 40 potential disinfectant inactivators, *Zentralbl. Bakteriol. Hyg. (B) 167*:528–534 (1978).
47. D. A. A. Mossel and H. Beerens, Studies on the inhibitory properties of sodium thioglycollate on the germination of wet spores of clostridia, *J. Hyg. (Camb.) 66*:269–272 (1968).
48. H. R. Hibbert and R. Spencer, An investigation of the inhibitory properties of sodium thioglycollate in media for the recovery of bacterial spores, *J. Hyg. (Camb.) 68*:131–135 (1970).
49. A. Kayser and G. van der Ploeg, Growth inhibition of staphylococci by sodium thioglycollate, *J. Appl. Bacteriol. 28*:286–293 (1965).
50. H. H. Laycock and B. A. Mulley, Application of the Ferguson principle to the

antibacterial activity of mono- and multi-component solutions of quaternary ammonium surface-active agents, *J. Pharm. Pharmacol. 22*:157–62 (1970).

51. C. Lark and K. G. Lark, The effects of D-amino acids on *Alcaligenes faecalis*, *Can. J. Microbiol. 5*:369–379 (1959).
52. P. B. Cowles, The disinfectant concentration exponent, *Yale J. Biol. Med. 12*:697–704 (1939).
53. F. W. Tilley, An experimental study of the relation between concentration of disinfectants and time required for disinfection, *J. Bacteriol. 38*:499–510 (1939).
54. W. B. Hugo and S. P. Denyer, The concentration exponent of disinfectants and preservatives (biocides), *Preservatives in the Food, Pharmaceutical and Environmental Industries* (R. G. Board, M. C. Allwood, and J. G. Banks, eds.), Blackwell Scientific Publications, Boston, 1987, pp. 281–291.
55. J. M. Prickett and D. D. Rawal, Membrane filtration method technique for testing disinfectants, *Lab. Pract. 21*:425–428 (1972).
56. J. Prince, C. E. A. Deverill, and G. A. J. Ayliffe, A membrane filter technique for testing disinfectants, *J. Clin. Pathol. 28*:71–76 (1975).
57. D. D. Rawal, Microbiological evaluation of disinfectants by membrane filtration capacity test, *Microbios Lett. 3*:213–215 (1976).
58. C. O. Chiori, R. Hambleton, and G. J. Rigby, The inhibition of spores of *Bacillus subtilis* by centrimede retained on washed membrane filters and on the washed spores, *J. Appl. Bacteriol. 28*:322–330 (1965).
59. N. D. Harris and J. P. Richards, The failure of phenol treated *Escherichia coli* to grow on membrane filters, *J. Appl. Bacteriol. 21*:86–93 (1958).
60. M. Van Ooteghem and H. Herbots, The adsorption of preservatives on membrane filters, *Pharm. Acta Helv. 44*:610–619 (1969).
61. *United States Pharmacopeia*, Microbiological tests, *USP XXII/NSF XVII*. United States Pharmacopeial Convention, Rockville, MD, 1990, pp. 1478–1488.
62. American Society for Testing Materials (ASTM), Standard practices for evaluating inactivators of antimicrobial agents used in disinfectant, sanitizers, antiseptic, or preserved products, *Am. Soc. Testing Mat.* E. 1054-91 (1991).
63. S. V. W. Sutton, The importance of neutralizer evaluations in the microbicidal testing of preservatives and disinfecting solutions, *Int. Contact Lens Clinics 19*:167–173 (1992).
64. B. Ray, Methods to detect stressed microorganisms. *J. Food Protect. 42*:346–355 (1979).
65. P. Gilbert, The revival of micro-organisms sublethally injured by chemical inhibitors, *The Revival of Injured Microbes* (M. H. E. Andrew, and A. D. Russell, eds.), Academic Press, London, 1984, pp. 175–197.
66. W. B. Hugo, Survival of microbes exposed to chemical stress, 26th Symposium of the Society for General Microbiology (T. G. R. Gray and J. R. Postgate, eds.), Cambridge University Press, Cambridge, 1976, pp. 383–413.
67. D. A. A. Mossel and J. E. L. Corry, Detection and enumeration of sublethally injured pathogenic and index bacteria in foods and water processed for safety, *Alimenta 16* (Special Issue on Microbiology):19–34 (1977).
68. D. A. A. Mossel and P. Van Netten, Harmful effects of selective media on stressed

micro-organisms: Nature and remedies, *The Revival of Injured Microbes* (M. H. E. Andrew, and A. D. Russell, eds.), Academic Press, London, 1984, pp. 329–369.
69. A. A. El-Banna and A. Hurst, Survival in foods of *Staphylococcus aureus* grown under optimal and stressed conditions and the effect of some food preservatives, *Can. J. Microbiol. 29*:297–302 (1983).
70. A. D. Russell, Factors influencing the activity of antimicrobial agents, *Microbios 10*:151–174 (1974).
71. S. F. Bloomfield, Methods for assessing antimicrobial activity, *Mechanisms of Action of Chemical Biocides: Their Study and Exploitation* (S. P. Denyer, and W. B. Hugo, eds.), Blackwell Scientific Publications, Boston, 1991, pp. 1–22.
72. USP ⟨51⟩ Antimicrobial Preservatives—Effectiveness, United States Pharmacopeia/ The National Formulary (USP), United States Pharmacopeial Convention, Inc., 1995, pp 1681.
73. American Society for Testing Materials (ASTM), PE testing, *Am. Soc. Testing Mat.* E 640-78, 1978.
74. P. Gilbert, M. R. W. Brown, and J. W. Costerton, Inocula for antimicrobial sensitivity testing: A critical review, *J. Antimicrob. Chemother. 20*:147–154 (1987).
75. H. Chick and C. J. Martin, The principles involved in the standardization of disinfectants and the influence of organic matter upon germicidal value, *J. Hyg. (Camb.) 8*:654–697 (1908).
76. Food and Drug Administration, Draft testing guidelines for class III soft (hydrophilic) contact lens solutions, FDA, Washington, DC, July 15, 1985.
77. A. Cremieux and J. Fleurette, Methods of testing disinfectants, *Disinfection, Sterilization and Preservation*, 3rd ed. (S. S. Block, ed.), Lea & Febiger, Philadelphia, 1983.
78. J. P. Calabrase and G. K. Bissonnette, Improved membrane filtration method incorporation catalase and sodium pyruvate for detection of chlorine-stressed coliform bacteria, *Appl. Environ. Microbiol. 56*:3558–3564 (1990).
79. J. P. Calabrese and G. K. Bissonnette, Improved detection of acid mine water stressed coliform bacteria on media containing catalase and sodium pyruvate, *Can. J. Microbiol. 36*:544–550 (1990).
80. A. W. Hoadley, Effect of injury on the recovery of bacteria on membrane filters, *Membrane Filtration: Application, Techniques, and Problems* (B. J. Dutka, ed.), Marcel Dekker, New York, 1981, pp. 413–450.
81. J. H. Zar, *Biostatistical Analysis*, Prentice-Hall, Englewood Cliffs, NJ (1984).
82. J. Arbuthnott, An argument for Divine Providence taken from the constant regularity in the births of both sexes, *Philo. Trans. 27*:186–190 (1710).
83. Q. McNemar, Note on the sampling error of the difference between correlated proportions or percentages, *Psychometrica 12*:153–157 (1947).
84. D. M. Ilstrup, Statistical methods in microbiology, *Clin. Microbiol. Rev. 3*:219–226 (1990).
85. F. J. Anscombe, The transformation of Poisson, binomial, and negative binomial data, *Biometrika 35*:246–254 (1948).

4

Antiseptics and Their Role in Infection Control

Joseph M. Ascenzi
Johnson & Johnson Medical, Inc.
Arlington, Texas

I. INTRODUCTION

The positive influence of antiseptics in the control of nosocomial infections has long been debated. Studies that definitely show antiseptics do aid in the control of hospital-acquired infections are almost nonexistent. Because infection rates are often low to begin with, such studies often require large numbers of patients and, as such, are (a) difficult to control, (b) must be carried out over a long time period, and (c) are expensive. In addition to these factors, since antisepsis is a routine procedure, it may not be ethical to withhold these procedures from patients as a control group. There are several in vitro and clinical trials comparing the various products that show the positive influence of antiseptics and, by deduction, the information indicates that the proper use of antiseptics in a variety of situations can be helpful in controlling cross-infection of patients by health care personnel. The first demonstration of the link between handwashing and reductions in postoperative infections was made by Semmelweis [1]. It is still believed that most hospital-acquired infections are transmitted by health care workers who fail to wash their hands adequately or with the required frequency [2]. However, the link between handwashing and nosocomial infections continues to be debated [3,4]. Larson [3] argues that the effect of handwashing is direct and not simply due to an increase in general cleanliness among those who wash their hands, nor is it because of increased health status of the host so that they are more likely to resist infection, as some have speculated. A significant correlation between frequency of handwashing and the presence of harmful microorganisms has been made [5].

II. SKIN ANTISEPSIS

Skin cells (squames), through shedding, are the main unit for the transfer of microorganisms. Baquero et al. [6] describe mechanical, electrostatic, and irritative effects of handwashing and the interaction of antimicrobials in this process. In the absence of antimicrobials, the mechanical agitation associated with handwashing may have negative consequences. It results in the release of skin squames colonized with microorganisms, clumps of bacteria, and isolated bacteria. This mechanical action results in exposure of the new noncolonized epithelial cells that then become colonized, continuing the process. Maki [7] states that there is a certain frequency beyond which more washing may not be helpful because of the shedding of bacteria and skin squames with associated bacteria.

The use of soaps and soaps containing antiseptics may modify the electrostatic interaction of bacteria and skin, which is important in the attachment of the predominantly negatively charged bacteria to the skin surface. It has also been proposed that the more damage to the skin from detergents and chemicals, the less shedding occurs, since with severe trauma to the skin, the stratum corneum can be so severely damaged that there are less squames available for shedding [8].

Probably the most overlooked property of continued handwashing on the microbiology of the skin is the irritative effect; that is, the effect of the handwashing agent on the integrity of the skin. Irritation causes ecological changes in the skin (i.e., changes in temperature, composition of skin fluids, and cell turnover). These changes, in turn, may affect the bacterial flora of the skin, leading to colonization by specific types of organisms. Maki [7] found an increase in the number of gram-negative bacilli.

Weather may also play an important role in the microflora of the skin. Seasonal changes in the microflora have been noted, with a decrease in the proportion of cocci and an increase in the gram-negative bacteria and fungi on hands in summer, as compared with winter. Cold weather is also associated with an increase in the degree of shedding, presumably owing to colder, drier conditions.

These combined nonspecific effects could encourage the spread of infection by the shedding of skin squames and bacteria [6]. The number of clustered infections that can be related to iatrogenic causes range from 2% to as much as 15.5% [9–11], suggesting that sometimes when the rate is low, the cause may not be related to personnel, whereas when the rate is high, the cause may be related to health care personnel and their inattention to protocol. Studies in intensive care units indicate that the rate of compliance with handwashing is less than 50% [4,12,13]. Compliance has increased when the staff is under observation, undergoes education, or is given feedback [3,15]. In a controlled study on handwashing in an intensive care unit, there was little evidence that infection rates were unrelated to handwashing [4]; however, the study was of relatively short duration and included only nurses, not physicians. These two parameters limit the power of the conclusions that can be drawn from the data. Maki [7] was able to demonstrate reduced infection rates with handwashing; however, it was suggested that the agent used for handwashing might be as important as the frequency. Table 1 summarizes the study design and results of studies from those of Semmelweis to current ones. In addition to those listed, a study by Frappier-Davignon and associates [16] demonstrated a significant difference in staphylococcal infections among isolated infants in a nursery that had a program of using medicated soap (hexachlorophene), as opposed to a nursery in which no program existed. However, in nurseries in which the infants were not isolated, the difference in infection rates between nurseries with and without handwashing programs was not as obvious.

III. CLINICAL USES FOR TOPICAL ANTIMICROBIALS

A. Surgical Handscrubs

Surgical handscrubs are used with the intention of reducing both the resident microflora and the transient pathogens, with the goal of preventing the transfer of bacteria from the surgeon to the operation site and to ensure that even if the surgeon's gloves are torn, the transfer of bacteria to the surgical site will be minimal [17]. Most persons agree that surgical scrubs should be an antimicrobial possessing the properties of broad-spectrum, rapid activity and, under some circumstances, residual activity [18,19]. Because of the effects from irritation, as noted earlier, the scrubs should also be as nonirritating as possible, yet have the desired antimicrobial properties. Surgical handscrubs have generally been evaluated using the glove-juice method, as described in Chapter 2. Recent findings have suggested that the primary source of bacteria recovered in the glove juice protocol are from the subungual spaces [20,21]. For a surgical handscrub to be effective, it must be capable of penetrating this space and eradicating the microflora. Techniques that have been developed to determine the effect of handscrub solutions without the contribution of the subungual space have shown that these several solutions were not effective in eradicating the subungual bacteria. Hands on which

Table 1 Prospective Studies of the Influence of Handwashing on Infections

			Elements of experimental design				
Year published	Investigator	Setting	Intervention?	Comparison group?	Randomization?	Blinding?	Significant results
1861	Semmelweis	Maternity Hosp. (Vienna)	Yes	Sequential	No	No	Reduced mortality from puerperal fever
1977	Casewell	Critical care unit (Great Britain)	Yes	Sequential	No	No	Reduced nosocomial infection rates caused by endemic *Klebsiella* strains
1981	Black	Day care centers (U.S.)	Yes	Concurrent	No	No	Reduced incidence of diarrhea
1982	Khan	Village (Bangladesh)	Yes	Concurrent	Yes	No	Reduced incidence of shigellosis
1982	Maki	Critical care unit (U.S.)	Yes	Crossover design	No	No	Reduced incidence of nosocomial infections
1984	Massanari	Critical care unit (U.S.)	Yes	Crossover design	No	No	Reduced incidence of nosocomial infections
Unpublished	Shahid	Village (Bangladesh)	Yes	Concurrent	No	No	Reduced incidence of diarrhea

Source: Ref. 3.

the subungual spaces were sealed with cyanoacrylate glue and scrubbed with either a 4% chlorhexidine gluconate (CHG) solution or a 10% polyvinyl pyrilidone–iodine (PVP-I) solution showed a 4-$\log_{10}$ reduction in recovered organisms, whereas when the same study was performed on hands without sealed subungual spaces, there was less than a 1-$\log_{10}$ reduction [21]. A comparison of surgical wound infection rates between surgeons who scrubbed with PVP-I and those who scrubbed with hexachlorophene showed no difference in the infection rate [22].

B. Health Care Personnel Handwashing

Health care personnel handwash solutions are those intended for use between patient contact. It is well documented that hospital patients become colonized with gram-negative bacteria [23–25] and also that most hands of health care workers are major routes for transmission of infection [26]. The purpose of these solutions used for in-between patient contact is to kill the pathogenic microorganisms that have been transferred to hands of health care practitioners, from either patients or the surrounding environment and, thereby, to eliminate the possibility of transfer to other patients. Some of these formulations may have residual effect for continued inactivation of microorganisms; however, because of the limited amount of time most practitioners have between patients, the primary intent of these solutions is to rapidly kill transient organisms [17]. Ideally, these solutions should be active in 15–30 s against the target microorganisms, the amount of time for a normal handwash. Because rapidity of kill is of prime concern, many of the conventional germicidal soaps and surgical scrub solutions are inadequate for this particular use [17]. Those formulations that are alcohol-based seem to provide the rapid rate of kill necessary for short wash times.

Alcoholic chlorhexidine solutions (usually containing 0.5% chlorhexidine in a 70% alcohol base) have both residual and cumulative effects that are effective on skin as well as latex gloves. The use of skin disinfectants for the elimination of bacteria from contaminated gloves so that they can be reused, with out the threat of cross-contamination of patients, has been suggested [27]. Several antimicrobial agents (chlorine, chlorhexidine gluconate, 70% ethanol) and nonmedicated soap are effective in removing *Staphylococcus aureus* from latex gloves [28]. The use of water alone did not suffice in the removal of the test organism from latex gloves; however, it was effective with vinyl gloves, a finding not easily explained.

In an 8-month crossover study comparing alcohol (70% isopropanol, IPA)–soap wash procedure with a CHG wash procedure, the data in all analyzed comparisons showed that the CHG wash procedure provided a nonsignificant trend over IPA–soap [29]. Differences in infection rates were seen in intensive care units among the cardiovascular, medical, and surgical patients. Fewer gastrointestinal, urinary tract, IV catheter, skin and soft tissue, bacteremia, and "other" infections were seen in the CHG group. None of the differences could be explained by characteristics of the patient population or by the types of organisms isolated from health care personnel's hands. One factor that may have affected the outcome was the increased amount of CHG used over that of IPA–soap by the personnel involved in the study, since it has been demonstrated that there is a dose–response effect with CHG and alcohol. The greater the amount used, the more the reduction in organisms on the skin, which then may affect the chance for transmission of organisms from health care worker to patient [3].

C. Use of Whole-Body Wash Solutions

The use of whole-body washes and shower solutions for the prevention of nosocomial infections, similar to that for handscrubs, is a debated subject. The evidence is confusing and contradictory [30]. Much of the confusion in the literature may be due to inconsistency in the protocols that have been used. Studies have consisted of one, two, or several showers or baths, preoperative, or preoperative and postoperative. Also, general washing or showering with nonmedicated soaps may actually increase the bacterial load on certain parts of the body by redistributing organisms from areas of high bacterial counts (e.g., perineum, axilla) [31].

Bruun [32] determined that preoperative bathing with hexachlorophene reduced the skin colonization with *S. aureus* by 30% and also concomitantly found a reduction in the postoperative infections attributable to staphylococci from 8 to 2%. A comparison of patients who did not shower before surgery with those who showered with soap, or with a soap containing the hexachlorophene, indicated that the infection rates were 2.3, 2.1, and 1.3%, respectively. This indicates the value of washing before surgery with an antimicrobial. As part of the same study, surgical wound infection rates decreased from 2% (251 of 12,849) to 1.2% (22 of 1,810) in patients who were washed with a PVP-I sponge before surgery as part of their preparation [22]. In a study involving vasectomy patients, the assessment of preoperative bathing with an antimicrobial soap was made. Among 94 patients in the study, there was a combined infection rate of 32.9%, and no significant difference between the group who had no shower, the group who used an antimicrobial soap (Hibiscrub), and the group who used a nonantimicrobial soap [33]. Although the group using the antimicrobial shower protocol had a much reduced skin bacterial count, the infection rates were the same, a seemingly contradictory condition. The reduction in bacterial colonization of skin after use of a chlorhexidine-containing soap (such as Hibiscrub) is consistent with the excellent antimicrobial activity of this chemical.

Brandberg and Andersson [34] also demonstrated the persistent effect of Hibiscrub on the skin microflora after a single shower bath. The effect was evident for up to a week after the shower. Patients undergoing vascular surgery in this study showered with Hibiscrub between three and eight times before surgery, with a concomitant reduction in the infection rates from 17.5 to 8%. Leigh et al. [35] found that a single wash with Hibiscrub reduced the number of *S. aureus* on the skin, yet there was no reduction in the infection rates. Likewise, Aylife et al. [36], with more than 5000 patients using a single preoperative bath with Hibiscrub, also were unable to find a reduction in the infection rate. The comparison of infection rates in patients assigned to groups who used bar soap, chlorhexidine, or a placebo showed that the group who used a chlorhexidine wash twice in a 24-h period before surgery had an overall reduced infection rate, a reduced infection rate in clean surgery, and a reduced *S. aureus* infection rate [36].

In a slightly different use for topical antimicrobial solutions, a study was undertaken to determine the effect of using chlorhexidine gluconate on the transmission of group B streptococci during labor and delivery. A prospective, randomized, double-blind, placebo-controlled study with 4483 subjects comparing chlorhexidine gluconate prophylaxis with a saline placebo was carried out by the Swedish Chlorhexidine Study Group [37]. Colonization of expectant mothers with *Streptococcus agalactiae* is associated with early neonatal morbidity owing to the aspiration of the organism during labor and delivery. A significant decrease in admission of neonates to the special care neonatal units was observed with mothers who were carriers and were treated with vaginal washes containing CHG before delivery, over placebo-treated mothers. Treatment proved to be significantly effective in reducing admission to special care units for neonates born to all mothers (carrier or noncarrier) [38].

IV. RESISTANCE TO ANTISEPTIC AGENTS

Microorganisms have the ability to adapt to a variety of conditions and, eventually, with prolonged exposure, resistant microorganisms are found that are resistant to virtually any antimicrobial agent. Organisms have been known to grow in disinfectant solutions and can survive and become resistant to antibiotics. For this reason, antiseptics must be protected against contamination either by high enough levels of active ingredient to preclude bacterial survival, or by the inclusion of effective preservative systems.

The mechanisms of resistance to antibiotics are well understood and fall into one of three categories: (a) altered targets, (b) production of detoxifying enzymes, and (c) decreased ability to take up the compound. The resistance to many of the antibiotics is genetically linked either as a plasmid- or chromosomal-linkage. With the exception of heavy metal antiseptics (mercurials and silver salts), there is little evidence for the genetic linkage of microbial resistance to antiseptics. There is some evidence that suggests an association between plasmid-linked antibiotic resistance and resistance to disinfectants [39]. Studies with strains of *Pseudomonas aeruginosa*, with and without the R_{TEM} plasmid (codes for β-lactamase), showed no difference in susceptibility to glutaraldehyde, cetrimide, chlorhexidine, chlorxylenol, or Lysol [40]. It appears that organisms that become resistant to various chemicals, as a result of their heavy use, revert to normal resistance after removal of the active ingredient [41].

Given what is known about the resistance of spores to chemical agents, because of the impermeability of the spore coat, it is likely that other vegetative cells that develop such resistance may do so by providing a barrier to entry of the chemical into the cell. This is thought to be why mycobacteria are resistant to most aqueous agents; the high degree of lipid in the cell wall provides a barrier to aqueous materials [42]. Information suggests that resistance of *P. aeruginosa* and *P. cepecia* to certain chemicals is associated with changes in the outer cell membrane [43].

REFERENCES

1. I. P. Semmelweis, The etiology, the concept, and the prophylaxis of childbed fever, C. A. Hartleben's Verlag-Expedition, Pest, Hungary, 1861.
2. F. D. Daschner, The transmission of infections in hospitals by staff carriers, methods of prevention and control, *Infect. Control 6*:97–99 (1985).
3. E. A. Larson, A casual link between handwashing and risk of infection? Examination of the evidence, *Infect. Control Hosp. Epidemiol. 9*:28–36 (1988).
4. B. Simmons, J. Bryant, K. Neiman, L. Spencer, and K. Arheart, The role of handwashing in prevention of endemic intensive care unit infections, *Infect. Control Hosp. Epidemiol. 11*:589–594 (1990).
5. D. G. Maki and J. A. Hecht, Comparative study of handwashing with chlorhexidine, povidone-iodine, and non-germicidal soap for prevention of nosocomial infection, *Clin. Res. 30*:303A (1982).
6. F. Baquero, C. Patron, R. Canton, and M. M. Ferrer, Laboratory and in-vitro testing of skin antiseptics: A prediction for in-vivo activity, *J. Hosp. Infect. 18B*:5–12 (1991).
7. D. G. Maki and J. A. Hecht, Comparative study of handwashing with chlorhexidine, povidone-iodine and non-germicidal soap for prevention of nosocomial infection, *Clin. Rev. 30*:303A (1982).
8. E. Larson, J. J. Leydon, K. J. McGinley, G. L. Grove, and G. H. Talbot, Physiologic and microbiologic changes in skin related to frequent handwashing, *Infect. Control 7*:59–63 (1986).
9. D. R. Schaberg, R. W. Haley, A. K. Highsmith, R. L. Andersson, and J. E. McGowen, Nosocomial bacteriuria, a prospective study of case clustering and antimicrobial resistance, *Ann. Intern. Med. 93*:420–424 (1980).
10. R. P. Wenzel, R. L. Thompson, S. M. Landry, B. S. Russell, P. J. Miller, S. Ponce de Leon, and G. B. Miller, Hospital-acquired infections in intensive care unit patients: An overview with emphasis in community hospitals, *Infect. Control 4*:371–375 (1983).
11. R. W. Haley, J. H. Tenny, J. O. Garner, and J. V. Bennett, How frequent are outbreaks of nosocomial infection in community hospitals? *Infect. Control 6*:223–226 (1985).

12. R. K. Albert and F. Condie, Handwashing patterns in medical intensive care units. *N. Engl. J. Med. 304*:1465–1466 (1981).
13. G. A. Preston, E. L. Larson, and W. E. Stamm, The effect of private isolation rooms on care practices, colonization, and infection in an intensive care unit, *Am. J. Med. 70*:641–645 (1980).
14. L. M. Kaplan and M. McGuckin, Increasing handwashing compliance with more accessible sinks, *Infect. Control 7*:408–410 (1986).
15. P. M. Dubbert, J. Dolce, W. Richter, M. Miller, and S. W. Chapman, Increasing ICU staff handwashing: Effect of education and group feedback. *Infect. Control Hosp. Epidemiol. 11*:191–193 (1990).
16. L. Frappier-Davignon, A. Frappier, and J. St-Pierre, Staphylococcal infection in hospital nurseries, *Can. Med. Assoc. J. 81*:531–536 (1951).
17. G. Reybrouck, Handwashing and hand disinfection, *J. Hosp. Infect. 8*:5–23 (1986).
18. G. A. J. Ayliffe, J. R. Babb, J. G. Davies, and H. A. Lilly, Hand disinfection: A comparison of various agents in laboratory and ward studies, *J. Hosp. Infect. 11*:226–243 (1988).
19. G. A. J. Ayliffe, Standardization of disinfection testing, *J. Hosp. Infect. 13*:211–216 (1989).
20. K. J. McGinley, E. L. Larson, and J. J. Leyden, Composition and density of microflora in the subungual space of the hand, *J. Clin. Microbiol. 26*:950–953 (1988).
21. J. J. Leyden, K. J. McGinley, G. G. Kates, and K. B. Myung, Contribution of the subungual bacteria to the recovery of bacteria in the glove juice test, *Infect. Control Hosp. Epidemiol. 10*:451–454 (1989).
22. P. J. E. Cruse and R. Foord, A five-year prospective study of 23,649 surgical wounds, *Arch. Surg. 107*:206–209 (1973).
23. M. E. McBride, W. E. Duncan, and J. M. Knox, Physiologic and environmental control of gram-negative bacteria on skin, *Br. J. Dermatol. 93*:191–199 (1975).
24. H. D. Rose and J. B. Babcock, Colonization of intensive care unit patients with gram-negative bacilli, *Am. J. Epidemiol. 101*:495–501 (1975).
25. E. A. Larson, Persistent carriage of gram-negative bacteria on hands, *Am. J. Infect. Control. 9*:112–119 (1981).
26. T. M. Bauer, E. Ofner, H. M. Just, H. Just, and F. D. Daschener, An epidemiological study assessing the relative importance of airborne and direct contact transmission of microorganisms in medical intensive care unit, *J. Hosp. Infect. 15*:301–309 (1990).
27. S. Mehtar, A. Tsakris, D. Castro, and F. Mayet, The effect of disinfectants on perforated gloves, *J. Hosp. Infect. 18*:191–200 (1991).
28. M. Best and M. E. Kennedy, Effectiveness of handwashing agents in eliminating *Staphylococcus aureus* from gloved hands, *J. Appl. Bacteriol. 73*:63–66 (1992).
29. B. N. Doebbeling, G. L. Stanley, C. T. Sheetz, M. A. Pfaller, A. K. Houston, L. A. N. Li, and R. P. Wenzel, Comparative efficacy of alternative hand washing agents in reducing nosocomial infections in intensive care units, *N. Engl. J. Med. 337*:88–93 (1992).
30. D. J. Byrne, A. Napier, and A. Cuschieri, Rationalizing whole body disinfection, *J. Hosp. Infect. 15*:183–187 (1990).

31. J. Davies, J. R. Babb, and G. A. J. Ayliffe, The effect on the skin flora of bathing with antiseptic solutions, *J. Antimicrob. Chemother. 3*:473–481 (1977).
32. J. N. Bruun, Post-operative wound infection. Predisposing factors and the effect in the dissemination of staphylococci, *Acta Med. Scand. 514*:3–89 (1970).
33. P. E. Randall, L. Ganguli, and R. W. Marcuson, Wound infection following vasectomy, *Br. J. Urol. 55*:564–567 (1983).
34. A. Brandberg and I. Andersson, Whole body disinfection by shower-bath with chlorhexidine soap, *R. Soc. Med. Int. Congr. Symp. 23*:65–70 (1979).
35. D. A. Leigh, J. L. Stronge, J. Marriner, and J. Sedgwick, Total body bathing with 'Hibiscrub' (chlorhexidine) in surgical patients: A controlled trial, *J. Hosp. Infect. 4*:229–235 (1983).
36. G. A. J. Ayliffe, M. F. Noy, J. R. Babb, J. G. Davies, and J. Jackson, A comparison of preoperative bathing with chlorhexidine–detergent and nonmedicated soap in the prevention of wound infection, *J. Hosp. Infect. 4*:237–244 (1983).
37. J. L. Hayek, J. M. Emerson, and A. M. N. Gardner, A placebo-controlled trial of the effect of two preoperative baths or showers with chlorhexidine detergent on postoperative wound infection rates, *J. Hosp. Infect. 10*:165–172 (1987).
38. L. G. Burman, P. Christensen, K. Christensen, B. Fryklund, A. M. Helgesson, N. W. Svenningsen, and K. Tullus, Prevention of excess neonatal morbidity associated with group B *Streptococci* by vaginal chlorhexidine disinfection during labour, *Lancet 340*:65–69 (1992).
39. A. D. Russell, Bacterial resistance to disinfectants, *J. Hosp. Infect. 7*:213–225 (1986).
40. A. D. Russell, Comparative resistance of R^+ and other strains of *Pseudomonas aeruginosa* to non-antibiotic antibacterial agents, *Lancet 2*:332 (1972).
41. A. D. Russell, The role of plasmids in bacterial resistance to antiseptics, disinfectants and preservatives, *J. Hosp. Infect. 6*:9–19 (1985).
42. V. Jahlier and H. Nikaido, Permeability barrier to hydrophilic solutes in mycobacteria, *J. Bacteriol. 172*:1418–1423 (1990).
43. D. J. Stickler and B. Thomas, Intrinsic resistance to nonantibiotic antibacterial agents, *Principles and Practices of Disinfection, Preservation and Sterilisation* (A. D. Russell, W. B. Hugo, and G. A. J. Ayliffe, eds.), Blackwell Scientific Publications, Oxford, 1982, pp. 186–198.

5

Disinfectant Use in Dentistry

JAMES A. COTTONE
The University of Texas Health Science Center at San Antonio
San Antonio, Texas

JOHN A. MOLINARI
University of Detroit Mercy School of Dentistry
Detroit, Michigan

I. INTRODUCTION

Dental treatment presents a formidable infection control problem as the dentist and dental team performs procedures in an environment with a full array of commensal microorganisms. The challenge increases with the use of many small, sharp instruments, an abundance of supplies, and several types of dental handpieces in this environment. More complications arise with relatively extensive surface contamination, resulting when dentist and dental team reach for many items in a nonsterile environment, open drawers and cabinets, activate the handpiece bur to rotate at thousands of revolutions per minute in a field of blood and saliva, and move the patient to different areas of the office where other procedures are performed.

For centuries, dental practitioners pursued their work with little attention to the possible infectious hazards inherent in the procedures, until the realization that potentially lethal infections, such as hepatitis B virus (HBV) [1] and human immunodeficiency virus (HIV) infection [2] could possibly be transmitted in this environment. The availability of various Centers for Disease Control and Prevention (CDC) [3,4], American Dental Association (ADA), and other organizations' guidelines and recommendations from 1978 [5] to the present [6] assisted the profession in increasing their use of a variety of infection control procedures. The greatest advances have been in the use the personal protective equipment; however, a substantial increase in the use of heat sterilization has caused a corresponding decrease in the use of chemicals for immersion disinfection and sterilization [7,8]. These same guidelines and recommendations have also caused an increase in the appropriate use of chemicals for surface cleaning and disinfection.

The selection of an efficacious immersion or surface disinfectant is confusing for many dental professionals, partly because of occasional exaggerated manufacturer claims and misleading reports in the literature. Additionally, most dental practitioners have not been trained to differentiate accurate from inaccurate or distorted claims and reports. Therefore, the actual performance capabilities of individual agents may be obscured. Confusion also arises when dentists and other dental treatment providers are not aware of guidelines and requirements for the selection of appropriate chemicals.

In an attempt to assist the dental profession with understanding the evaluation of products, the ADA convened a group to develop a policy for ADA acceptance of chemical agents for disinfection or sterilization in the mid 1980s. The resultant criteria can be found in Table 1. In the fall of 1991, these criteria were reevaluated because of several issues, including the lack of reproducibility of results [9], the validity of the Environmental Protection Agency's (EPA) pass–fail performance standards, problems in company claims concerning acceptance of multiple tuberculocidal testing methods, and the lack of an independent laboratory to verify test results [10]. Concerns about the applicability of laboratory test results to actual use conditions in dentistry have been limited and have resulted in problematic findings [11–13]. Although these are valid concerns, we feel that the

TABLE 1 Former Criteria for ADA Acceptance of Surface Disinfectants

1. EPA registration as a hospital level disinfectant
2. Tuberculocidal action
3. Viricidal efficacy
   a. Lipophilic virus
   b. Hydrophilic virus

Source: Ref. 17.

ADA acceptance program continued to assist the dental profession with product evaluation; however, it was disbanded.

Dentistry has struggled with an easy way to classify sterilants and disinfectants. A standard system of classification for chemical sterilants and disinfectants proposed by Spaulding in 1972 [14] was originally developed for classifying hospital instruments according to their use and degree of contamination (Table 2). Patient care items and equipment placed into one of three categories (critical, semicritical, and noncritical) often overlap, depending on how the instrument is used or what procedures are performed. In addition, the three levels of disinfection defined (high, intermediate, and low [15]) do apply to disinfectants in dentistry; however, low-level disinfectants function as cleaners only in dentistry. At best, only with difficulty can this system be adapted to dental instruments and equipment.

II. MOST COMMONLY USED DISINFECTANTS AND STERILANTS

Infection control in dental treatment facilities requires utilization of disinfectants in several forms: surface disinfectants, immersion disinfectants, and immersion sterilants. Each category is distinguished by different needs, and care must be exercised to separate the categories.

A. Immersion Disinfectants and Sterilants

At present, the group of chemical agents used for immersion disinfection and sterilization continues to be the 2% alkaline glutaraldehydes [16]. There is also use for 3.2% glutaraldehydes as well as acid and alkaline glutaraldehydes. Examples of these chemical agents used in dentistry are shown in Table 3 along with their EPA registration numbers.

TABLE 2 Spaulding Classification of Inanimate Objects

Critical: sterilization required
Items that penetrate or touch broken skin or mucous membranes, such as needles, scalpels, surgical instruments, mirrors, and dental explorers.
Semicritical: sterilization or high-level disinfection required
Items that touch mucous membranes, but do not enter sterile body areas, such as amalgam condensers, handpieces, and ultrasonic scalers.
Noncritical: tuberculocidal intermediate-level disinfection
Surfaces that do not touch mucous membranes, such as countertops, light handles, and chair surfaces.

Source: Ref. 7.

TABLE 3 Representative Table of Chemical Agents

Office Sterilization and Asepsis Procedures (OSAP) Research Foundation Guide to Chemical Agents for Disinfection and/or Sterilization					
Products (EPA Reg no.)	Chemical classification	TB Directions (test)	Hydrophilic virucide	Sterilant	Sterilant reuse (days)
Immersion only					
Multicide Plus (1043-36)	1% phenylphenol, 5% benzyl chlorophenol, soap	1:32, 20 min, 20C (AOAC)	No	No	None
Banicide (15136-1)	Glutaraldehyde 2% acidic	20 min (FS), 20C (Quant)	Yes	FS, 5 h, 40C	None
Sterall, Wavicide 01				FS, 10 h, 21C	30
Cidex Plus (7078-14)	Glutaraldehyde 3.2% alkaline	FS, 20 min, 25C (Quant)	Yes	FS, 10 h, 20–25C	28
CoeCide XL Plus (46781-4) Metricide Plus 30	Gluraraldehyde 3.2% alkaline	FS, 20 min, 25C (AOAC)	Yes	FS, 10 h, 20–25C	30
Cidex 7 (7078-1)	Glutaraldehyde 2% alkaline	FS, 90 min, 25C (Quant)	Yes	FS, 10 h, 20–25C	28

Baxter/Omnicide (46851-2)	Glutaraldehyde 2% alkaline	FS, 45 min, 20C (AOAC)	Yes	FS, 10 h, 20 C	28
Omnicide					
ProCide					
CoeCide XL (46781-2)	Glutaraldehyde 2% alkaline	FS, 90 min, 25C (AOAC)	Yes	FS, 10 h, 20–25C	30
Maxicide					
Metricide					
Protec-Top					
Surface only					
Bleach (5.25%)	Chlorine compounds	1:10, 10 min, 20C	No	No	None
Lysol Spray (777-53)	0.1% phenylphenol, 79% ethanol	10 min, 20C (AOAC)	Yes	No	None
Coe Spray-The Pump (334-417)	0.216% phenylphenol, 0.054% *t*-amylphenol, 66% ethanol	10 min, 20C (AOAC)	Yes	No	None
ProSpray (46581-5)	0.28% phenylphenol 0.03% benzyl chlorophenol	10 min, 20C (AOAC)	Yes	No	None

Table 3 Continued

Office Sterilization and Asepsis Procedures (OSAP) Research Foundation Guide to Chemical Agents for Disinfection and/or Sterilization					
Products (EPA Reg no.)	Chemical classification	TB Directions (test)	Hydrophilic virucide	Sterilant	Sterilant reuse (days)
Surface/immersion					
Biocide, Surf-A-Cide (4959-16)	Iodophor (1.75% titrable iodine)	1:213, 10–25 min, 20C (AOAC)	Yes	No	None
Iodofive (1677-22)	Iodophor (1.75% titrable iodine)	1:213, 5 min, 20C (AOAC) (10 min hydrophilic viruses)	Yes	No	None
Omni II, Vital Defense-D (46851-1)	9.0% phenylphenol 1% benzyl chlorophenol	1:32, 10 min, 20C (AOAC)	Yes	No	None
Asepti-phene 128 (303-87)	4.28% phenylphenol	1:128, (AOAC)	No	No	None

Source: Other products available. Listing does not imply endorsement, recommendation or warranty. 20C = 68F, 25C = 77F. Purchasers are legally required to consult the package insert for changes in formulation and recommended product uses. Check materials compatibility before use on dental equipment. Updated tables are available from the OSAP Research Foundation 800/243-1233. February 1994, OSAP Research Foundation.

Even though high-level disinfectants are capable of sterilization of immersed items, these chemicals are often misused. Instead of immersing items in the solution for the required interval for sterilization (i.e., 10 h at room temperature), dental personnel may use only a 20- to 30-min exposure and rinse the "sterilized" materials in nonsterile, potable water. These items are at best disinfected and represent one of the most abused aspects of infection control in dentistry. Additionally, the terms "use life" and "reuse life" are often misused and used interchangeably. This can be problematic with a product that has a 28- to 30-day use life, but cannot be reused for multiple batches of instruments and, therefore, has no reuse life.

B. Surface Disinfectants

All surface disinfectants used in dentistry must meet the requirements of an intermediate-level disinfectant, particularly inactivation of *Mycobacterium tuberculosis*. Additionally, these products must also inactivate small, nonlipid virus (hydrophilic) as well as medium-sized, lipid-coated virus (lipophilic) [17]. Available product formulations include surface solutions and sprays, each with a specific rationale for use. Unfortunately, product misuse is common and has led to numerous reactions [18] and equipment failures [19].

Surface disinfectants that meet efficacy requirements and are now often used in dentistry include iodophors, chlorine-containing agents, and complex synthetic phenolics [7,16]. The surfactant properties of iodophors make them excellent cleaning agents before disinfection, and newer iodophor commercial formulations have received EPA-approved tuberculocidal activity within 5 min of exposure. Fresh solutions must be prepared daily to ensure the tuberculocidal activity required in dentistry [16].

Accepted chlorine-containing compounds in common use in dentistry are primarily hypochlorite solutions (see Table 3). Diluted bleach (1:100) in water was shown in the 1970s to be useful as a disinfectant, especially in areas considered to have been contaminated with hepatitis viruses [20]. The CDC has recommended the use of 500 ppm (0.05%) sodium hypochlorite as an effective agent in destroying hepatitis B viruses [21]. Because this chemical is unstable, fresh solutions must be prepared daily. Despite its effectiveness as a disinfectant, this chlorine-releasing preparation has some obvious disadvantages. It is corrosive to metals, irritating to skin and other tissues, and destroys many fabrics.

In the mid-1980s, a new class of phenolic compounds was approved by the EPA as surface disinfectants. These contain more than one phenolic agent. Currently, most products contain two, but newer formulations have three phenols as the active compounds. After appropriate dilution in water, these phenolics act in a synergistic manner to offer a broad antimicrobial spectrum, including tuberculocidal and hydrophilic virucidal activity. These products are often used in dentistry, as they are also good surface cleaners and are effective in the presence of detergents [7].

III. SUMMARY

Although heat sterilization is the primary method to prepare instruments for reuse in virtually all dental offices, many commercial products are recognized as being effective for use in dentistry as disinfectants or sterilants. Some of these products will only disinfect and do not sterilize, some have an odor that is unpleasant, others stain or bleach surfaces, and some of the formulations are corrosive to metallic surfaces. The EPA and the OSAP Research Foundation currently list iodophors, synthetic phenolics, and chlorine-containing compounds as acceptable intermediate hospital-level surface disinfectants with both lipophilic and hydrophilic virucidal activity. Note that alcohols and quaternary ammonia compounds should not be used as disinfectants in a dental office [5–7]. For liquid sterilants, most dental offices currently use a 2–3.2% glutaraldehyde solution, although usually in a time frame for chemical disinfection. Therefore, it is essential that before purchasing a chemical sterilant or disinfectant and a surface disinfectant for routine use in dentistry, the practitioner, dental hygienist, assistant, and laboratory technician be provided with as much information as possible. This will then allow subsequent decisions to be based on appropriate efficacy criteria, and reduce the potential for product misuse.

REFERENCES

1. J. A. Cottone, Hepatitis B virus infection in the dental profession, *J. Am. Dent. Assoc. 110*:617–621 (1985).
2. Centers for Disease Control (CDC), Update: Transmission of HIV infection during an invasive dental procedure—Florida, *MMWR 40*(23):377–381 (1991).
3. Centers for Disease Control (CDC), Recommended infection control practices for dentistry, *MMWR 41*:(RR-8)1–12 (1993).
4. Centers for Disease Control (CDC), Recommendations for prevention of HIV transmission in health-care settings, *MMWR 36*(2S):2S–18S (1987).
5. American Dental Association Council on Dental Therapeutics, Quaternary ammonium compounds not acceptable for disinfection of instruments and environmental surfaces in dentistry, *J. Am. Dent. Assoc. 97*:855–856 (1978).
6. American Dental Association, Infection control recommendations for the dental office and the dental laboratory, *J. Am. Dent. Assoc.* (Suppl.):1–8 (1992).
7. J. A. Cottone, G. T. Terezhalmy, and J. A. Molinari, *Practical Infection Control in Dentistry*, Lea & Febiger, Philadelphia, 1991.
8. L. P. Samaranayake, F. Scheutz, and J. A. Cottone, *Infection Control for the Dental Team*, Munksgaard, Copenhagen, 1991.
9. W. A. Rutala, and E. C. Cole, Ineffectiveness of hospital disinfectants against bacteria: A collaborative study, *Infect. Control. 8*:501–506 (1987).
10. Disinfectants—EPA lacks assurance they work. U. S. General Accounting Office, Gaithersburg, MD, GAO/RCED-90-139, 1990.
11. Robinson, et al., Log reductions for three glutaraldehyde-based disinfectants during actual clinical use, *Appl. Environ. Microbiol. 54*:158 (1988).

12. R. P. Christensen, R. A. Robinson, D. F. Robinson, et al., Antimicrobial activity of environmental surface disinfection in the absence and presence of bioburden, *J. Am. Dent. Assoc. 119*:493–505 (1989).
13. Letters to the editor, Surface disinfectants, *J. Am. Dent. Assoc. 120*:10, 12, 14 (1990).
14. E. H. Spaulding, Chemical disinfection and antisepsis in the hospital, *J. Hosp. Res. 9*:5–31 (1972).
15. E. H. Spaulding, *Chemical Disinfection of Medical and Surgical Materials*, Lea & Febiger, Philadelphia, 1968, pp. 517–531.
16. J. A. Cottone and J. A. Molinari, State-of-the-art infection control in dentistry, *J. Am. Dent. Assoc. 122*:33–41 (1991).
17. American Dental Association Council on Dental Therapeutics, Provisions for acceptance of products by the Council on Dental Therapeutics, *J. Am. Dent. Assoc.* October, 1986.
18. Centers for Disease Control (CDC), Chlorine gas toxicity from mixture of bleach with other cleaning products—California, *MMWR 40*(36):619–621, 629 (1991).
19. Surface disinfectants. Adec inter-office correspondence. August 9, 1990.
20. H. Kobayashi, Susceptibility of hepatitis B virus to disinfectants or heat, *J. Clin. Microbiol. 20*:214–216 (1984).
21. W. W. Bond, M. S. Favero, N. J. Petersen, and J. W. Ebert, Inactivation of hepatitis B virus by intermediate-to-high level disinfectant chemicals, *J. Clin. Microbiol. 18*:535–538 (1983).

6

Contact Lens Disinfectants

Michael J. Miller
Bausch & Lomb, Inc.
Rochester, New York

I. THE HISTORY OF CONTACT LENS CARE PRODUCTS

The use of contact lenses has gained popularity over the years, with over 26 million individuals in the United States enjoying the comfort and convenience of lens wear. Contact lenses are used for therapeutic, visual, and cosmetic purposes, and are medical devices that require proper care for safe and successful use. Contact lens care products, such as disinfecting solutions, have been influenced by the evolution of contact lens materials. The introduction of new polymers for daily and extended wearing of contact lenses subsequently prompted the industry to develop new lens care regimens.

A. Hard (Polymethylmethacrylate) Contact Lenses

In 1948, the original hard contact lens was introduced. The lens consisted of polymethylmethacrylate (PMMA), a hydrophobic polymer that resisted binding of proteins, lipids, and other contaminants on the lens surface. As a result, cleaners were not used as part of a lens maintenance program. In addition, disinfecting these lenses was not considered to be a critical part of lens wear, since the surface provided a poor environment for the survival of microorganisms. As the PMMA lens-wearing population increased, so did an apparent increase in discomfort and microbial contamination during lens storage. The hydrophobic nature of the lens contributed to discomfort when the lens was initially inserted into the eye. Wetting solutions were developed to improve wearing comfort (by providing a "comfortable" layer between the lens and eyelids and cornea) and lens wettability (the solution facilitated the adherence of the tear film to the lens surface). In addition, the development of storage or soaking solutions provided a safe method for storing lenses when not in use. These solutions were preserved with antimicrobial agents, such as benzalkonium chloride and thimerosal, which protected the soaking solution against microbial contamination.

The wearing of PMMA lenses was not without complications. Oxygen transmissibility to the cornea was significantly reduced by the lens polymer, causing redness, blurred vision, corneal edema, and distortion. Many patients experienced discomfort because of the hard lens surface and fitting characteristics. In addition, epithelial abrasions occurred in some patients whose lenses did not fit properly.

B. Soft (Hydrogel) Contact Lenses

In 1971, the first soft contact lenses were introduced, and patients dissatisfied with hard lenses began wearing the new, hydrogel lens. The hydrogel polymer (Soflens) was composed of approximately 38% water, and provided increased oxygen permeability, less epithelial abrasion, and improved wearing comfort. Although soft lenses offered significant advantages for comfort and ocular safety, a greater amount of care and maintenance was required for these lenses compared with PMMA lenses. Shortly after the introduction of hydrogels, the contact lens industry rapidly developed a variety of lens care products; specifically, cleaners, disinfectants, and saline storage solutions that would support the care and maintenance of these new devices. Today, soft contact lenses are composed of ionic and nonionic polymers, with water contents ranging from 36 to 79%, and are worn either daily or continuously (overnight) for extended periods (hence, the name, "extended-wear" lenses).

C. Rigid Gas-Permeable Contact Lenses

Although soft contact lenses were a major advantage over PMMA lenses, the early soft lenses were unsuccessful in correcting astigmatism. In the 1980s, contact lenses were developed that were rigid enough to correct astigmatism, while allowing adequate oxygen permeability, hence the name "rigid gas-permeable" (RGP).

II. THE IMPORTANCE OF CLEANING, RINSING, AND DISINFECTING CONTACT LENSES

Most complications associated with contact lens wear (dry- or red eye syndrome, conjunctivitis, excessive tearing, surface deposition, or other) have no long-term visual significance and usually resolve when the lens is removed, replaced, or the lens care regimen is changed. Microbial keratitis (corneal infection), however, is a serious complication that may result in partial or permanent visual damage from corneal scarring. Many instances of ocular infection have been attributed to improper hygiene and noncompliance with recommended lens care procedures [5,18,45,74]. Bowden et al. [5] found that 23 of 24 patients were not compliant in their lens care regimen. Other investigators have demonstrated that between 40 and 74% of normal patients were noncompliant [9,12]. In one study, patients did not wash their hands before lens manipulation, disinfected their lenses less than once a month, or used no disinfection system at all [45]. Proper care of contact lenses requires the wearer to follow a precise lens care regimen, which normally includes cleaning, rinsing, and disinfection; the importance of each are presented in the following.

A. Cleaning

Routine cleaning is an important step in the proper care of contact lenses; however, it is often the most neglected. One study reported that a clean, unused lens when placed on the eye is 50% coated with a macromolecular tear film within 30 min, and is 90% coated by 8 h [20]. Failure to clean lenses may lead to the accumulation of surface deposits, increased irritation, discomfort, and the adherence of microorganisms. Deposits may reduce the visual acuity of the wearer and may cause corneal injury owing to reduced oxygen transmission across the lens [67]. Finally, bacteria that colonize soft contact lenses preferentially adhere to areas of lens deposits [21,35], and some studies conclude that tear coatings are a major factor facilitating this adherence [7,8,63]. Most regimens recommend the use of a daily surfactant cleaner (with finger rubbing) for the removal of loosely bound contaminants, such as lipid, protein, cosmetic, and other environmental deposits. In addition, surfactant cleaning may reduce the level of microbial contaminants by 1–4 logs (1 log is equivalent to a 90% reduction of a microbial population) [27,51,59].

B. Rinsing

In most regimens, the lens is rinsed with a saline or similar solution immediately following the cleaning step. Rinsing the lens helps remove surface deposits released during cleaning and may also aid in the removal of microorganisms (one study demonstrated that rinsing removed 99.9% of surface-bound microorganisms [59]).

C. Disinfection

Normally, the surface of a contact lens can be expected to carry a few microorganisms that are part of the relatively harmless flora in the conjunctival cul-de-sac [68]. These may include coagulase-negative (predominating flora) and coagulase-positive staphylococci, micrococci, α-hemolytic streptococci, corynebacteria, and other gram-positive and gram-negative bacilli [28,40]. Microorganisms not considered to be part of the normal ocular flora have been implicated in causing contact lens-related infectious keratitis. The presence of potential pathogens in the ocular environment, however, does not necessarily mean that an infection will occur. Relatively few patients ever develop an eye infection, for the eye contains an elaborate and effective natural defense system. The tear film washes microorganisms from the corneal surface, the eyelids help to push away debris during blinking, and tear components, such as lysozyme, lactoferrin, and immunoglobulins, possess antimicrobial activity. Most importantly, the corneal epithelium serves as a barrier in protecting the eye, since most microorganisms cannot penetrate this ocular tissue.

It is well established that epithelial trauma is a necessary prerequisite for bacterial corneal ulceration, since infection begins with bacteria adhering to the edges of injured epithelial tissue [55,64]. Corneal epithelial breakdown may be related to recent contact lens manipulation, such as the insertion and removal of lenses [2,22], surface deposition [67], long-term anoxic stress (a decrease in the oxygen supply to the cornea [26,45,63]), or to hypersensitivity reactions to chemical agents [81]. Corneal damage coupled with the potential that the contact lens may introduce microorganisms into the eye from contaminated sources, makes the lens wearer more susceptible to ocular infection. Contact lens disinfection systems significantly reduce large numbers of potentially harmful microorganisms present on a lens to a safe level before the lens is placed on the eye. Therefore, the proper use of disinfection systems will reduce the risk of contact lens-related infectious keratitis.

III. MICROORGANISMS AND RISK OF INFECTION

Where do ocular pathogens come from? Most are ubiquitous in nature, residing in moist environments such as water, soil, and vegetation. *Pseudomonas aeruginosa*, the major etiological agent of contact lens-related keratitis, has been isolated from contaminated sinks and faucets, distilled water, chlorinated swimming pools, and hot tubs. These environments may serve as sources of contamination of other items, such as contact lenses, care systems, and lens cases.

A. Mode of Transmission

Numerous studies have associated contact lens-related infectious keratitis with the use of contaminated lens care systems [5,38,39,77,78]. Donzis et al. [15] determined that 50% of asymptomatic patients possessed contaminated lens care products and that 100% of bottles containing home-prepared saline contained gram-negative bacilli. Microbial contamination is most prevalent in contact lens cases [15,79]. Microbial populations may be found in lens cases in which nutrients from lens deposits or unwashed fingers have accumulated [77]. In addition, microorganisms can survive and replicate in lens cases with residual solution, especially nonpreserved saline, in as few as 8 h of exposure [80]. Contaminated cases may provide reservoirs of potential ocular pathogens that may be transported to the eye by the lens, since microorganisms can readily attach to the surface of hydrogel and rigid gas-permeable lenses [17,42–44]. Such transportation is supported by studies demonstrating that organisms isolated from contaminated lens care systems are the same strains cultured from corneal ulcers and other ocular infections [38,39]. The predisposition of the eye by a number of clinical factors, such as anoxia, or by accidental trauma during insertion of a lens or during lens wear, may permit microorganisms adhering to the lens to initiate infection.

B. Extended-Wear

The use of extended-wear soft contact lenses, as compared with conventional daily-wear soft contact lenses, increases the risk of corneal ulcers [13,54,57,61]. This is primarily due to overnight anoxic stress and corneal injury. One study [57] suggested that the risk of ulcerative keratitis increases with each consecutive night of lens wear. Patients who wear extended-wear lenses overnight increase the risk 10–15 times when compared with patients who remove their daily-wear lenses at night [54,57]. Additionally, patients who wear daily-wear lenses overnight have a risk nine times higher than patients who do not [57]. Estimates of microbial corneal ulcer incidence rates are relatively low; recent studies have reported 18.2:10,000–20.9:10,000 patients per year wearing cosmetic extended-wear soft lenses, and 4.1:10,000–5.2:10,000 patients per year wearing cosmetic daily-wear lenses [36,54].

C. Microorganisms Involved in Contact Lens-Related Infectious Keratitis

1. Bacteria

Most lens-associated ocular infections are caused by bacteria. Of these, *P. aeruginosa* has been the major etiologic agent. Corneal ulcers caused by this gram-negative bacterium are usually centrally located, with large epithelial defects, stromal melting, and mucoid material clinging to the lesion. *Serratia mar-*

cescens corneal infections are not as devastating as *P. aeruginosa*; however, *S. marcescens* has become more resistant to chemical preservatives such as benzalkonium chloride, chlorhexidine, and polyquaternium-1 [1,23,39]. Ulcers caused by gram-positive bacteria, such as *Staphylococcus* spp., are usually smaller and less purulent than ulcers caused by gram-negative bacteria. Table 1 lists the most commonly associated bacteria with lens-related microbial keratitis.

2. Fungi

Several studies have noted fungal invasion (penetration) of hydrogel contact lenses [6,60,78,82]; however, corneal infections associated with contact lenses are rare (only 3–4% of reported lens-associated infections are caused by fungi [58,75,78]). Fungal contamination of soft lenses is most often related to improper hygiene, poor cleaning and disinfection practices, or to accidental contamination from the environment [78]. Fungi most commonly associated with keratitis and lens invasion are listed in Table 2. The appearance of corneal ulcers may be similar to fungal growth on lenses or other substrata, with feathery borders, raised infiltrates, or satellite lesions.

Fungi colonizing the lens surface may appear as white, yellow, brown, or gray deposits, ranging in size from less than 1 to greater than 3 mm. Within 24–48 h, conidia (spores) germinate and produce hyphal elements; the hyphae are able to penetrate the lens and "corkscrew" through a hydrogel polymer. Several in vitro studies have suggested that fungal penetration may result in physical and metabolic degradation of the lens [60,82]. Penetration from the anterior to the posterior surface may occur in fewer than 7 days. This behavior of fungi is clinically relevant for extended-wear patients, because growth of certain fungi is enhanced in higher water content lenses [60,82].

3. Acanthamoeba

Acanthamoeba are free-living protozoa that are uniquitous in nature. They are common inhabitants of chlorinated waters (swimming pools and hot tubs), salt and freshwater estuaries, tap water, distilled water, soil, and air.

TABLE 1 Bacteria Most Associated With Contact Lens-Related Keratitis

Gram-negative	Gram-postitive
Pseudomonas aeruginosa	*Staphylococcus aureus*
Serratia marcescens	*S. epidermidis*
	Streptococcus pneumoniae

TABLE 2 Fungi Most Associated With Contact Lens-Related Keratitis and Hydrogel Lens Invasion

Fungi most associated with lens-related keratitis	Fungal invasion of hydrogel lenses
Candida	*Acremonium*
Curvularia	*Aspergillus*
Fusarium	*Aureobasidium*
	Beauveria
	Bipolaris
	Cladosporium
	Exophiala
	Paecilomyces
	Penicillium

The first documented cases of contact lens-associated acanthamoeba keratitis occurred in the 1980s. In one study, 82% of the 62 cases reported from 1984 to 1986 were associated with contact lens wear [46]. Both forms of the amoeba (trophozoite and cyst) have been cultured from soft contact lenses [11,31,34], soft contact lens cases [48,49,76], distilled water bottles [30], and saline solutions made with distilled water and salt tablets [34,62]. In addition, infections have been associated with the use of tap water and saliva [10,47]. The organisms cultured most often have been *A. castellani* and *A. polyphaga*.

Contaminated lens cases provide an ideal environment for the survival of acanthamoeba (16), and lenses soaking in this environment may harbor amoebae [29] that may be transported to the eye. Although corneal ulcers caused by these protozoa are severe, infection is rare (one study estimated a 0.01% infection incidence rate [62]).

4. Viruses

The most common viral ocular pathogens are adenovirus and herpes simplex. Both viruses are occasionally encountered in patients wearing contact lenses, but neither have been etiologically linked to any form of lens wear. Human immunodeficiency virus (HIV), the etiological agent of AIDS, has been isolated from a variety of ocular sources, including the tears, conjunctiva, iris, retina, vitreous, and cornea. In addition, HIV has been isolated from high–water-content lenses of patients tested with AIDS or AIDS-related complex [66]. Nonetheless, there have been no documented cases suggesting that the virus can be transmitted by contact lens wear.

IV. ANTIMICROBIAL ACTIVITY OF CONTACT LENS SYSTEMS

Contact lens solutions usually contain several "excipient" ingredients that may support the survival and growth of a variety of microorganisms. Contaminated preparations may allow the growth of potential pathogens to concentrations in excess of 10^6 colony forming units (CFU)/ml. This is why contact lens solutions should be preserved with active agents that will prevent the growth of microbes that may gain access into product containers. With multidose containers (the majority of contact lens solutions), it is necessary that the product be adequately preserved such that any microbial contamination that is introduced into the preparation during use is eliminated before the next dose is utilized [19]. In addition to being preserved, disinfecting solutions require rapid and efficacious agents that will reduce numbers of potential ocular pathogens on a lens to a safe level before the lens is placed on the eye. The response of microorganisms to active ingredients in disinfectants and preserved solutions (e.g., cleaners and saline solutions) will depend on a number of factors, including the microorganism, the active ingredient and its corresponding concentration, and the duration of cellular exposure to the active agent.

A considerable amount of work has been performed studying the effects of chemical and physical agents on microorganisms; however, the precise mechanism of many inhibitory compounds remains unclear [56]. Antimicrobial agents normally interfere with one or more critical cellular processes; these may include (a) action on the cytoplasmic membrane, (b) inhibiting cell wall synthesis, (c) inhibiting protein synthesis, and (d) activity against nucleic acids. Details on the possible mode of action of contact lens disinfectants will be discussed in the following sections.

A. Thermal Disinfection

Thermal disinfection was the first method used to disinfect hydrogel lenses. In the original method, patients prepared a saline solution by dissolving salt tablets in distilled water; the resulting diluent was used to suspend the lenses in a storage case. The case was placed into a larger heating unit, which disinfected the lenses at a temperature of 95°C for 20 min. Modern heating units are portable and compact, and provide a temperature of 80°C for 10 min (an FDA requirement) to disinfect soft lenses. Furthermore, either preserved or unpreserved saline solutions may be used as the suspending diluent. Thermal disinfection acts on microorganisms by denaturing their cellular components (e.g., proteins and enzymes), rupturing the plasma membrane causing leakage of low molecular weight material, and damaging RNA and DNA [56]. The advantages of thermal disinfection include short cycle times, efficacy against a variety of vegetative microorganisms, and low risk of ocular reaction when unpreserved saline is used. On the other hand, thermal disinfection may not be compatible with all lens types, such as RGP and high–water-content lenses. Repeated heating may change the physical parameters of lenses with greater than 40% water content, or lenses containing poly(vinyl pyrrolidone). Lens discoloration caused by thermal denatured protein deposits or saline components (e.g., sorbic acid, thimerosal, polyvinyl alcohol) may also result. Finally, denatured proteins may cause an allergic reaction (e.g., giant papillary conjunctivitis) in some individuals who are predisposed toward autoimmune phenomena.

B. Chemical Disinfection

In an attempt to provide a more convenient regimen, and be less destructive to lenses than thermal disinfection, chemical systems were introduced. These contain antimicrobial agents that interact with microorganisms; each agent has a unique chemical structure and reactive group(s) that induce varying degrees of microbial damage. The most commonly used agents in contact lens disinfecting systems are described in the following sections.

1. Hydrogen Peroxide

Hydrogen peroxide (H_2O_2) has successfully been used as a contact lens disinfectant for many years; it is a rapid microbicidal agent, effective against a wide range of organisms, including bacteria, yeasts, molds, viruses, spores, and protozoa (a more detailed discussion on H_2O_2 may be found in Chap. 9 of this volume).

The H_2O_2 molecule spontaneously decomposes to form free radicals which, in turn, break down into water and oxygen vapor. Some of the intermediate species, such as superoxide ion and hydroxyl radical, are short-lived, but highly reactive with organic matter. These species can attack membrane lipids, DNA, and other essential cell components of microorganisms [4]. Although the decomposition of H_2O_2 is essential to its antimicrobial activity, currently available contact lens care systems control this rate by adding stabilizers to a 3% solution. Consequently, this provides an adequate soak time (e.g., 10–20 min) for lens disinfection.

Following disinfection, it is necessary that residual H_2O_2 be neutralized before the lens is placed on the eye; otherwise, extensive damage to the ocular surface may result [69]. Neutralization may be accomplished through catalysis, chemical reaction, or dilution. In catalytic neutralization, the natural decomposition of H_2O_2 to water and oxygen is accelerated by a component in the neutralization system. Examples include the use of a platinum-coated disk or catalase, a peroxidase enzyme. Reactive neutralization accelerates the decomposition of H_2O_2 by reaction with another chemical agent, such as sodium thiosulfite or sodium pyruvate. Finally, serially diluting the disinfecting solution with an appropriate saline will reduce the concentration of residual H_2O_2 to an acceptable physiological level. Further information on neutralization and neutralizers may be found in Chapter 3 of this book.

To summarize, H_2O_2 disinfection systems are rapid, are efficacious against a broad spectrum of microorganisms, and produce nontoxic decomposition products. There are, however, drawbacks to this lens care system. Neutralization of residual peroxide following disinfection is an absolute necessity; omitting this critical step could result in permanent damage to the cornea and surrounding tissue (it is not uncommon for patients, who might be confused by the two-step disinfection regimen, to place non-neutralized lenses onto the eye). Finally, neutralized H_2O_2 solutions do not offer continuous antimicrobial activity; care must be taken to avoid contaminating these systems if lenses are stored for extended periods [24,79].

Table 3 contains a list of current H_2O_2 disinfecting systems and their corresponding neutralization agents.

2. Thimerosal

Thimerosal is a compound composed of organic mercury and thiosalicylic acid:

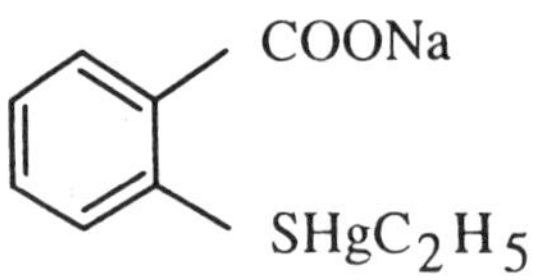

Thimerosal (ethylmercuric thiosalicylate)

It is a broad-spectrum antimicrobial agent (with activity against bacteria, yeast fungi, protozoa, and viruses), and its actions against bacteria are predominantly bacteriostatic, rather than bactericidal. Thimerosal works synergistically with other antimicrobial compounds, thereby increasing its microbial efficacy. For example, the first chemical contact lens-disinfecting system employed thimerosal, chlorhexidine gluconate, and EDTA. The organic mercurial interacts with cellular thiol groups normally present in many proteins, cell walls, membranes, enzymes, and ribosomes. In addition, the compound binds to amides, amines, amino acids, sulfur-free proteins, carboxylates, phosphates, purines, pyrimidines, phenols, imidazoles, indoles, and nucleotides [25]. These interactions may account for membrane permeability changes, inhibition of enzyme functions, suppression of catabolic metabolism, and loss of other critical cellular processes.

As a result of the widespread use of mercurial compounds, there is a large population who are sensitive and allergic to thimerosal. In fact, the reported frequency of reactions may be as high as 40% [50]. Lens wearers diagnosed as being sensitive to this agent are usually recommended to make use of another

TABLE 3 Disinfection Systems Employing Stabilized, 3% Hydrogen Peroxide (H_2O_2)[a]

Product name	Manufacturer	Neutralizer formulation(s)
Oxysept Disinfecting/ Neutralizing System	Allergan Optical	Catalase solution (520 units/mL), EDTA (0.1%) Catalase tablet
UltraCare System	Allergan Optical	Catalase tablet
Hydrogen Peroxide System	Charter	Rinse/soak with saline solution, sorbic acid (0.1%), EDTA (0.1%)
AOSept Disinfection/ Neutralization System	CIBA Vision	Platinum-coated catalytic disk
Lensept Disinfection System	CIBA Vision	Catalase solution, sorbic acid, EDTA
Consept Cleaning and Disinfecting System	Paragon Vision Sciences	Sodium thiosulfate solution (0.5%), chlorhexidine gluconate (0.001%), EDTA (0.1%) Aerosol sodium thiosulfate solution (0.5%)
MiraSept Disinfecting/ Neutralizing System	Alcon Laboratories	Sodium pyruvate solution, EDTA

[a]For use with soft (hydrogel) contact lenses.

chemical lens care system. Table 4 contains a list of current mercurial-based disinfecting systems.

3. Quaternary Ammonium Compounds

Quaternary ammonium compounds are water-soluble cationic surface-active agents that are bactericidal against gram-positive bacteria, and to a lesser extent, gram-negative organisms (*Pseudomonas* and *Serratia* species are especially resistant to these compounds [1,41]). Furthermore, these compounds are less active against fungi, and solutions are occasionally formulated with additional agents (e.g., thimerosal) to increase microbial efficacy. Effects on the cytoplasmic membrane are the primary mode of action by these compounds. Alterations in the membrane can lead to leakage of metabolites and coenzymes, resulting in a loss of enzymatic activity [41]. In addition, the active compounds can denature proteins, dissociate enzymes, and may affect metabolic reactions.

The first-generation quaternary ammonium compound is benzalkonium chloride; its structure follows:

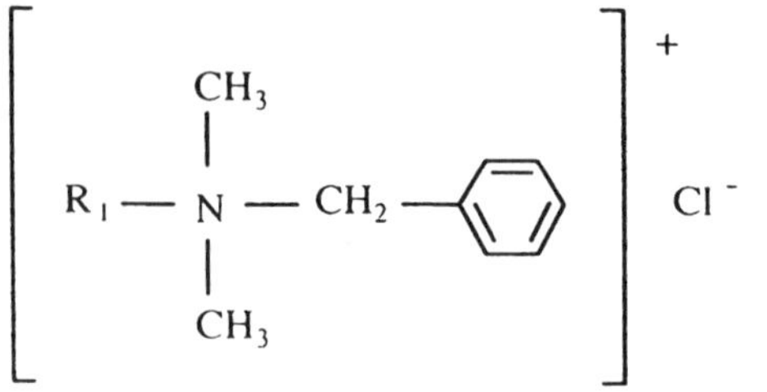

Benzalkonium chloride

$R_1 = C_{12}$ (40%), C_{14} (50%), C_{16} (10%)
$R_1 = C_{12}$ (5%), C_{14} (60%), C_{16} (30%), C_{18} (5%)

The contact lens industry first used benzalkonium chloride when developing soaking solutions for PMMA lenses. Later, when soft lenses were introduced, the industry recognized the need for soft lens disinfection systems, and simply suggested the use of PMMA solutions for soft lenses. Unfortunately, many of these solutions, particularly those preserved with benzalkonium chloride, were incompatible with hydrogel material. Benzalkonium chloride significantly binds to soft lenses, which causes an increased incidence of ocular irritation in many patients. This is why benzalkonium chloride is no longer used in soft contact lens care products.

A newer class of biocides, known as polymeric quaternaries, are less toxic than benzalkonium chloride. One compound, polyquaternium 1 (Polyquad), has been incorporated into several preserved contact lens solutions, and its structure is illustrated:

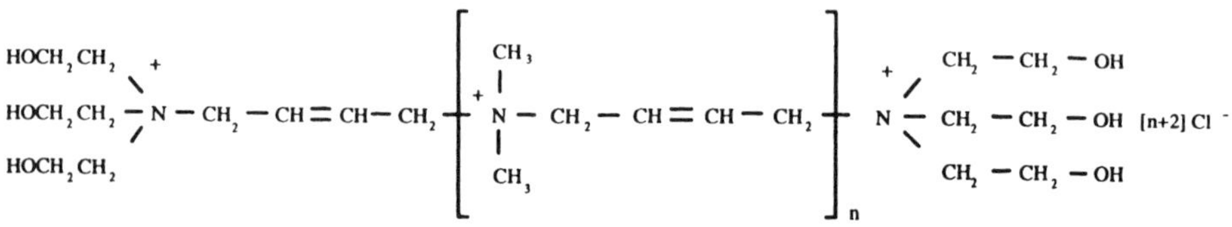

Polyquaternium 1 plus triethanolamine hydrochloride

This compound is used in two current disinfecting formulations: one in a citrate buffer system, and the other, a borate buffer. The latter solution is contraindicated for use with soft contact lenses having high water contents and polymers containing methacrylic acid (lenses may concentrate polyquaternium 1 by adsorption to levels that are toxic to the cornea [50]). Table 5 contains a list of disinfecting systems employing quaternary ammonium compounds as the active ingredient.

TABLE 4 Disinfection Sytems Employing Thimerosal

Product name	Manufacturer	Antimicrobial agent(s)	Approved lens type(s)[a]
Hydrocare Cleaning and Disinfecting Solution	Allergan Optical	Tris (2-hydroxyethyl) tallow ammonium chloride (0.013%), thimerosal (0.002%), and bis (2-hydroxyethyl) tallow ammonium chloride	Soft
Disinfecting Solution	Bausch & Lomb	Thimerosal (0.001%), chlorhexidine (0.005%), EDTA (0.1%)	Soft
Soaclens Wetting & Soaking Solution	Alcon Laboratories	Thimerosal (0.004%), EDTA (0.1%)	PMMA

[a]Soft (hydrogel); PMMA (polymethylmethacrylate).

TABLE 5 Disinfection Systems Employing Quaternary Ammonium Compounds

Product name	Manufacturer	Antimicrobial agent(s)	Approved lens type(s)[a]
OptiFree Rinsing, Disinfection and Storage Solution; Opti-One Multi-Purpose Solution	Alcon Laboratories	Polyquad (polyquaternium 1, 0.001%), EDTA (0.05%), in a citrate buffer system	Soft
Opti-Soft Disinfecting Solution	Alcon Laboratories	Polyquad (polyquaternium 1, 0.001%), EDTA (0.1%), in a borate buffer system	Soft[b]
Total	Allergan Optical	Benzalkonium chloride (0.004%), EDTA (0.127%)	PMMA
Wet-N-Soak Plus Wetting and Soaking Solution	Allergan Optical	Benzalkonium chloride (0.003%), EDTA (0.127%)	PMMA, SA, FSA
Soaking Solution	Lobob	Benzalkonium chloride (0.01%), EDTA (0.25%)	PMMA
Sereine Soaking/Cleaning	Optikem Int'l	Benzalkonium chloride (0.01%), EDTA (0.1%)	PMMA
Sereine Wetting/Soaking	Optikem Int'l	Benzalkonium chloride (0.01%), EDTA (0.1%)	SA
Cleaning and Soaking Solution	Paragon Vision Sciences	Benzalkonium chloride (0.01%), EDTA (0.2%)	SA, FSA

[a]Soft hydrogel; PMMA, polymethylmethacrylate; SA, siloxane acrylate rigid gas-permeable; FSA, fluoro-siloxane acrylate rigid gas-permeable.
[b]Soft lenses with 45% or less water content except CSI lens; 70–74% water content lenses except Permalens.

4. Chlorhexidine

Chlorhexidine is a cationic bis(biguanide) of the following formula:

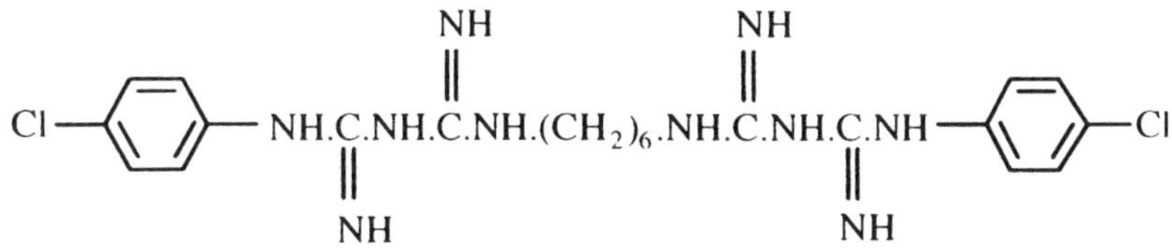

Chlorhexidine (1,6-di[4′-chlorophenyldiguanido]hexane)

This active agent is insoluble in water; however, contact lens disinfecting solutions have made use of the more soluble derivative, chlorhexidine digluconate. Chlorhexidine is bacteriostatic and bactericidal against a variety of organisms, notably gram-positive bacteria. It is less active against gram-negative bacteria, especially *Serratia* species [1,23,50]. Chlorhexidine is also active against yeasts, molds, protozoa, and certain lipophilic viruses (e.g., herpes, HIV, influenza). The mechanism of action of this agent against microorganisms, particularly bacteria, appears to consist of a series of events [14]. The sequence is thought to be as follows: (a) adsorption to the negatively charged bacterial surface; (b) overcoming cell wall exclusion mechanisms; (c) attraction towards the cytoplasmic membrane; (d) leakage of low molecular weight components (e.g., potassium ions) and inhibition of specific membrane-bound enzymes; (e) precipitation of the cytoplasm through complex formation with phosphated components (e.g., ATP and nucleotides). As the concentration of chlorhexidine increases, so does the leakage of higher molecular weight components, such as nucleotides.

Chlorhexidine is capable of binding to the surface of hydrogel polymers (although considerably less than benzalkonium chloride), particularly in areas of adsorbed tear components. This may lead to the formation of a yellowish discoloration, as well as increased ocular irritation. The symptoms of hypersensitivity to this agent are indistinguishable from those provoked by mercurial or quaternary ammonium compounds [32]. Table 6 contains a list of disinfecting systems employing chlorhexidine as the active ingredient. See Chapter 11 in this volume for additional information concerning chlorhexidine.

5. Polymeric Antimicrobial Agents

One class of antimicrobial compounds contain active agents within polymeric structural groups. These polymers are usually water-soluble, and their activity does not depend on molecular hydrolysis or other degradative reactions [37].

When it was determined that bis(biguanides), such as chlorhexidine, were more efficacious than monomeric biguanides, there became an interest in developing polymeric biguanides for use as novel antimicrobial agents. The resulting

TABLE 6 Disinfection Systems Employing Chlorhexidine

Product name	Manufacturer	Antimicrobial agent(s)	Approved lens type(s)[a]
Flex-Care Rinsing Disinfecting & Storage Solution	Alcon Laboratories	Chlorhexidine gluconate (0.005%), EDTA (0.1%)	Soft, SA
Disinfecting Solution	Bausch & Lomb	Chlorhexidine gluconate (0.005%), thimerosal (0.001%), EDTA (0.1%)	Soft
Soft Mate Disinfecting Solution	Paragon Vision Sciences	Chlorhexidine gluconate (0.005%), EDTA (0.1%)	Soft
Wetting and Soaking Solution	Bausch & Lomb	Chlorhexidine gluconate (0.006%), EDTA (0.05%)	PMMA, SA, FSA
Barnes Hind Comfort Care G.P. Wetting & Soaking Solution	Paragon Vision Sciences	Chlorhexidine gluconate (0.005%), EDTA (0.02%)	SA, FSA
Barnes Hind Gas Permeable Wetting and Soaking Solution	Paragon Vision Sciences	Chlorhexidine gluconate (0.005%), EDTA (0.02%)	SA, FSA
Boston Conditioning Solution	Polymer Technology	Chlorhexidine gluconate (0.006%), EDTA (0.05%)	SA

[a]Soft, hydrogel; PMMA, polymethylmethacrylate; SA, siloxane acrylate rigid gas-permeable; FSA, fluoro-siloxane acrylate rigid gas-permeable.

polymer was polyhexamethylene biguanide (PHMB) hydrochloride (otherwise known as polyaminopropyl biguanide [PAPB]), and its structure follows:

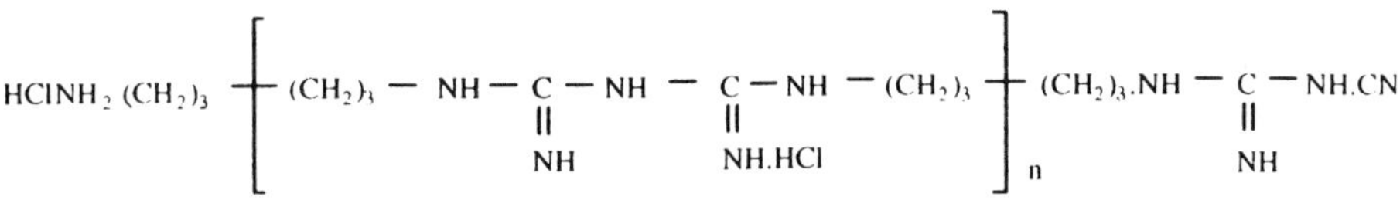

Polyhexamethylene biguanide (PHMB) hydrochloride

Although various terminal groups and molecular weights are associated with PHMB, contact lens disinfectants employ polymers terminated by amino groups, and have molecular weights in the 1000–3000 range.

This agent is bactericidal, fungicidal, virucidal, and amebicidal. The mechanism of action of PHMB is similar to other cationic compounds previously discussed; namely, quaternary ammonium compounds and chlorhexidine: cell death is primarily due to the irreversible loss of cellular components because of cytoplasmic membrane damage.

The use of PHMB as a contact lens disinfectant has several advantages over all other antimicrobial actives previously discussed. The molecule possesses high antimicrobial activity, even when used at low concentrations. It exhibits a lower binding affinity to soft contact lenses and, more importantly, offers the lowest ocular toxicity of all agents currently used in contact lens chemical disinfection systems (see Table 7 for a description of disinfecting systems employing PHMB).

6. Ethylenediaminetetraacetate

Ethylenediaminetetraacetic acid (edetic acid; EDTA) is a chelating agent that interacts with cations associated with the outer membrane of gram-negative

Table 7 Disinfection Systems Employing Polyhexamethylene Biguanide (PHMB) [Also Known as Polyaminopropyl Biguanide (PAPB)]

Product name	Manufacturer	Antimicrobial agent(s)	Approved lens type(s)[a]
ReNu Multi-Purpose Solution	Bausch & Lomb	DYMED (PAPB, 0.00005%), EDTA (0.1%)	Soft
Boston Advance Conditioning Solution	Polymer Technology	PAPB (0.0015%), EDTA (0.05%)	SA
Complete Multi-Purpose Solution	Allergan Optical	TrisChem (PHMB; 0.0001%), EDTA (0.05%)	Soft

[a]Soft, hydrogel; SA, siloxane acrylate rigid gas-permeable.

bacteria [56]. This compound causes the release of lipopolysaccharide and causes lysis in these types of bacteria, particularly *P. aeruginosa*. Most importantly, EDTA can potentiate the effects of chemically unrelated antimicrobial compounds, such as benzalkonium chloride and chlorhexidine. On the other hand, EDTA has no effect on gram-positive bacteria, yeasts, or molds, nor does it increase these organisms' sensitivity to other agents. Many chemical contact lens-disinfecting systems employ EDTA in their formulations; these may be listed in the tables located throughout this chapter.

7. Alcohols

Alcohols have been used for centuries as disinfectants and antiseptics (see Chap. 10 in this volume for additional information). They have a bactericidal, rather than a bacteriostatic effect on vegetative organisms; the mechanism of action include protein denaturation, interference with metabolism (e.g., inhibition of enzymes and metabolites required for cell division), and cellular lysis [33]. In addition, alcohols have demonstrated activity against fungi and viruses. Alcohols have traditionally been used to disinfect inanimate surfaces and for the removal of transient or pathogenic organisms on the skin. Isopropyl and benzyl alcohols are used in current contact lens disinfecting systems (Table 8). Isopropyl alcohol has produced toxic reactions when applied to the skin; therefore, it is important that alcohol compounds be removed from contact lenses before placing them on the eye.

TABLE 8 Disinfection Systems Employing Alcohol

Product name	Manufacturer	Antimicrobial agent(s)	Approved lens type(s)[a]
Cleaning, Disinfecting, Storage Solution	Lobob	Benzyl alcohol (0.1%), EDTA (0.5%)	SA, FSA
de-STAT 3	Sherman Pharmaceuticals	Benzyl alcohol (0.1%), EDTA (0.5%)	PMMA, SA, FSA
de-STAT 4	Sherman Pharmaceuticals	Benzyl alcohol (0.3%), EDTA (0.5%)	PMMA, SA, FSA
Quick Care Starting Solution	CIBA Vision	Isopropyl alcohol (16%)	Soft

[a]Soft, hydrogel; PMMA, polymethylmethacrylate; SA, siloxane acrylate rigid gas-permeable; FSA, fluoro-siloxane acrylate rigid gas-permeable.

V. EVALUATING SOFT CONTACT LENS DISINFECTION SYSTEMS

The U. S. Food and Drug Administration (FDA) has classified contact lenses as class II (lenses used for daily wear [amended as of April, 1994]) or class III (lenses used for extended wear), and all contact lens solutions and heat disinfection units as class III [71]. Class II devices generally are tested against performance standards to assure reasonable safety and efficacy before these devices are commercially distributed. Class III devices require extensive preclinical and clinical testing because (a) insufficient information exists to establish a performance standard, or (b) the device presents a potential unreasonable risk of illness or injury (e.g., corneal injury resulting from extended wearing of contact lenses). Although contact lens solutions were originally labeled class III, the FDA is currently considering reclassifying these products as class II.

The Division of Ophthalmic Devices of the Center for Devices and Radiological Health (FDA) has developed a guidance document [70] for manufacturers of soft contact lens solutions who are performing preclinical and clinical testing required for a Premarket Approval Application (PMA) or supplement, or an Investigation Device Exemption (IDE). The document outlines microbiology and toxicity requirements that must be met to support the safety and effectiveness of a contact lens solution. Microbiological testing evaluates (a) sterilization; (b) shelf life; and (c) disinfection characteristics of the solution. Safety testing assesses the potential of the solution to produce an (a) acute oral toxicity, (b) acute systemic toxicity, (c) acute ocular irritation and cytotoxicity, or (d) allergic response caused by sensitization (refer to the guideline [70] for additional information, protocols, and test requirements).

A. Sterilization Requirements

The manufacturer must demonstrate that the contact lens solution will be sterile when shipped for distribution or use:

1. Sterilization cycles must be validated.
2. Product sterility must be confirmed according to the current *United States Pharmacopeia* (*USP*) [73].
3. Preservative systems in lens care solutions must meet the current *USP* Antimicrobial Preservatives—Effectiveness Test with a significant rechallenge added for multiple dose containers. The FDA modifications of the *USP* test include (a) an initial inoculation concentration equal to 1×10^6 CFU/ml; (b) a rechallenge of the original inoculated tubes with 1×10^5 CFU/ml after the *USP*-required 14-day incubation period; and (c) incubation for an additional 14 days. The acceptance criteria for the rechallenge test are the same as those described in the *USP* procedure for the original challenge [72].

B. Shelf Life-Testing Requirements

1. The manufacturer must demonstrate that the contact lens solution will remain sterile for the recommended shelf life as packaged; samples are tested for sterility as described in the previous section.
2. For any preserved solution, the antimicrobial preservative effectiveness must be demonstrated for the proposed shelf life; samples are tested as described earlier.

C. Disinfection Requirements

A manufacturer must demonstrate that in the hands of the consumer, a representative lens from each polymer group can be effectively disinfected by the recommended lens care system. The following tests are required for chemical disinfection systems: (a) "contribution of elements" test; (b) D-value determinations; and (c) multi-item microbial challenge test (a practical consumer-use type evaluation of the process employing bacterial, fungal, and viral strains). Tests required for thermal disinfection units include (a) submitting time–temperature curves demonstrating consistency and uniformity of heating performance (units must reach and maintain a temperature of at least 80°C for 10 min), and (b) performing a multi-item microbial challenge test against a single bacterial strain.

1. Contribution of Elements Test

If a regimen other than simple exposure to heat or a chemical is considered as a disinfection system, its various parts must be evaluated for their contribution to microbicidal activity or physical–chemical removal of microorganisms. One part of the system must be amenable to D-value determination (a *D-value* is defined as the time required to reduce a microbial population by 1 log [90%]). A theoretical model follows:

Cleaning step	Disinfection step	Rinsing step
Evaluate log reduction	Evaluate D-values	Evaluate log reduction

To evaluate log reductions for the cleaner (or rinse), 20 contact lenses are separately inoculated with at least two challenge organisms (see next section for description) suspended in a proteinaceous and particulate soil (heat-killed cells of *Saccharomyces cerevisiae* in heat-inactivated serum; this is referred to as an "organic load" and is intended to mimic deposits on lenses from ocular secretions). Both sides of the lens are inoculated with a known volume (e.g., 0.01 ml/side), and the final concentration of each microbial challenge is between 0.5 and 2.0×10^6 CFU/lens. Following a 3- to 10-min–contact period the cleaner (or rinse) is applied as directed for patient use. Each lens is then transferred into a known volume of phosphate buffer and mixed for 30 s (if the cleaner [or rinse] contains a preservative, a suitable neutralizer is also added to the buffer). Plate counts are performed, and the number of viable organisms are compared with initial inoculum levels.

2. D-Value Determinations

The D-value provides a reliable estimation of the time required to kill any specified level of cells so that the probability of surviving organisms after exposure to a disinfecting agent can be determined. The test is performed with inoculated lenses or by introducing microorganisms directly into the disinfecting solution being evaluated. The FDA recommends three methods for determining D-values [70]. The first, and most commonly used, is the Stumbo or endpoint method [65]. Briefly, microorganisms are exposed to a disinfecting agent for a specified amount of time, and the number of survivors determined. The D-value is calculated using the following formula:

$$\text{D-value} = \frac{\text{time of exposure}}{(\log N_0 - \log N_t)}$$

where

N_0 = number of organisms at the initial time point
N_t = number of organisms at the end of the exposure time point

An averaging of the multiple endpoint D-values is required in the second method, whereas linear regression analysis of multiple data points is used in the third. These methods will provide similar D-values as long as microbial death follows first-order kinetics. On the other hand, if kill kinetics are nonlinear, then differences in observed D-values for a particular disinfection system may result [27,52,53,65]. Finally, the FDA requires that disinfection cycles be calculated by using the number of D-values to bring the level of contamination to log 0 (one organism), and then adding an additional three D-values to yield the total disinfection time. This calculation must be based on the highest D-value found among the following six test organisms:

Staphylococcus epidermidis	ATCC 17917
Pseudomonas aeruginosa	ATCC 15442
Serratia marcescens	ATCC 14041
Candida albicans	ATCC 10231
Aspergillus fumigatus	ATCC 10894
Herpes simplex virus	ATCC VR260

With an initial challenge population equal to 10^6 the resulting disinfection efficacy corresponds to a 9-log reduction.

3. Multi-item Microbial Challenge Test for Chemical Systems

The multi-item microbial challenge test (MIMCT) evaluates the bactericidal, fungicidal, and virucidal effectiveness of a disinfecting solution or regimen. The test incorporates a specific soft contact lens, or lenses representing the four polymer groups, if the system will be labeled for use with different lens types. To perform the test, 20 contact lenses are inoculated with 10^6 microorganisms in an organic load, as previously described (see Sec. V.C.1). Each of the challenge organisms listed in the previous section (V.C.2) are individually tested against the disinfection system being evaluated (refer to the FDA guideline [70] for a detailed protocol on testing herpes simplex virus). Inoculated lenses are then subjected to the disinfection regimen as directed for patient use. At the end of the disinfection regimen, each of the previously inoculated lenses are aseptically transferred into neutralizing broth media (a neutralizer is used to inactivate any of the antimicrobial agent that may be carried over; otherwise, an overestimation of microbial kill may result). When applicable, the solution from which the lens is removed (e.g., within a lens case) is cultured separately. All lenses and corresponding solutions are incubated for 14 days; bacterial cultures are usually incubated at 30–35°C, and yeast and mold cultures at 20–25°C. For the test to be acceptable, there should be no growth in the neutralizing media for either the test lenses or the corresponding solutions at the end of the specified incubation time.

Because a neutralizer is used during testing, it is imperative that two aspects of the neutralizer be tested: (a) the inherent toxicity of the neutralizer to each of the challenge organisms employed in the test, and (b) the efficacy of the neutralizer in counteracting the antimicrobial effect of the disinfection system.

The toxicity of a neutralizer is evaluated by comparing the response of each challenge organism with the neutralizer and with a nonneutralized, buffer solution. Each solution is inoculated with 10^5 organisms per milliliter, and viable counts are determined at specified intervals (e.g., 1, 10, 30, and 60 min). Neutralizer efficacy is assessed from comparisons of the response of each challenge organism with the neutralizer, the neutralizer–disinfectant solution (e.g., 1.0 ml disinfectant in 9.0 ml neutralizing solution), and a nonneutralized, buffer solution. Each solution is inoculated with 10^6 organisms per milliliter, and viable counts determined at specified intervals (e.g., 3, 10, and 30 min).

4. Multi-item Microbial Challenge Test for Thermal Systems

The procedure used to test thermal disinfection systems is similar to the procedure described for chemical systems, except that only one challenge organism is required (*Streptococcus faecalis*, Ward's strain 85W 1100).

5. Future Chemical Disinfection System Requirements

Recently, an *American National Standards Institute* (ANSI) committee was created to participate in developing an ISO standard entitled "Performance of Products for Disinfection of Contact Lenses" [3]. The standard's acceptance criteria depends on which protocol is utilized; the "stand-alone" challenge or the "disinfection regimen" procedure.

The stand-alone test challenges a disinfecting product with a standard inoculum of the following microorganisms:

Staphylococcus aureus	ATCC 6538
Pseudomonas aeruginosa	ATCC 9027
Serratia marcescens	ATCC 13880
Candida albicans	ATCC 10231
Fusarium solani	ATCC 36031

Products are inoculated with a suspension of test organisms sufficient to provide a final concentration of 1.0×10^5–1×10^6 per milliliter. Viable counts are determined at 25, 50, 75, and 100% of the minimum recommended disinfection time for all organisms and, in addition, not less than four times the minimum disinfection time for yeast and mold. Where overnight disinfection is recommended, disinfection time is taken to be 8 h. Additionally, validated measures are taken to inactivate or remove residual antimicrobial agents during the culturing and counting of survivors. The primary acceptance criteria for bacteria requires at least a 3-log (99.9%) reduction within the minimum recommended disinfection period; for yeasts and molds a 1-log (90%) reduction is required, with no increase at not less than four times the minimum recommended disinfection time.

Products failing to meet the foregoing criteria may be evaluated by the regimen procedure, provided there is a combined log reduction of not less than 5 for all three bacteria. In addition, the minimum acceptable log reduction for any single bacterium is 1. Finally, stasis for the yeast and mold must be observed. The regimen procedure employs inoculated contact lenses and is similar to the multiitem microbial challenge previously described. Fewer than 10 CFU of bacteria, yeasts, or molds recovered from the lens–storage solution combination is acceptable.

At the time of this writing, the FDA was considering adopting these requirements for approval of products to be marked as chemical disinfecting systems in the United States [the foregoing procedures and acceptance criteria are in draft form and final requirements may change].

VI. CONCLUSION

Disinfection systems are required to effectively reduce the number of microorganisms on a contact lens before the lens is placed on the eye. However, the chemical properties of antimicrobial agents used in these systems may also make them toxic to ocular tissue. The ideal solution, therefore, is one in which a delicate balance between antimicrobial efficacy and ocular toxicity exists. The development of novel antimicrobial compounds, such as PHMB, has brought us closer toward achieving this goal.

Finally, the ultimate responsibility for safe and effective lens wear relies on the consumer. Although disinfection systems are proved to be safe and effective, the lens wearer must comply with all recommended directions for use, use adequate hygiene during lens handling, and seek the advise of their practitioner at the earliest sign of ocular redness or discomfort.

ACKNOWLEDGMENTS

The author gratefully acknowledges Dr. Donald G. Ahearn for his review of, and contributions to, this manuscript. In addition, the author wishes to thank Samantha, Max, and Susan for their patience and understanding.

REFERENCES

1. D. G. Ahearn, C. A. Penley, and L. A. Wilson, Growth and survival of *Serratia marcescens* in hard contact lens wetting solutions, *CLAO J. 10*:172–174 (1984).
2. E. Alfonso, S. Mandelbaum, M. J. Fox, and R. K. Forster, Ulcerative keratitis associated with contact lens wear, *Am. J. Ophthalmol. 101*:429–433 (1986).
3. American National Standards Institute, Performance of products for disinfection of contact lenses (draft). Z80 Subcommittee for Contact Lenses. American National Standards Institute, 1994.
4. S. S. Block, Peroxygen compounds, *Disinfection, Sterilization, and Preservation*, 4th ed. (S. S. Block, ed.), Lea & Febiger, Malvern, PA, 1991, pp. 167–181.
5. F. W. Bowden, E. J. Cohen, J. J. Arentsen, and P. R. Laibson, Patterns of lens care practices and lens product contamination in contact lens associated microbial keratitis, *CLAO J. 15*:49–54 (1989).
6. J. R. Buffington, R. B. Simmons, L. A. Wilson, and D. G. Ahearn, Effect of polymer differences on fungal penetration of hydrogel lenses, *J. Ind. Microbiol. 3*:29–32 (1988).
7. S. I. Butrus, S. A. Klotz, and R. P. Misra, The adherence of *Pseudomonas aeruginosa* to soft contact lenses, *Ophthalmology 94*:1310–1314 (1987).
8. S. I. Butrus and S. A. Klotz, Contact lens surface deposits increase the adhesion of *Pseudomonas aeruginosa*, *Curr. Eye Res. 9*:717–724 (1990).
9. M. W. Chun and B. A. Weissman, Compliance in contact lens wear, *Am. J. Optom. Physiol. Opt. 64*:274–276 (1987).
10. E. J. Cohen, H. W. Buchanan, P. A. Laughrea, L. P. Adams, P. G. Galentine, G. S. Visvesvara, R. Folberg, J. J. Arentsen, and P. R. Laibson, Diagnosis and management of *Acanthamoeba* keratitis, *Am. J. Ophthalmol. 100*:389–395 (1985).
11. E. J. Cohen, C. J. Parlato, J. J. Arentsen, G. I. Genvert, R. C. Eagle, Jr., M. R. Wieland, and P. R. Laibson, Medical and surgical treatment of *Acanthamoeba* keratitis, *Am. J. Ophthalmol. 103*:615–625 (1987).
12. M. J. Collins and L. G. Carney, Patient compliance and its influence on contact lens wearing problems, *Am. J. Optom. Physiol. Opt. 63*:952–956 (1986).

13. J. K. G. Dart, F. Stapleton, and D. Minassian, Contact lenses and other risk factors in microbial keratitis, *Lancet 338*:650–653 (1991).
14. G. W. Denton, Chlorhexidine, *Disinfection, Sterilization, and Preservation*, 4th ed. (S. S. Block, ed.), Lea & Febiger, Malvern, PA, 1991, pp. 274–289.
15. P. B. Donzis, B. J. Mondino, B. A. Weissman, and D. A. Bruckner, Microbial contamination of contact lens care systems, *Am. J. Ophthalmol. 104*:325–333 (1987).
16. P. B. Donzis, B. J. Mondino, and B. A. Weissman, Microbial analysis of contact lens care systems contaminated with *Acanthamoeba*, *Am. J. Ophthalmol. 108*:53–56 (1989).
17. J. A. Duran, M. F. Refojo, I. K. Gipson, and K. R. Kenyon, *Pseudomonas* attachment to new hydrogel contact lenses, *Arch. Ophthalmol. 105*:106–109 (1987).
18. N. Efron, A. Wohl, N. G. Toma, L. W. J. Jones, and R. Lowe, *Pseudomonas* corneal ulcers associated with daily wear of disposable hydrogel contact lenses, *ICLC 18*: 46–51 (1991).
19. R. A. Fassihi, Preservation of medicines against microbial contamination, *Disinfection, Sterilization, and Preservation*, 4th ed. (S. S. Block, ed.), Lea & Febiger, Malvern, PA, 1991, pp. 871–886.
20. S. A. Fowler and M. R. Allansmith, Evolution of soft contact lens coatings, *Arch. Ophthalmol. 98*:95–99 (1980).
21. S. A. Fowler, J. V. Greiner, and M. R. Allansmith, Attachment of bacteria to soft contact lenses, *Arch. Ophthalmol. 97*:659–660 (1979).
22. P. G. Galentine, E. J. Cohen, P. R. Laibson, C. P. Adams, R. Michaude, and J. J. Arentsen, Corneal ulcers associated with contact lens wear, *Arch. Ophthalmol. 102*:891–894 (1984).
23. P. A. Gandhi, A. D. Sawant, L. A. Wilson, and D. G. Ahearn, Adaptation and growth of *Serratia marcescens* in contact lens disinfectant solutions containing chlorhexidine gluconate, *Appl. Environ. Microbiol. 59*:183–188 (1993).
24. G. Garrigue, S. Bremont, A. M. Burguiere, and J. D. Poveda, Bactericidal activity of six contact lens care solutions—a comparative study, *Contactologia 17F*:1–10 (1995).
25. N. Grier, Mercurials—inorganic and organic, *Disinfection, Sterilization, and Preservation*, 3rd ed. (S. S. Block, ed.), Lea & Febiger, Philadelphia, PA, 1983, pp. 346–374.
26. B. A. Holden, D. F. Sweeney, A. Vannas, K. T. Nilsson, and N. Efron, Effects of long-term extended contact lens wear on the human cornea, *Invest. Ophthalmol. Vis. Sci. 26*:1489–1501 (1985).
27. R. D. Houlsby, M. Ghajar, and G. Chavez, Microbial evaluation of soft contact lens disinfection solutions, *J. Am. Optom. Assoc. 55*:205–211 (1984).
28. G. Hovding, The conjunctival and contact lens bacterial flora during lens wear, *Acta Ophthalmol. 59*:387–401 (1981).
29. T. John, D. Desai, and D. Sahm, Adherence of *Acanthamoeba castellani* cysts and trophozoites to unworn soft contact lenses, *Am. J. Ophthalmol. 108*:658–664 (1989).
30. K. J. Johns, D. M. O'Day, and S. S. Feman, Chorioretinitis in the contralateral eye of a patient with *Acanthamoeba* keratitis, *Ophthalmology 95*:635–639 (1988).
31. K. J. Johns, W. S. Head, C. M. Parrish, T. E. Williams, R. D. Robinson, and D. M. O'Day, Examination of hydrophilic contact lenses with light microscopy to aid in the diagnosis of *Acanthamoeba* keratitis, *Am. J. Ophthalmol. 108*:329–331 (1989).

32. J. Larke, ed., *The Eye in Contact Lens Wear*, Butterworths, London, 1986.
33. E. L. Larson and H. E. Morton, Alcohols, *Disinfection, Sterilization, and Preservation*, 4th ed. (S. S. Block, ed.), Lea & Febiger, Malvern, PA, 1991, pp. 191–203.
34. T. D. Lindquist, N. A. Sher, and D. J. Doughman, Clinical signs and medical therapy of early *Acanthamoeba* keratitis, *Arch. Ophthalmol. 106*:73–77 (1988).
35. S. Liotet, D. Guillaumin, P. Cochet, V. N. Warnel, and H. D. Cao, The genesis of organic deposits on soft contact lenses, *CLAO J. 9*:49–56 (1983).
36. S. MacRae, C. Herman, D. Stulting, R. Lippman, D. Whipple, E. Cohen, D. Egan, C. P. Wilkinson, C. Scott, R. Smith, and D. Phillips, Corneal ulcer and adverse reaction rates in premarket contact lens studies, *Am. J. Ophthalmol. 111*:457–465 (1991).
37. O. W. May, Polymeric antimicrobial agents, *Disinfection, Sterilization, and Preservation*, 4th ed. (S. S. Block, ed.), Lea & Febiger, Malvern, PA, 1991, pp. 322–333.
38. M. S. Mayo, W. L. Cook, R. L. Schlitzer, M. A. Ward, L. A. Wilson, and D. G. Ahearn, Antibiograms, serotypes, and plasmid profiles of *Pseudomonas aeruginosa* associated with corneal ulcers and contact lens wear, *J. Clin. Microbiol. 24*:372–376 (1986).
39. M. S. Mayo, R. L. Schlitzer, M. A. Ward, L. A. Wilson, and D. G. Ahearn, Association of *Pseudomonas* and *Serratia* corneal ulcers with use of contaminated solutions, *J. Clin. Microbiol. 25*:1398–1400 (1987).
40. M. E. McBride, Evaluation of microbial flora of the eye during wear of soft contact lenses, *Appl. Environ. Microbiol. 37*:233–236 (1979).
41. J. J. Merianos, Quaternary ammonium antimicrobial compounds, *Disinfection, Sterilization, and Preservation*, 4th ed. (S. S. Block, ed.), Lea & Febiger, Malvern, PA, 1991, pp. 225–255.
42. M. J. Miller, and D. G. Ahearn, Adherence of *Pseudomonas aeruginosa* to hydrophilic contact lenses and other substrata, *J. Clin. Microbiol. 25*:1392–1397 (1987).
43. M. J. Miller, L. A. Wilson, and D. G. Ahearn, Effects of protein, mucin, and human tears on adherence of *Pseudomonas aeruginosa* to hydrophilic contact lenses, *J. Clin. Microbiol. 26*:513–517 (1988).
44. M. J. Miller, L. A. Wilson, and D. G. Ahearn, Adherence of *Pseudomonas aeruginosa* to rigid gas-permeable contact lenses, *Arch. Ophthalmol. 109*:1447–1448 (1991).
45. B. J. Mondino, B. A. Weissman, M. D. Farb, and T. H. Pettit, Corneal ulcers associated with daily wear and extended wear contact lenses, *Am. J. Ophthalmol. 102*: 58–65 (1986).
46. M. B. Moore, J. P. McCulley, C. Newton, L. M. Cobo, G. N. Foulks, D. M. O'Day, K. J. Johns, W. T. Driebe, L. A. Wilson, R. J. Epstein, and D. J. Doughman, *Acanthamoeba* keratitis. A growing problem in soft and hard contact lens wearers. *Ophthalmology 94*:1654–1661 (1987).
47. M. B. Moore, *Acanthamoeba* keratitis, *Arch. Ophthalmol. 106*:1181–1183 (1988).
48. M. B. Moore, J. P. McCulley, H. E. Kaufman, and J. B. Robin, Radial keratoneuritis as a presenting sign in *Acanthamoeba* keratitis, *Ophthalmology 93*:1310–1315 (1986).
49. M. B. Moore, J. P. McCulley, M. Luckenbach, H. Gelender, C. Newton, M. B. McDonald, and G. S. Visvesvara, *Acanthamoeba* keratitis associated with soft contact lenses, *Am. J. Ophthalmol. 100*:396–403 (1985).
50. J. F. Morgan, Soft contact lens hygiene—clinical experience, *Contact Lenses. The CLAO Guide to Basic Science and Clinical Practice*, Vol 2, 2nd ed. (O. H. Dabezies, Jr., ed.), Little, Brown, & Co., Boston, 1988, pp. 40A.1–18.

51. C. A. Penley, C. Llabres, L. A. Wilson, and D. G. Ahearn, Efficacy of hydrogen peroxide disinfection systems for soft contact lenses contaminated with fungi, *CLAO J. 11*:65–68 (1985).
52. C. A. Penley, D. G. Ahearn, R. L. Schlitzer, and L. A. Wilson, Laboratory evaluation of chemical disinfection of soft contact lenses, *Contact Intraocul. Lens Med. J. 7*:101–110 (1981).
53. C. A. Penley, D. G. Ahearn, R. L. Schlitzer, and L. A. Wilson, Laboratory evaluation of chemical disinfection of soft contact lenses. II. Fungi as challenge organisms, *Contact Intraocul. Lens Med. J. 7*:196–204 (1981).
54. E. C. Poggio, R. J. Glynn, O. D. Schein, J. M. Seddon, and K. R. Kenyon, The incidence of ulcerative keratitis among users of daily-wear and extended-wear soft contact lenses, *N. Engl. J. Med. 321*:779–783 (1989).
55. R. Ramphal, M. T. McNiece, and F. P. Polack, Adherence of *Pseudomonas aeruginosa* to the injured cornea: A step in the pathogenesis of corneal infections, *Ann. Ophthalmol. 13*:421–425 (1981).
56. A. D. Russell, Principles of antimicrobial activity, *Disinfection, Sterilization, and Preservation*, 4th ed. (S. S. Block, ed.), Lea & Febiger, Malvern, PA, 1991, pp. 27–58.
57. O. D. Schein, R. J. Glynn, E. C. Poggio, J. M. Seddon, K. R. Kenyon, and The Microbial Keratitis Study Group, The relative risk of ulcerative keratitis among users of daily-wear and extended-wear soft contact lenses, *N. Engl. J. Med. 321*:773–778 (1989).
58. O. D. Schein, L. D. Ormerod, E. Barraquer, E. Alfonso, K. M. Egan, B. G. Paton, and K. R. Kenyon, Microbiology of contact lens-related keratitis, *Cornea 8*:281–285 (1986).
59. K. L. Shih, J. Hu, and M. J. Sibley, The microbial benefit of cleaning and rinsing contact lenses, *ICLC 12*:235–242 (1985).
60. R. B. Simmons, J. R. Buffington, M. Ward, L. A. Wilson, and D. G. Ahearn, Morphology and ultrastructure of fungi in extended-wear soft contact lenses, *J. Clin. Microbiol. 24*:21–25 (1986).
61. F. Stapleton, J. K. G. Dart, and D. Minassian, Risk factors with contact lens related suppurative keratitis, *CLAO J. 19*:204–210 (1993).
62. J. K. Stehr-Green, T. M. Bailey, and G. S. Visvesvara, The epidemiology of *Acanthamoeba* keratitis in the United States, *Am. J. Ophthalmol. 107*:331–336 (1989).
63. G. A. Stern and Z. S. Zam, The pathogenesis of contact lens-associated *Pseudomonas aeruginosa* corneal ulceration, I: the effect of contact lens coatings on adherence of *Pseudomonas aeruginosa* to soft contact lenses, *Cornea 5*:41–45 (1986).
64. G. A. Stern, A. Lubniewski, and C. Allen, The interaction between *Pseudomonas aeruginosa* and the corneal epithelium, *Arch. Ophthalmol. 103*:1221–1225 (1985).
65. S. V. W. Sutton, R. J. Franco, D. A. Porter, M. F. Mowrey-McKee, S. C. Busschaert, J. F. Hamberger, and D. W. Proud, D-value determinations are an inappropriate measure of disinfecting activity of common contact lens disinfecting solutions, *Appl. Environ. Microbiol. 57*:2021–2026 (1991).
66. T. Tervo, J. Lähdevitra, A. Vaheri, S. L. Valle, and J. Suni, Recovery of HTLV-III from contact lenses, *Lancet 1*:379–380 (1986).
67. R. C. Tripathi, B. J. Tripathi, and M. Ruben, The pathology of soft contact lens spoilage, *Ophthalmology 87*:365–380 (1980).

68. R. C. Tripathi and B. J. Tripathi, Soft lens spoilage: Nature, etiology, and pathogenesis, *Ophthalmol. Forum 2*:80–91 (1984).
69. B. J. Tripathi and R. C. Tripathi, Hydrogen peroxide damage to human corneal epithelial cells in vitro. Implications for contact lens disinfecting systems, *Arch. Ophthalmol. 107*:1516–1519 (1989).
70. United States Food and Drug Administration, Testing guidelines for class III soft (hydrophilic) contact lens solutions (draft), U. S. Food and Drug Administration, Washington, DC, 1985.
71. United States Food and Drug Administration, Part 886—Ophthalmic devices, *Code of Federal Regulations 21*. U. S. Food and Drug Administration, Washington, DC, 1992, pp. 369–389.
72. United States Pharmacopeial Convention, Microbiological tests. Antimicrobial preservatives—effectiveness, *United States Pharmacopeia 23. National Formulary 18*, United States Pharmacopeial Convention, Rockville, MD, 1994, p. 1681.
73. United States Pharmacopeial Convention, Sterility tests, *United States Pharmacopeia 23. National Formulary 18*, United States Pharmacopeial Convention, Rockville, MD, 1994, p. 1681.
74. M. A. Ward and M. J. Miller, The microbiology of contact lens wear, *Contact Lens Forum 13*:25–29 (1988).
75. K. R. Wilhelmus, N. M. Robinson, R. A. Font, M. B. Hamill, and D. B. Jones, Fungal keratitis in contact lens wearers, *Am. J. Ophthalmol. 106*:708–714 (1988).
76. K. R. Wilhelmus, M. S. Osato, R. L. Font, N. M. Robinson, and D. B. Jones, Rapid diagnosis of *Acanthamoeba* keratitis using calcofluor white, *Arch. Ophthalmol. 104*:1309–1312 (1986).
77. L. A. Wilson, R. L. Schlitzer, and D. G. Ahearn, *Pseudomonas* corneal ulcers associated with soft contact lens wear, *Am. J. Ophthalmol. 92*:546–554 (1981).
78. L. A. Wilson and D. G. Ahearn, Association of fungi with extended wear soft contact lenses, *Am. J. Ophthalmol. 101*:434–436 (1986).
79. L. A. Wilson, A. D. Sawant, R. B. Simmons, and D. G. Ahearn, Microbial contamination of contact lens storage cases and solutions, *Am. J. Ophthalmol. 110*:193–198 (1990).
80. L. A. Wilson, A. D. Sawant, and D. G. Ahearn, Comparative efficacies of soft contact lens disinfectant solutions against microbial films in lens cases, *Arch. Ophthalmol. 109*:1155–1157 (1991).
81. L. A. Wilson, J. McNatt, and R. Reitschel, Delayed hypersensitivity to thimerosal in soft contact lens wearers, *Ophthalmology 88*:804–808 (1981).
82. T. Yamaguchi, A. Hubbard, A. Fukushima, T. Kimura, and H. E. Kaufman, Fungus growth on soft contact lenses with different water contents, *Ann. Ophthalmol. 11*:1731–1735 (1984).

7

Glutaraldehyde-Based Disinfectants

Joseph M. Ascenzi
Johnson & Johnson Medical, Inc.
Arlington, Texas

I. INTRODUCTION

Aldehydes, especially glutaraldehyde and formaldehyde, have been used as antimicrobial agents for the past several decades. Both have exhibited excellent antimicrobial properties, specifically in their use as sporicidal agents in sterilization. Pepper and Lieberman [1] describe the use of glutaraldehyde as a replacement for formaldehyde, and Stonehill and associates [2] showed the rapid sporicidal activity achieved with a 2% glutaraldehyde formulation, which became the prototype of the several glutaraldehyde formulations currently available for use as a disinfectant and cold sterilization solution.

Formaldehyde and glutaraldehyde are the best-known aldehydes in terms of practical antimicrobial compounds, but there are several other aldehydes that exhibit antimicrobial activity, including benzaldehyde and substituted benzaldehydes, succinaldehyde, malonaldehyde, glyoxal, and several of the α,β-unsaturated aldehydes. Of all the aldehydes that have been studied, more information is available on glutaraldehyde in terms of activity and safety. Several of the α,β-unsaturated aldehydes (e.g., cinnamaldehyde) have been used in the flavor and food industry, but little has been done to take advantage of their antimicrobial activity and low toxicity. Products containing combinations of glutaraldehyde and other aldehydes, especially formaldehyde, have been marketed primarily in European countries, but because of the uncertainty of formaldehyde's safety, these products have not been sold in the United States.

Because of the substantial antimicrobial activity and the broad-spectrum activities associated with glutaraldehyde, it has become the chemical of choice for rapid disinfection of medical and dental equipment. However, because of problems with irritating vapors produced by glutaraldehyde, much effort has gone into reformulation of products with reduced aldehyde content, yet equivalent or increased activity. A review of the available information on aldehydes and their antimicrobial activity is presented. The wealth of information available is concerned with the activity of glutaraldehyde, since this has been the aldehyde of choice for use in chemical disinfectants.

II. CHEMICAL NATURE OF ALDEHYDES AND RELATIONSHIP TO ANTIMICROBIAL ACTIVITY

Aldehydes have the general structure shown in Figure 1, where R = alkyl, benzyl, or carbonyl moiety. The simplest monoaldehyde (other than formaldehyde)—acetaldehyde (CH_2-CHO)—is a highly reactive compound, but does not have good antimicrobial activity, possibly because it is easily metabolized to its corresponding acid, acetic acid, and becomes incorporated into cell constituents. The corresponding dialdehyde, glyoxal (CHO−CHO) does show substantially more activity than the monoaldehyde. [In general, the dialdehydes do show substantially greater antimicrobial activity than the corresponding monoaldehydes.] Egyud [3] showed the relative activity of low concentrations of a homologous series of saturated monoaldehydes on inhibition of cell division of *Escherichia*

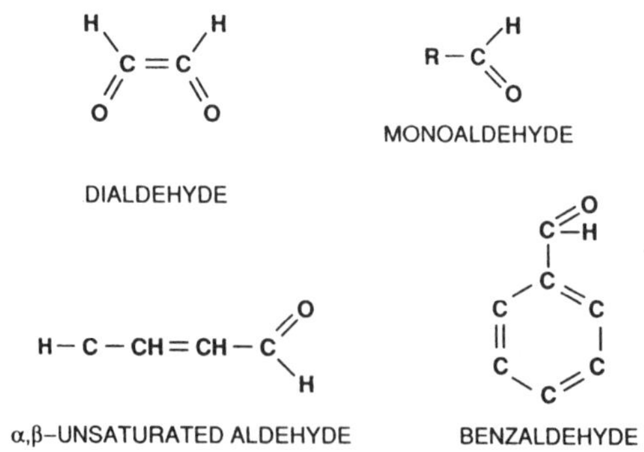

FIGURE 1 General chemical structure of aldehydes.

coli, in which formaldehyde and acetaldehyde exhibited equal and the greatest activity, followed by *n*-butyraldehyde and *n*-valeraldehyde, *n*-heptaldehyde, and propionaldehyde. The inhibitory effect was transient for all the compounds except formaldehyde and acetaldehyde. When examining sporicidal activity of aldehydes, the same trend is seen (i.e., the longer-chain compounds have less activity, with the exception of glutaraldehyde; Table 1). Power and Russell [4] demonstrated that 2% glutaraldehyde was much more active against *Bacillus subtilis* spores than 8% formaldehyde, 10% glyoxal, 10% butyraldehyde, or a formulation containing formaldehyde and butane-1,4-dial-2,5-dimethoxytetrahydrofuran.

Although not used to a great degree in disinfectants or chemical sterilants, the aromatic aldehydes show great promise for this purpose. The general structure of the aromatic aldehydes is presented in Figure 1. [For a description of the theoretical basis for the reactivity of the carbonyl carbon on aromatic aldehydes, the reader is referred to Ref. 5]. The activity of the carbonyl carbon is essential for activity of these compounds, as demonstrated by the fact that the corresponding acids and alcohols do not exhibit equal or greater activity [5–7]. The activity of benzaldehyde and its substituted derivatives seems to be related to their solubility in lipids, as demonstrated by the fact that those substituted benzaldehydes that have higher partition coefficients in cyclohexane–sodium borate have greater antimicrobial activity against *Pseudomonas aeruginosa*. The activity is pH-dependent and is greater at acidic and alkaline pH than at a neutral pH [5,7].

Data presented in an exhaustive study of benzaldehyde compounds suggest that the compounds are bacteriostatic, rather than bactericidal. 4-Isopropylbenzaldehyde, however, was particularly effective against *Candida albicans* and *Trichophyton mentagrophytes*. Addition of a hydroxy function to this compound did not increase the activity, as was expected [5]. α,β-Unsaturated aldehydes,

TABLE 1 Relative Sporicidal Activity of Aldehyde

Corresponding acid	Aldehyde	No. of carbons	Ratio of aldehyde groups to carbons	Sporicidal activity
Formaic	Methanal (formaldehyde)	1	1	Good
Oxalic	Ethanedial (glyoxal)	2	1	Good
Malonic	Propanedial (malonaldehyde)	3	0.69	Slight
Succinic	Butanedial (succinaldehyde)	4	0.5	Slight
Glutaric	Pentanedial (glutaraldehyde)	5	0.4	Excellent
Adipic	Hexanedial	6	0.33	Slight
Pimelic	Heptanedial	7	0.28	Poor
Suberic	Octanedial	8	0.25	Poor

Source: Ref. 41.

similar to the benzaldehyde compounds, are active antimicrobials, but this trait has not been used in disinfectants, or in chemical sterilants. This is particularly interesting, because many of these chemicals are naturally occurring, and others have been safely used in the food industry as flavors (cinnamaldehyde, vanillin, and others). The activity of these compounds, again, is directly related to the activity of the carbonyl group, which is affected by the substituent groups found on the backbone. The simplest compound in this series, acrolein (CH_2=CH−CHO), is highly reactive with proteins [8,9] and has antimicrobial activity [unpublished data]; however, it is also very toxic; therefore, it would not be useful in applications that involve human contact, such as disinfection of medical instruments. For a complete summary on the toxic effects of acrolein and other aldehydes, the reader is referred to the monograph by Schaunstein et al. [10].

III. MECHANISM OF ACTION

The activity of aldehydes is related to the reactivity of the aldehyde group, whether it is a saturated, unsaturated, or aromatic aldehyde. The reactivity of the aldehyde group can be changed by the other groups associated with the molecule. Aldehydes tend to become hydrated when in aqueous solutions and in equilibrium with the free aldehyde form, which is the active species. Because glutaraldehyde has been in use as a tissue fixative, tanning agent, and disinfectant for many years, much of the data gathered on the biocidal activity of aldehydes have been for glutaraldehyde. Figure 2 depicts the species available when glutaraldehyde is in an aqueous system [11]. This distribution of hydrates is similar for other types of aldehydes in solution. With glutaraldehyde, the monomeric free aldehyde molecule is in equilibrium with the cyclic hemiacetal and the acetal polymer, with smaller quantities of the mono- and dihydrates. This equilibrium appears to be somewhat temperature-sensitive, which may account for the changes in activity with temperature, since, at higher temperature, there is a shift to the more active monomeric form.

The primary reactive species for aldehydes are amines and sulfhydral groups. The molecules most likely to react with aldehydes would be proteins and nucleic acids. The involvement of these molecules indicates that aldehydes will interfere with enzymatic activity and with synthesis and function of the nucleic acids, which would then control the other biochemical and physiological events of the cell. Formaldehyde is known to react with amino groups [12] and bring about intermolecular crosslinking with phenolic or indole residues of protein amino groups [13]. The active site of specific aldehydes in killing bacteria and fungi or in inactivation of viruses has not been determined, and in all likelihood, there is no one site for the action of these compounds, but more probably, they act in a nonspecific manner, attacking more than one essential target molecule. Although formaldehyde will react with both RNA and DNA, the reaction with RNA is much

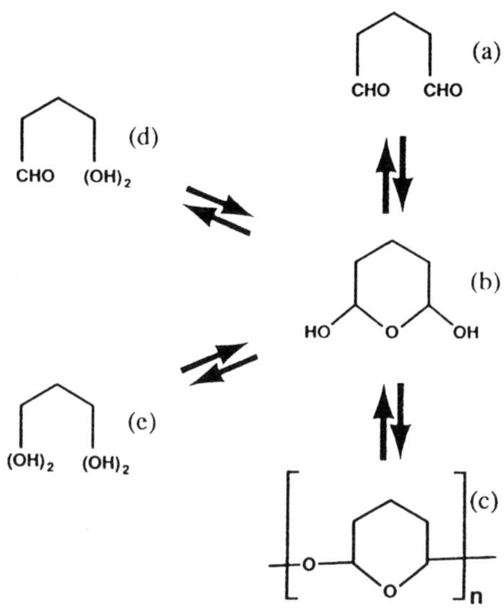

FIGURE 2 Glutaraldehyde species available in aqueous solutions: (a) monomer, (b) cyclic hemiacetal, (c) acetal-like polymer, (d) monohydrate, (e) dihydrate. (From Ref. 11.)

stronger [14]. Power and Russell [15] demonstrated a correlation between uptake of glutaraldehyde by cells and their sensitivity to the biocidal action of the molecule. Formaldehyde at high concentrations causes inter- and intramolecular crosslinking of proteins. Glutaraldehyde produces the same effect, but in a more efficient manner because of the two aldehyde groups on the same molecule. Some evidence exists suggesting that the hydrophobicity on hydrophilicity of the aldehyde molecule is important in the penetration of the compound into the cell [5,7]. Jalier and Nikaido [16] have demonstrated permeability barriers to hydrophilic solutes in mycobacteria, which are particularly resistant to aqueous glutaraldehyde. Without good penetration of the compound into the cell, little or no antimicrobial activity is likely to occur or, at best, the activity is less effective because of the action on less essential compounds located in the cell wall region. Other evidence exists suggesting that, at least with *Bacillus subtilis* spores, interaction of the spore coat with glutaraldehyde causes the spore surface to become more hydrophobic, thereby resulting in less penetration of glutaraldehyde into the spore [15].

Munton and Russell [17] presented evidence that there is interaction of the cell membrane with glutaraldehyde. Protoplasts of *Bacillus megaterium* resist lysis after being treated with glutaraldehyde, suggesting a crosslinking and, thereby, strengthening of the cell membrane. Further evidence demonstrates that dehydrogenases located in the cell membrane region are inactivated by treatment with glutaraldehyde [18]. Treatment of *Micrococcus lysodeikticus* cells with glutaraldehyde [19] prevented the release of selective enzymes from the periplasmic space (i.e., the space between the cell wall and cell membrane), indicating an interaction of glutaraldyde with the outer surface of the cell wall. Similar reactivity of formaldehyde with the cell surface of *E. coli* has been demonstrated, thereby "sealing" the surface of the cell [20,21]. Interaction of aldehydes at this level may cause death by inhibiting the uptake of essential nutrients, such as carbon and nitrogen sources, into the cell. Hughes and Thurman [22] demonstrated that when peptidoglycan, isolated from *B. subtilis*, was exposed to glutaraldehyde, 30–50% of the ϵ-NH_2 groups were bound, and D-alanine residues in cell wall teichoic acids react with glutaraldehyde, resulting in intramolecular bonds between teichoic acid chains. They also determined that peptidoglycan, when exposed to glutaraldehyde, is more resistant to the enzymatic activity of lysozyme, a finding supported by studies with *Pseudomonas aeruginosa* [23] and *Staphylococcus aureus* [24]. The work of McGucken and Woodside [25] demonstrated the immediate and complete cessation of protein, RNA, and DNA synthesis in *E. coli* cells treated with 100 μg/ml glutaraldehyde. The authors attributed this action to inhibition of uptake of precursor molecules, rather than to direct effects on incorporation into macromolecules. This view is supported by evidence of interaction of glutaraldehyde with cell walls. Inactivation of several enzymes located in the periplasmic space has been reported, suggesting the ability of glutaraldehyde to penetrate the cell wall complex to the periplasmic region of the outer cell surface [26,27]. It is likely that the degree of penetration is related to the concentration of glutaraldehyde. Comparison of transport of [^{14}C]-α-aminoisobutyric acid in glutaraldehyde-treated and untreated *E. coli* mureinoplasts (cells without the outer lipoprotein layer) shows a 50% reduction in transport in treated versus untreated cells [27]. Evidence of the importance of lipid solubility of aldehydes to their bactericidal activity has been demonstrated for benzaldehyde. The bactericidal activity of substituted benzaldehyde compounds is directed related to their partition coefficient in cyclohexane–0.5 *M* aqueous sodium borate [7].

The effect of pH on the activity of glutaraldehyde has been extensively studied and reported. The reaction of glutaraldehyde with protein increases significantly as the pH is increased from 4 to 9 [28]. The activity of benzaldehyde is known to be greater at pH 9–10 than at 6–8, either because of the effect of pH on the bacterial cell or because condensation reactions of aldehydes are catalyzed at alkaline pH. The effect of sodium bicarbonate, as the pH-increasing agent, is more an effect on the cell than on modification of the glutaraldehyde molecule [29]. Munton and Russell [30] showed that both acidic and alkaline glutaraldehyde react with the cell wall to different extents, with alkaline glutaraldehyde showing more immediate reactivity. Glutaraldehyde uptake isotherms for alkaline glutaraldehyde show initial and secondary increases, suggesting penetration of the glutaraldehyde, presumably aided by the bicarbonate, into the reactive sites. Cells exposed to acidic glutaraldehyde show a fairly uniform uptake over a wide concentration range, suggesting no penetration into the cell [31].

The role of metal ions in the activity of aldehydes has been examined. The influence of metal ions on the activity of salicylaldehyde has been demonstrated, with its increased activity in their presence [32]. Chelation of the metal ions by salicylaldehyde seems to be the mechanism by which the increased activity results. Methylation of salicylaldehyde to yield 2-methoxybenzaldehyde (which has no chelation properties) results in no increased effect of cations [7]. The potentiation of glutaraldehyde by cations (Mg, Ca, or other) has also been demonstrated [33].

With the limited amount of information concerning mechanism of action of aldehydes, the data for glutaraldehyde seems to point to an interaction with the cell wall and membrane components, which effectively reduces uptake of nutrients and interaction with the membrane–wall components (especially enzymes) that are essential for cell survival.

IV. MICROBIOLOGICAL ACTIVITY

The available data on activity of aldehydes against microorganisms remains tied to the requirements for registration of a product by governmental entities. All registered hospital disinfectants in the United States must be effective against *S. aureus*, *P. aeruginosa* and *Salmonella choleraesuis*, the model organisms for efficacy. Because of the limited "required" data for registration of a product, there is often little else available, since manufacturers are not required to determine the limits or scope of effectiveness of their product. Much of the data available on efficacy against bacteria, fungi, and viruses, other than those required for registration, is generated for experimental purposes and does not necessarily indicate the effectiveness of a chemical entity in a disinfectant formulation under actual in-use conditions. The biological activity of chemicals as well as products that have been studied under scientifically controlled conditions will be examined. Data published in manufacturers' literature is not included.

Several proprietary formulations with glutaraldehyde as the active ingredient have been marketed for use in medical, dental, and veterinary applications. Most formulations claim to be similar in activity because of the glutaraldehyde concentration, whereas others claiming to be glutaraldehyde-based disinfectants have as the major component phenol or a substituted phenolic derivative that then acts as the primary germicide in the formulation. Lack of regulation in the labeling of such formulations as primarily phenolics has further confused the users about what a glutaraldehyde-based formulation is. In an attempt to discuss the activity of these formulations relative to bactericidal, fungicidal, virucidal, sporicidal, and mycobactericidal activity.

Glutaraldehyde, in the standard 2–3% formulation, is active against a wide range of microorganisms; gram-positive, gram-negative, and mycobacteria; viruses, and spores. The degree of activity is usually associated with a variety of variables, such as temperature, glutaraldehyde concentration at the time of use, pH, presence of inorganic ions, dilution from use, and age of the solution. A study by Gelinas et al. [34] demonstrated the effect of temperature on disinfectant activity, with glutaraldehyde being the chemical most affected by temperature. The effect of temperature on glutaraldehyde activity was also demonstrated by Ascenzi and associates [35]. Because product claims are generated under artificial conditions, the activity of a solution in actual use is always subject to some question. Also, because data generated from actual use of the product under real conditions are not required for registration, the claims for microbiological activity that are listed on the product may not hold up under real-use conditions. Before we examine the efficacy of commercially available formulations, an examination of the information that is available on the activity of aldehyde solutions under experimental conditions follows.

A. Bactericidal Activity

Several publications describe the effectiveness of glutaraldehyde against both gram-positive and gram-negative bacteria. Every hospital disinfectant on the market must have, as a minimum, a claim of bactericidal effectiveness against gram-positive (*S. aureus*) and gram-negative (*Salmonella choleraesuis* and *Pseudomonas aeruginosa*) bacteria. However, these are the only organisms that a registered product must be effective against under artificial laboratory conditions. Work by Munton and Russell [18,30,36,37] has demonstrated the mechanism of action of glutaraldehyde against the gram-negative *Escherichia coli*. The effectiveness of low concentrations of glutaraldehyde (0.5%) in reducing the number of *Salmonella typhimurium* bacteria on poultry carcasses was demonstrated by Bailey et al. [38] and, as a result, there is an increased shelf life of these carcasses because of their reduced bacterial loads [39]. The rapid activity of 2% glutaraldehyde against *E. coli*, *P. aeruginosa*, *Serratia marcescens*, *Proteus vulgaris*, and *Klebsiella pneumoniae*, with complete kill in less than 1 min, and against the spores of the gram-negative bacteria *Clostridium perfringens* and *C. tetani* in less than 3 h has been reported [40].

Other aldehydes have been effective under laboratory conditions against gram-negative organisms; for example, benzaldehyde and substituted benzaldehydes against *P. aeruginosa*, *Proteus vulgaris*, and *K. pneumoniae* [6,7,41], and *o*-methoxycinnamaldehyde against *E. coli*, *Proteus vulgaris*, *P. aeruginosa*, and spores of *C. perfringens* and *C. botulinum* [42]. To some degree alkyl, mono-, di-, and trihydroxy, and chloro-substituted benzaldehydes show inhibitory activity against *P. aeruginsa*, *Proteus vulgaris*, and *K. pneumoniae* [5]. None of these three groups of aldehydes have been formulated into a commercial product. Formaldehyde, which has been formulated into several commercially available disinfectants outside the United States, has been effective against the gram-negative bacteria *Enterobacter aerogenes* [43] and *E. coli* [44].

The activity of aldehydes against gram-positive bacteria, similar to that for gram-negative bacteria, shows a wide spectrum across many genera. Again, because glutaraldehyde is the predominant aldehyde in commercially available products, there are substantially more data for this chemical's activity than for the other aldehydes.

Glutaraldehyde is active against *S. aureus* [33,40,45] *Micrococcus lysodeikticus* [19,40], and *Streptococcus pyogenes* [40]. Other aldehydes, such as formaldehyde, succinaldehyde, or glyoxal, all show activity against gram-positive bacteria.

B. Mycobactericidal Activity

The resistance of mycobacteria to chemical agents has been considered to be intermediate between that of other vegetative microorganisms and that of spores [46,47]. This has been the subject of much recent debate, especially for the activity of glutaraldehyde-based disinfectants. The basis for this resistance to chemical agents may be the hydrophobic nature of the cell wall complex of these organisms. The barrier to hydrophilic solutes by mycobacteria has been demonstrated [16]. For many years, 2% glutaraldehyde was considered to be mycobactericidal in a relatively short time period, as evidenced by the 10-min tuberculocidal claims contained on most hospital disinfectants. Substantiation of these claims with improved, more-accurate methods indicates that this is not true [48]. It generally requires longer than 30-min exposure of *Mycobacterium bovis* BCG to 2% glutaraldehyde at 20°C for a significant reduction in viable population. The same is true for *M. tuberculosis* H37Rv [49]. The innate resistance of mycobacteria to formaldehyde has also been demonstrated, with distilled water-grown *M. chelonei* surviving an exposure to 8% formaldehyde for longer than 120 min at 25°C [50], whereas laboratory strains of *M. chelonei* and *M. fortuitum* both required 20 min for a 4-log inactivation of cells. The same strain of water-grown *M. chelonei* required from 5 to more than 60 min for inactivation by 2% glutaraldehyde. One strain showed only a 2-log decrease after exposure to 2% alkaline glutaraldehyde for 60 min at 25°C. At lower concentrations of glutaraldehyde (0.2%), survivors of water-grown *M. chelonei* isolates were found at 96 h. These data on the "atypical" mycobacteria are substantiated by Collins [51,52].

The rapidly growing mycobacteria (*M. marinum* and *M. fortuitum*) seem to be more susceptible to the activity of glutaraldehyde, suggesting a relation with metabolic activity. No correlation between sensitivity to glutaraldehyde and the presence of pigmentation or the degree of virulence was found [52]. In general, the relative resistance of organisms in suspension and those dried on surfaces was the same, and the rate of inactivation was the same, regardless of the surface type or suspension [49]. These data suggest that careful consideration should be given to choosing a model for effectiveness of chemical agents against *M. tuberculosis*. *Mycobacterium smegmatis* was, and still is, used in the AOAC presumptive test for efficacy testing [53]. Given the data of Collins [49] and Best et al. [54,55], the use of *M. smegmatis* is not recommended. Studies by van Klingeren and Pullen [56] suggests that *M. terrae* is a good model for *M. tuberculosis* when measuring the activity of chemical agents.

C. Spores

The effectiveness of disinfectant solutions against spore-forming bacteria has become synonymous with cold chemical sterilization. Spores have long been considered to be the most resistant form of life. The more recent discovery of chemical resistance of some prions puts this conventional wisdom in doubt, however. Aldehydes have been the chemicals of choice for irradication of spores since the work of Pepper and Lieberman [1] in which the pH dependence of aldehyde activity was demonstrated with the sporicidal activity of glyoxal and glutaraldehyde. Glutaraldehyde is effective against spores of both genera, *Bacillus* and *Clostridium*.

The addition of bicarbonate to 1% alcoholic glyoxal or to 1% glutaraldehyde solutions provided substantial sporicidal activity, compared with no detectable activity with bicarbonate itself. The sporicidal activity of glutaraldehyde was tenfold greater than that of glyoxal. The effect of bicarbonate was on the integrity of the spore coat. Pretreatment of spores with bicarbonate rendered them as sensitive to acidic glutaraldehyde as they were to glutaraldehyde brought to alkaline pH with bicarbonate. The same effect was not evident with pretreatment with NaOH, suggesting something specific about the interaction of $NaHCO_3$ with the spore [4], and the effect is not one of pH, but rather, it is a chemical interaction of the spore with bicarbonate. Bicarbonate may aid in opening up the spore coat for better penetration of the glutaraldehyde. Pretreatment of spores with chemicals known to specifically remove selective components of the spore coat renders the spore susceptible to glutaraldehyde; the greater the amount of spore coat removed, the more susceptible it is to the activity of glutaraldehyde.

Although the sporicidal activity of glutaraldehyde has been accepted for over two decades, evidence that much of the activity attributed to this agent may be due to a lack of recovery of dormant spores [57]. Treatment of spores with NaOH after they were treated with glutaraldehyde "revived" a portion of the spores that were otherwise considered killed. This treatment was specific for glutaraldehyde and the same response was not shown by formaldehyde-treated spores. Just as bicarbonate is thought to open the spore coat for better penetration, it is hypothesized that posttreatment with NaOH of glutaraldehyde-exposed spores may provide better access of the germicide to the germicidal receptor of the spore. Because this is an extreme condition to bring about the revival of the spores, it is not cause for concern from a practical point of view. Spores on medical equipment that are rendered "killed" with glutaraldehyde are not likely to encounter the extreme alkaline conditions that would result in their outgrowth, and the infection of a patient.

The use of ultrasonics in augmenting the activity of chemicals has been effective in increasing the activity of glutaraldehyde against spores. Synergism between glutaraldehyde and ultrasonics at both low and high frequencies has been demonstrated [58] and verified [59]. Rendering instruments free of organisms can be accomplished much more easily with ultrasonics because of the physical removal achieved with the sonic energy. Removal of spores from medical instruments has never been deemed necessary when used on nonsterile areas of the body (e.g., colonoscope, bronchoscope, and such). However, in immunocompromised and aged patients this is not always valid.

Clostridium difficile has been recognized as a major infection-control problem, especially in aged patients and in conjunction with certain antibiotic therapies [60]. Data presented in Tables 2 and 3 indicate the in vitro susceptibility of this spore-former to glutaraldehyde with short periods of exposure: 5 min for 2% glutaraldehyde and 30 min for 0.2% glutaraldehyde. This is in comparison with the longer exposure times and higher glutaraldehyde concentrations required for killing *Bacillus* spores [61]. A study evaluating the performance of 2% glutaraldehyde against *C. difficile* in situ corroborates the in vitro data. Endoscopes contaminated with *C. difficile* were effectively disinfected in 5 min [62].

Because of the generally longer exposure time required for inactivation of spores than for vegetative bacteria, fungi, and viruses, one must keep in mind the sterility of the solution used for disinfection. Since spores can remain viable in glutaraldehyde solutions, as they can in most chemical disinfectants, cross-contamination of instruments can occur. Because more spore-forming organisms of the *Bacillus* species do not cause serious infections, and because those of the genus *Clostridium* seem to be more susceptible to chemical sterilization, this is not a major concern, but should be remembered for the special situations, especially for the immunocompromised patient.

D. Fungi

Most glutaraldehyde-based products available commercially have claims indicating effectiveness against fungi. In reality, this claim is based on the activity of the active ingredient against the fungus *Trichophyton mentagrophytes*. Although a

TABLE 2 Sporicidal Activity of Glutaraldehyde Against *Bacillus subtilis*, *B. stearothermophilus* and *Clostridium difficile* (Broth Recovery Method)[a]

Test organism and disinfectant	Challenge (spores/ml)	Number of broths positive on subculture at							Spore media	Recovery media
		2 min	5 min	10 min	30 min	1 h	2 h	3 h		
B. subtilis	5×10^6		5/5	5/5	5/5	5/5	3/5	0/5	NA	2 NB
(2% glutaraldehyde)	5×10^3		5/5	5/5	5/5	5/5	0/5	0/5	NA	2 NB
	5×10^1		1/5	0/5	1/5	0/5	0/5	0/5	NA	2 NB
	8×10^5		1/5	3/5	1/5	1/5	0/5	0/5	NA	2 NB
B. stearothermophilus	8×10^3		0/5	0/5	0/5	0/5	0/5	0/5	NA	2 NB
(2% glutaraldehyde)	5×10^2		0/5	0/5	0/5	0/5	0/5	0/5	NA	2 NB
	1×10^5		1/5	0/5	0/5				FAB	FAB
	7×10^6	0/5	0/5	0/5	0/5				EL	TNT
	7×10^4	0/5	0/5	0/5	0/5				EL	TNT
C. difficile	2×10^5	0/5	0/5	0/5	0/5				TG	TNT
(2% glutaraldehyde)	2×10^4	0/5	0/5	0/5	0/5				TG	TNT
	2×10^2	0/5	0/5	0/5	0/5				TG	TNT
	2×10^1	0/5	0/5	0/5	0/5				TG	TNT
	6×10^3	0/5	0/5	0/5	0/5				TG	TG
	1×10^5	0/5	1/5	0/5	0/5				FAB	FAB
C. difficile	3×10^5	0/5	0/5	0/5	0/5				TG	TNT
(0–2% glutaraldehyde)	3×10^4	0/5	0/5	0/5	0/5				TG	TNT
	3×10^3	0/5	0/5	0/5	0/5				TG	TNT
	6×10^3	5/5	3/5	1/5	0/5				TG	TG

[a]NA, nutrient agar; 2 NB, double-strength nutrient broth; FAB, fastidious anaerobe broth; EL, Ellners broth; TNT, tryptic nitrate/thiol broth; TG, thioglycollate broth.
Source: Ref. 61.

TABLE 3 Sporicidal Activity of Glutaraldehyde Against *C. difficile* (plate recovery method)

Glutaraldehyde concentration (%)	Spore challenge	Number of surviving spores at 2 min	5 min	10 min	15 min	20 min	25 min	30 min
2	2×10^3	400	0	0	0	0	0	0
1	1×10^4	500	900	0	0	0	0	0
0–2	2×10^3	400	400	400	100	0	0	0
0–1	1×10^4	4000	1400	2900	1000	2100	700	400

Source: Ref. 61.

pathogenic fungus, it does not represent a cross section of the genera of fungi that are potentially infectious to humans. Because of diversity in the fungi (e.g., molds and yeasts), and even the diversity that exists in the molds and yeasts separately, it is important to identify the specific organisms that are susceptible to the action of aldehydes, specifically glutaraldehyde.

Glutaraldehyde has been tested both in vivo and in vitro for its effectiveness against fungi. Once again, the effect of pH can be seen from the studies of Gorman and Scott [63]. Half-percent acidic glutaraldehyde was less effective at inhibiting mycelial growth and spore production of *Aspergillus niger* than was alkaline glutaraldehyde. The pH effect was also true for inhibition of *Saccharomyces cerivissiae* for which the *minimal inhibitory concentration* (MIC) was five times higher at an acidic pH than at a slightly alkaline pH. Sonacide, an acidic glutaraldehyde, however, was effective in 15 min at 25°C against a wide variety of fungi, except for *A. niger*. An increased exposure time of 30 min was necessary to completely kill *A. niger* [64]. The effect of glutaraldehyde concentration on effectiveness can be seen in this study in that a glutaraldehyde–phenate solution, containing approximately 0.15% glutaraldehyde (with phenol), was not effective against either *A. niger* or *A. fumigatus* with a 90-min exposure. Table 4 shows the broad spectrum of activity against fungi, as measured by zone of inhibition. Several of the fungi are sensitive to low concentrations of this chemical, which is useful for topical application of glutaraldehyde.

In vivo activity of glutaraldehyde has been evaluated for treatment of onychomycoses [65]. A 10% solution of buffered glutaraldehyde was effective in eradicating *T. mentagrophytes*, and *Cephalosporium* and *Fusarium* species [66].

In an in-use type evaluation of glutaraldehyde, Isenberg et al. [67] demonstrated that glutaraldehyde solutions, reused as they are in normal hospital practice, maintained effectiveness against *Candida albicans*, whereas, glutaraldehyde–

Table 4 Inhibition of Fungi by Glutaraldehyde

Days incubation	3 d		7 d		12 d	
Glutaraldehyde (%)	0.1	1.0	0.1	1.0	0.1	1.0
Alternaria	—	11.0[a]				
Cephalosporium	8.0	11.0				
Cladosporium	—	10.0				
Fusarium	—	9.0				
Scopulariopsis brevicaulis	—	9.5				
Microsporium audouini	—	—	—	—	—	14.0
M. canis	—	—	—	12.5		
M. gypseum	7.5	12.5				
Epidermophyton floccosum	—	—	—	—	—	14.0
Trichophyton mentagrophytes	7.5	15.0				
T. rubrum	—	—	—	—	—	13.0
T. terrestre	9.5	12.0				
T. tonsurans	8.0	13.0				
T. violaceum	—	—	—	—	7.0	13.0
Candida albicans	—	11.0				
C. parapsilosis	—	11.0				
Geotrichum candidum	—	—	—	11.5		

[a]Zone of inhibition (mm).
Source: Ref. 66.

phenate solutions did not, presumably because of the lower concentration of glutaraldehyde in the latter solution.

Other aldehydes that have shown effectiveness against fungi are *o*-methoxycinnamaldehyde [42], benzaldehyde and substituted benzaldehyde derivatives [6,68], and *o*-phthalaldehyde [unpublished data].

E. Viruses

The role of viruses in nosocomial infections is well documented and, like bacteria and other microorganisms, they too can be spread through the use of improperly disinfected medical and dental instruments. The use glutaraldehyde-based disinfectants for disinfection of virus-contaminated equipment is substantiated by the data from manufacturers of products and by data in the literature. However, one should be aware of differences in the measurement of infectivity or antigenicity of disinfectant-treated viruses. Measurement of changes in the infectivity, rather than the antigenicity of disinfectant-treated virus particles is more meaningful.

The original work of Klein and Deforest [69] determined the differential activity of glutaraldehyde against the enveloped (lipophilic) and nonenveloped (hydrophilic) viruses, with the former being more susceptible. Poliovirus type 1, coxsackievirus type B1, and echovirus type 6 were more difficult to inactivate—requiring $> 1\%$ glutaraldehyde—than were the herpes simplex virus, vaccina virus, Asian influenza virus, and adenovirus. (Although adenovirus is classified as a nonenveloped virus, it does contain lipids adsorbed to its protein surface, but they are not considered an integral part of the virus structure, as they are with other enveloped viruses.) The enveloped viruses and adenovirus required only $\geqslant 0.2\%$ glutaraldehyde for inactivation in 1 min at room temperature. Other viruses that are susceptible to the antiviral activity of glutaraldehyde are coxsackievirus B1 [70] and B3 [71,72], human immunodeficiency virus type 1 [73–77], herpes simplex virus type 2 [77,78], yellow fever virus and influenza virus PR-8 [79], aphthovirus, pestivirus, and iridovirus [72], hepatitis B virus [80–83], poliovirus type 1 [70], echovirus 11 [70], and rotavirus [84]. Succinaldehyde and formaldehyde have also been effective against hepatitis B virus [85]. Exposure time, glutaraldehyde concentration, presence or absence and concentration of organic soil, and temperature are variables that have an effect on the virucidal activity of glutaraldehyde.

Organic soil, which adversly affects many other types of chemical agents, has little effect on the virucidal activity of glutaraldehyde [71,72,77,79,84]. Saitanu and Lund [71] have presented evidence that the inactivation of enterovirus by glutaraldehyde is faster in the presence of 10% calf serum and 20% mouse droppings than in the absence of organic soil. The effect of pH on glutaraldehyde activity was discussed earlier; however, data for inactivation of enterovirus indicate that the activity peaks at pH 7.4 (neutral) when tested at pH values ranging from 5 to 9, with a pronounced decrease in activity on the acidic side [71]. Data for poliovirus, however, are similar to those for spores, in that the activity is minimal at acidic pH and inactivation is greater at alkaline pH than at neutral pH [86].

The kinetics of inactivation of the enterovirus, coxsackievirus B3, which may or may not be the same for other viruses follows first-order kinetics with time, and for reduction in log units of active virus. However, there is a pronounced deviation when the reduction in virus runs over 2-$\log_{10}$ units of virus. This is similar to the tailing effect seen with mycobacteria and spores. Because of this deviation, extrapolation of the straight line decreases to an endpoint that suggests that "no-active-virus" may not be an appropriate method for determining time for complete kill [71]. The same first-order kinetics has been seen for poliovirus, without the tailing effect [86]. Concentration and exposure time are statistically significant factors in the effectiveness of glutaraldehyde against hepatitis A virus, with 0.5% glutaraldehyde reducing infectious particles by 3.5 $\log_{10}$ in 3 min [87].

Passagot et al. [87] suggest that the site for inactivation of hepatitis A virus is either the lysine or hydroxylysine moieties of the protein coat, the RNA, or both. Reaction with the protein coat would interfere with infectivity of the virus, whereas reaction with the nucleic acid would interfere with replication of the virus after penetration into the host cell. Data with RNA from poliovirus indicate that it is a poor reactive site for glutaraldehyde [86].

V. TOXICITY AND SAFETY

The excellent activity of aldehydes, and glutaraldehyde in particular, is associated with their ability to react with essential cell components. This reactivity is not selective and, therefore, these agents exhibit activity against human cells and have a certain degree of toxicity.

The most severe reaction noted among users of glutaraldehyde include respiratory and ocular irritation from vapors and sensitization. Oral median lethal dose (LD_{50}) data indicate that glutaraldehyde exhibits relatively low toxicity, whereas the acute dermal toxicity of the compound is five to ten times lower, with an LD_{50} of approximately 2500 mg/kg in the rabbit. Although not highly toxic by the dermal route, glutaraldehyde does cause allergic contact dermatitis [88]. However, the number of reports, considering the widespread use of the compound, suggests a relatively low incidence. Suringa [65] found no development of contact dermatitis in four patients treated for 6 weeks with glutaraldehyde for superficial onychomycosis. This may be due to the small population used in the study. The eliciting threshold concentration of glutaraldehyde in water that is necessary to induce sensitization in humans is 0.5%. Presumably, because of the ease with which aldehydes combine with amine groups of proteins, glutaraldehyde may produce an antigenic site, thereby stimulating the production of antibodies.

Studies on the teratogenic potential of glutaraldehyde have not indicated that it is a teratogen. Relatively high doses of Sonacide (2% acidic glutaraldehyde), equivalent to 100 mg/kg per day of glutaraldehyde, although toxic to mouse dams, were not judged to be teratogenic, nor was formaldehyde at doses as high as 185 mg/kg per day [89]. Along these same lines, the rate of spontaneous abortions, measured in hospital workers, showed that users of glutaraldehyde and formaldehyde did not have the statistically significant higher rates that were found in workers using ethylene oxide [90].

Mutagenicity and carcinogenicity of aldehydes have been questioned since the advent of widespread use of these chemicals for disinfection or sterilization. Glutaraldehyde has been tested several times in various systems for mutagenicity or mutagenic potential. Evaluation of glutaraldehyde by the Ames' procedure has produced equivocal information. In one study, the results indicated that glutaraldehyde was a weak mutagen in *Salmonella typhimurium* TA100 [91,92]. The data in these studies was not consistently reproducible, and the increase in the reversion rate was less that twofold over the controls. Other studies with *S. typhimurium* TA104 [93] and TA102 [94] demonstrated the mutagenicity of glutaraldehyde. However, these particular strains of *S. typhimurium* were developed specifically for sensitivity to mutations by aldehydes; therefore, they may give results that are exaggerated because of the sensitivity of the strain used. Studies with other strains of *S. typhimurium*, TA98 and TA100 [95,96], TA98, TA100, and TA1537 [96], and TA1535 and TA1538 [96], and with *E. coli* WP2 [90] failed to show that glutaraldehyde was a mutagen.

In other in vitro evaluations conducted by Union Carbide, no dose-related effects were found in the mutation frequency of Chinese Hamster Ovary (CHO) cells, either in the presence or absence of metabolic activation, nor was there a significant dose-related increase in sister-chromatid exchange in CHO cells. Finally, no significant increase in unscheduled DNA synthesis was noted as a function of glutaraldehyde concentration. Mutations in unscheduled DNA synthesis in hepatocytes is a measure of nonspecific DNA damage [96]. No damage to germ cells was found in either the sex-linked recessive lethal assay in *Drosophila melanogaster* [97,98] nor in the dominant lethal assay in mice. No reports of carcinogenicity have been made for glutaraldehyde.

REFERENCES

1. R. E. Pepper, and E. R. Lieberman, Dialdehyde alcoholic sporicidal composition, U.S. Patent 3016328, 1962.
2. A. A. Stonehill, S. Krop, and P. M. Borick, Buffered glutaraldehyde a new chemical sterilizing solution, *Am. J. Hosp. Pharm. 20*:458–465 (1965).
3. L. G. Egyud, Studies on cell division: The effect of aldehydes, ketones and α-keto-aldehydes on the proliferation of *Escherichia coli, Curr. Mod. Biol. 1*:14–20 (1967).
4. E. G. M. Power, and A. D. Russell, Sporicidal action of alkaline glutaraldehyde: Factors influencing activity and a comparison with other aldehydes, *J. Appl. Bacteriol. 69*:261–268 (1990).
5. D. Rehn, and H. Nolte, Zur antimikrobiellen Wirksamkeit substituierter aromatischer Aldehyde and Alkohole. *Zentrabl. Bakteriol. Hyg. Abt. I Orig. B 168*:507–516 (1979).
6. I. Leifertova, N. Hejtmankova, H. Hlava, J. Kudrnacova, and F. Santavy, Antifungal and antibacterial effects of phenolic substances. A study of the relationship between the biological activity and the constitution of the investigated compounds, *Acta Univ. Palacki. 74*:83–101 (1975).
7. D. E. Burton, K. Clarke, and G. W. Gray, The mechanism of the antibacterial action of phenols and salicylaldehyde. Part III. Substituted aldehydes, *J. Chem. Soc.* Nov., pp. 2458–2460 (1964).
8. H. D. Mayor, and L. E. Jordan, *J. Cell Biol. 18*:207 (1963).
9. N. Feder, and M. K. Wolf, Studies on nucleic acid metachromasy. II. Metachromatic and orthochromatic staining by toluidine blue of nucleic acids in tissue sections, *J. Cell Biol. 27*:327–336 (1965).
10. E. Schaunstein, H. Esterbauer, and H. Zollner, Aldehydes in biological systems, Pion., London, 1977.
11. S. P. Gorman, E. M. Scott, and A. D. Russell, Antimicrobial activity, uses and mechanisms of action of glutaraldehyde, *J. Appl. Bacteriol. 48*:161–190 (1980).

12. H. Fraenkel-Conat, Chemical modification of viral RNA. 1. Alkylating agents. *Biochim. Biophys. Acta 49*:169–180 (1961).
13. H. Fraenkel-Conat, M. Cooper, and H. S. Olcott, The reaction of formaldehyde with proteins, *J. Am. Chem. Soc. 67*:950–954 (1945).
14. M. Staehlin, Reaction of tobacco mosaic virus nucleic acid with formaldehyde, *Biochim. Biophys. Acta 29*:410–417 (1958).
15. E. G. M. Power, and A. D. Russell, Glutaraldehyde: Its uptake by sporing and non-sporing bacteria, rubber, plastic, and an endoscope, *J. Appl. Bacteriol. 67*:329–342 (1989).
16. V. Jarlier, and H. Nikaido, Permeability barrier to hydrophilic solutes, *J. Bacteriol. 172*:1418–1423 (1990).
17. T. J. Munton, and A. D. Russell, Effect of glutaraldehyde on protoplasts of *Bacillus megaterium*, *J. Gen. Microbiol. 63*:367–370 (1970).
18. T. J. Munton, and A. D. Russell, Effect of glutaraldehyde on cell viability, triphenyl tetrazolium reduction, oxygen uptake and β-galactosidase activity of *Escherichia coli*, *Appl. Microbiol. 26*:508–511 (1973).
19. D. J. Ellar, E. Munoz, and M. R. J. Salton, The effect of low concentrations of glutaraldehyde on *Micrococcus lysodeikticus* membranes, *Biochim. Biophys. Acta 225*:140–150 (1971).
20. H. E. Kubitschek, Constancy of uptake during the cell cycle in *Escherichia coli*, *Biophys. J. 8*:1401–1408 (1968).
21. S. Silver, P. Johnseine, and R. King, Manganese active transport in *Escherichia coli*, *J. Bacteriol. 104*:1299–1306 (1979).
22. R. C. Hughes, and P. F. Thompson, Cross linking of bacterial cell walls with glutaraldehyde, *Biochem. J. 119*:925–926 (1970).
23. A. D. Russell, and H. Haque, Inhibition of EDTA-lysozyme lysis of *Pseudomonas aeruginosa* by glutaraldehyde, *Microbios 13*:151–153 (1975).
24. A. D. Russell, and G. N. Vernon, Inhibition by glutaraldehyde of lysostaphin-induced lysis of *Staphylococcus aureus*, *Microbios 13*:147–149 (1975).
25. P. V. McGucken, and W. Woodside, Studies on the mode of action of glutaraldehyde on *Escherichia coli*, *J. Appl. Bacteriol. 36*:419–426 (1973).
26. B. K. Wetzel, S. S. Spicer, H. F. Dvorak, and L. A. Heppel, Cytochemical localization of certain phosphatases in *Escherichia coli*, *J. Bacteriol. 104*:529–542 (1970).
27. S. P. Gorman, and E. M. Scott, Transport capacity, alkaline phosphatase activity and protein content of glutaraldehyde-treated cell forms of *Escherichia coli*, *Microbios 19*:205–212 (1977).
28. D. Hopewood, C. R. Allen, and C. McCabe, The reactions between glutaraldehyde and various proteins: An investigation of their kinetics, *Histochem. J. 2*:137–150 (1970).
29. S. P. Gorman, and E. M. Scott, Effect of alkalination on the bacterial cell and glutaraldehyde molecule, *Microbios Lett. 6*:39–44 (1977).
30. T. J. Munton, and A. D. Russell, Interaction of glutaraldehyde with spheroplasts of *Escherichia coli*, *J. Appl. Bacteriol. 36*:211–217 (1973).
31. S. P. Gorman, and E. M. Scott, Uptake and media reactivity of glutaraldehyde solutions related to structure and biocidal activity, *Microbios Lett. 5*:163–169 (1977).

32. K. Clarke, R. A. Cowen, G. W. Gray, and E. H. Osborne, Mechanism of the antibacterial action of phenols and salicylaldehydes, *J. Chem. Soc.* pp. 168–173 (1963).
33. S. P. Gorman, and E. M. Scott, Effect of inorganic cations on the biocidal and cellular activity of glutaraldehyde, *J. Appl. Bacteriol. 47*:463–468 (1979).
34. P. Gelinas, J. Goulet, G. M. Tastayre, and G. A. Picard, Effect of temperature and contact time on the activity of eight disinfectants—a classification. *J. Food Protect. 47*:841–847 (1991).
35. J. M. Ascenzi, R. J. Ezzell, and T. M. Wendt, Evaluation of carriers used in the test methods of the AOAC, *Appl. Environ. Microbiol. 51*:91–95 (1986).
36. T. J. Munton, and A. D. Russell, Aspects of the action of glutaraldehyde on *Escherichia coli*, *J. Appl. Bacteriol. 33*:410–419 (1970).
37. T. J. Munton, and A. D. Russell, Effect of glutaraldehyde on the outer layers of *Escherichia coli*, *J. Appl. Bacteriol. 35*:193–199 (1972).
38. J. B. Bailey, J. E. Thompson, and N. A. Cox, Elimination of *Salmonella* and extension of shelf life with a glutaraldehyde product on broiler carcasses, *Poultry Sci. 56*:1346–1349 (1977).
39. M. G. Mast, and J. H. MacNeil, Use of glutaraldehyde as a disinfectant in immersion chilling of poultry, *Poultry Sci. 57*:681–684 (1978).
40. P. M. Borick, Chemical sterilizers (chemosterilizers), *Adv. Appl. Microbiol. 10*: 291–312 (1968).
41. R. M. G. Boucher, On biocidal mechanisms in the aldehyde series, *Can. J. Pharm. Sci. 10*:1–7 (1975).
42. S. Morozumi, Isolation, purification, and antibiotic activity of *o*-methoxycinnamaldehyde from cinnamon, *Appl. Environ. Microbiol. 36*:577–583 (1978).
43. K. Chan, and T. M. Lau, Effect of formaldehyde on oxygen uptake and β-galactosidase activity in *Enterobacter aerogenes*, *Microbios Lett. 10*:69–74 (1979).
44. W. A. Hamilton, Membrane active antibacterial compounds, *Inhibition and Destruction of the Microbial Cells* (W. B. Hugo, ed.), Academic Press, London, 1971, pp. 78–93.
45. S. D. Rubo, J. F. Gardner, and R. L. Webb, Biocidal activities of glutaraldehyde and related compounds, *J. Appl. Bacteriol. 30*:78–87 (1967).
46. B. Crowshaw, The destruction of mycobacteria, *Inhibition and destruction of the Microbial Cell* (W. B. Hugo, ed.), Academic Press, London, 1971, pp. 419–449.
47. A. D. Russell, Factors influencing the efficacy of antimicrobial agents, *Principles and Practices of Disinfection, Preservation, and Sterilization* (A. D. Russell, W. B. Hugo, and G. A. F. Ayliffe, eds.), Blackwell Scientific Publications, London, 1982, pp. 123.
48. J. M. Ascenzi, R. J. Ezzell, and T. M. Wendt, A more accurate method for measurement of tuberculocidal activity of disinfectants, *Appl. Environ. Microbiol. 53*:2189–2192 (1987).
49. F. M. Collins, Kinetics of tuberculocidal response by alkaline glutaraldehyde in solution and on an inert surface, *J. Appl. Bacteriol. 61*:87–93 (1986).
50. L. A. Carson, N. J. Peterson, M. S. Favero, and S. M. Aguero, Growth characteristics of atypical mycobacteria in water and their comparative resistance to disinfectants, *Appl. Environ. Microbiol. 3*:839–846 (1978).
51. F. M. Collins, and V. Montalbaine, Mycobactericidal activity of glutaraldehyde solutions, *J. Clin. Microbiol. 4*:408–412 (1976).

52. F. M. Collins, Bactericidal activity of alkaline glutaraldehyde solutions against a number of atypical mycobacteria, *J. Appl. Microbiol. 61*:247–251 (1986).
53. [AOAC] Association of Official Analytical Chemists, Official Methods of the Association of Official Analytical Chemists, Arlington, VA, 1984.
54. M. Best, S. A. Sattar, V. S. Springthorpe, and M. E. Kennedy, Comparative mycobactericidal efficacy of chemical disinfectants in suspension and carrier tests, *Appl. Environ. Microbiol. 54*:2856–2858 (1988).
55. M. Best, S. A. Sattar, V. S. Springthorpe, and M. E. Kennedy, Efficacy of selected disinfectants against *Mycobacterium tuberculosis*, *J. Clin. Microbiol. 28*:2234–2239 (1990).
56. B. van Klingeren, and W. Pullen, Comparative testing of disinfectants against *Mycobacterium tuberculosis* and *Mycobacterium terrae* in a quantitative suspension test, *J. Hosp. Infect. 10*:292–298 (1987).
57. E. G. M. Power, B. N. Dancer, and A. D. Russell, Emergence of resistance to glutaraldehyde in spores of *Bacillus subtilis* 168, *FEMS Microbiol. Lett. 50*:223–226 (1988).
58. G. Sierra, and R. M. G. Boucher, Ultrasonic synergistic effects in liquid-phase chemical sterilization, *Appl. Microbiol. 22*:160–164 (1971).
59. D. K. Warfield, S. Q. Bryington, Ultrasonic potentiation of the sporicidal activity of glutaraldehyde, *Oral Surg. Oral Med. Oral Pathol. 53*:342–346 (1982).
60. D. G. Teasley, D. N. Gerding, and M. M. Olson, Prospective randomized trial of metronidazole versus vancomycin for *Clostridium difficile*-associated diarrhea and colitis, *Lancet 2*:1043–1046 (1983).
61. A. Dyas, and B. C. Das, The activity of glutaraldehyde against *Clostridium difficile*, *J. Hosp. Infect. 6*;41–45 (1985).
62. C. E. Hughes, R. L. Gebhard, L. R. Peterson, and D. N. Gerding, Efficacy of routine fiberoptic endoscope cleaning and disinfection for killing *Clostridium difficile*, *Gastrointest. Endosc. 32*:7–9 (1986).
63. S. P. Gorman, and E. M. Scott, A quantitative evaluation of antifungal properties of glutaraldehyde, *J. Appl. Bacteriol. 43*:83–89 (1977).
64. B. Terleckyj, and D. A. Axler, Quantitative neutralization assay of fungicidal activity of disinfectants, *Antimicrob. Agents Chemother. 31*:794–798 (1987).
65. D. W. R. Suringa, Treatment of superficial onychomycoses with topically applied glutaraldehyde, *Arch. Dermatol. 102*:163–167 (1970).
66. N. Dabrowa, J. W. Landau, and V. D. Newcomer, Antifungal activity of glutaraldehyde in vitro, *Arch. Dermatol. 105*:555–557 (1972).
67. H. D. Isenberg, E. R. Giugliano, K. France, and P. Alperstein, Evaluation of three disinfectants after in-use stress, *J. Hosp. Infect. 11*:278–285 (1988).
68. D. Rehn, H. Nolte, and W. Zerling, Antimicrobial activity of substituted aromatic aldehydes, *Zentralbl. Bakteriol. Hyg. Abt. I Org. B 172*:508–519 (1981).
69. M. Klein, and A. Deforest, The inactivation of virus by germicides, *Proc. Chem. Spec. Manuf.* pp. 116–118 (1963).
70. H. K. Narang, and A. A. Codd, Action of commonly used disinfectants against enteroviruses, *J. Hosp. Infect. 4*:209–212 (1983).

71. K. Saitanu, and E. Lund, Inactivation of enterovirus by glutaraldehyde, *Appl. Microbiol. 29*:571–574 (1975).
72. H. R. Cunliffe, J. H. Blackwell, and J. S. Walker, Glutaraldehyde inactivation of exotic animal viruses in swine heart tissue, *Appl. Environ. Microbiol. 37*:1044–1046 (1979).
73. B. Spire, L. Montagnier, F. Barre-Sinoussi, and J. C. Chermann, (1984). Inactivation of lymphadenopathy associated virus by chemical disinfectants, *Lancet 2*:899–901 (1984).
74. P. J. V. Hanson, D. Gor, D. J. Jefferies, and J. V. Collins, Chemical inactivation of HIV on surfaces, *Br. Med. J. 298*:862–864 (1989).
75. P. J. V. Hanson, D. Gor, D. J. Jefferies, and J. V. Collins, Elimination of high titre HIV from fiberoptic endoscopes, *Gut 31*:657–660 (1990).
76. P. J. V. Hanson, D. Gor, J. R. Clarke, M. V. Chadwick, G. Nicholson, N. Shah, B. Gazzard, D. J. Jefferies, H. Gaya, and J. V. Collins, Contamination of endoscopes used in patients with AIDS, *Lancet 2*:86–88 (1989).
77. D. L. Prince, R. N. Prince, and N. Prince, Inactivation of human immunodeficiency virus type 1 and herpes simplex virus type 2 by commercial hospital disinfection, *Chem. Times Trends* Jan. pp. 13–16 (1990).
78. J. H. Blackwell, and J. H. S. Chen, Effects of various germicidal chemicals on HEp. 2 cell culture and herpes simplex virus, *J. Assoc. Off. Anal. Chem. 53*:1229–1236 (1970).
79. F. L. Sable, A. Hellman, and J. McDade, Glutaraldehyde inactivation of virus in tissue, *Appl. Microbiol. 17*:645–646 (1969).
80. U. Seefeld, G. Bansky, M. Jaeger, and M. Schmid, Prevention of hepatitis B virus transmission by the gastrointestinal fibrescope. Successful disinfection with an aldehyde liquid, *Endoscopy 13*:238–239 (1981).
81. K. Adler-Storthz, L. M. Sehulster, G. R. Dressman, F. B. Hollinger, and J. L. Melnick, Effect of alkaline glutaraldehyde on hepatitis B virus antigen. *Eur. J. Clin. Microbiol. 2*:316–320 (1983).
82. W. W. Bond, M. S. Favero, N. J. Petersen, and J. W. Ebert, Inactivation of hepatitis B virus by intermediate-to-high level disinfectant chemicals, *J. Clin. Microbiol. 18*:535–538 (1983).
83. H. Kobayashi, M. Tsuzuki, K. Koshimizu, H. Toyama, N. Yoshihara, T. Shikata, K. Abe, K. Mizuno, N. Otomo, and T. Oda, Susceptibility of hepatitis B virus to disinfectants or heat, *J. Clin. Microbiol. 20*:214–216 (1984).
84. S. A. Sattar, R. A. Raphael, H. Lochman, and V. S. Springhorpe, Rotavirus inactivation by chemical disinfectants and antiseptics used in hospitals, *Can. J. Microbiol. 29*:1464–1469 (1983).
85. O. Thraenhart, E. K. Kuwert, R. Dermietzel, N. Scheiermann, and F. Wendt, Influence of different disinfection conditions on the structure of the hepatitis B virus (Dane particle) as evaluated in the morphological alteration and disintegration test (MADT), *Zentralbl. Bakteriol. Hyg., Abt. I Orig. A. 242*:299–314 (1978).
86. J. Bailey, M. Chambon, H. Peigue-Lafeuille, H. Laveran, C. DeChamps, and D. Beytout, Activity of glutaraldehyde at low concentrations ($< 2\%$) against poliovirus

and its relevance to gastrointestinal endoscope disinfection procedures, *Appl. Environ. Microbiol. 57*:1156–1160 (1991).
87. J. Passagot, J. M. Crance, E. Biziagos, H. Laveran, F. Agbalika, and R. Deloince, Effects of glutaraldehyde on the antigenicity and infectivity of hepatitis A virus, *J. Virol. Methods 16*:21–28 (1987).
88. B. Ballantyne, and B. Berman, Dermal sensitizing potential of glutaraldehyde: A review and recent observations, *J. Cutan. Ocular Toxicol. 3*:251–262 (1984).
89. T. A. Marks, W. C. Worthy, and R. E. Staples, Influence of formaldehyde and Sonacide (potentiated acid glutaraldehyde) on embryo and fetal development in mice, *Teratology 22*:51–58 (1980).
90. K. Hemminki, K. Falk, and M. Vainio, Comparison of alkylation rates and mutagenicity of directly acting industrial and laboratory chemicals, *Arch. Toxicol. 46*:277–285 (1980).
91. National Toxicology Program, Mutagenicity testing results, *NTP Tech. Bull. 1*:9–11 (1981).
92. M. J. Wade, L. Freedman, and V. C. Ross, Biochemical and toxicological effects of glutaraldehyde, *FDA By-Lines 1*:14 (1982).
93. L. J. Marnett, H. K. Hurd, M. C. Hollstein, S. E. Levin, H. Esterbauer, and B. Ames, Naturally occurring carbonyl compounds are mutagens in *Salmonella* tester strain TA104, *Mutat. Res. 148*:25–34 (1985).
94. D. E. Levin, M. Holstein, M. F. Christman, E. A. Schwiers, and B. N. Ames, A new *Salmonella* tester strain (TA102) with A–T base pairs at the site of mutation detects oxidative mutagens, *Proc. Natl. Acad. Sci. USA 79*:7445–7449 (1982).
95. Y. Saski, and R. Endo, Mutagenicity of aldehydes in *Salmonella*, *Mutat. Res. 54*:251–252 (1978).
96. R. S. Slesinski, W. C. Hengler, A. J. Guzzie, and K. J. Wagner, Mutagenicity evaluation of glutaraldehyde in a battery of in vitro bacterial and mammalian test systems, *Food Cosmet. Toxicol. 21*:621–629 (1983).
97. M. Tamada, S. Saski, Y. Kadono, S. Kato, S. Amitani, Y. Ogasahara, T. Tamura, and N. Sato, Mutagenicity of glutaraldehyde in mice, *Bobkin Bobai 6*:62–69 (1978).
98. National Toxicology Program, Mutagenesis (*Drosophila melanogaster*) test results, *NTP Tech. Bull. 8*:11 (1982).

8

Chlorine and Iodine Formulations

Sally F. Bloomfield
King's College London
London, England

I. CHLORINE AND CHLORINE-RELEASING AGENTS

A. Introduction

Disinfection by chlorine-releasing agents (CRAs) has been used extensively for the last century. In 1798, bleaching powder was first made and was employed by Alcock, in 1827, as a deodorant and disinfectant. In 1847, Semmelweis used chlorinated lime to combat puerperal fever in his clinic in Vienna. Wide usage of chlorine as a disinfectant began during World War I, when Dakin [1] introduced sodium hypochlorite solution containing 0.45–0.5% w/v available chlorine ($AvCl_2$) for disinfection of open wounds. Up to 1963, chlorinated soda solution (Dakin's solution), was a standard product of the *British Pharmacopoeia.*

B. Structural, Chemical, and Physical Properties of Chlorine-Releasing Agents

Various antimicrobially active chlorine compounds are commercially available. The types of CRAs most frequently used in the health care profession are sodium hypochlorite and also the *N*-chloro compounds, mainly sodium dichloroisocyanurate, but also chloramine-T. All of these compounds release hypochlorous acid in aqueous solution, which is thought to be the active species. The chemical and physical properties of these agents is described in the following sections. Other CRAs used as disinfectants in various fields are reviewed elsewhere [2–4].

1. Chlorine and Hypochlorous Acid

Chlorine dissolves in water to give hypochlorous acid and hydrochloric acid:

$$Cl_2 + H_2O \leftrightharpoons HOCl + HCl$$

Hypochlorous acid is a weak acid, with a dissociation constant of 2.8×10^{-8}. The reaction is essentially complete in a few seconds at ordinary temperatures. In dilute solution, at pH levels above 4, the equilibrium is displaced to the right and very little chlorine exists in solution.

Hypochlorous acid undergoes ionization in water:

$$HOCl \leftrightharpoons H^+ + OCl^-$$

The relative amounts of hypochlorous acid and hypochlorite ion present in a solution of free available chlorine are a function of pH as shown in Table 1. It can be seen that below pH 6.0, HOCl dissociates poorly. Between pH 6.0 and 8.5, there is a sharp change from undissociated HOCl to almost complete dissociation. Thus, chlorine exists predominantly as hypochlorous acid (HOCl) at pH between 4 and 7, and above pH 9.0, hypochlorite ion (OCl^-) predominates.

2. Hypochlorites

Hypochlorites (sodium and calcium) are the oldest and most widely used chlorine compounds in the field of chemical disinfection. Sodium hypochlorite, rather than calcium hypochlorite, is now mostly used and is available commercially in

Table 1 Variation of the Percentage of Undissociated Hypochlorous Acid in Solutions of Various pH

pH	% HOCl	pH	% HOCl
4	100	8	23.3
5	99.7	9	2.9
6	96.8	10	0.30
7	75.2	11	0.030

solution at concentrations varying from 1 to 14% (w/v) available chlorine. Sodium hypochlorite (NaOCl), when dissolved in water, ionizes according to the equation

$$NaOCl + H_2O \leftrightharpoons Na^+ H_2O + OCl^-$$

Since NaOCl is a salt of a weak acid, the hypochlorite ions establish an equilibrium with HOCl, the concentration of the hypochlorite ion increasing with increasing pH:

$$OCl^- + H_2O \leftrightharpoons HOCl + OH^-$$

Solutions of sodium hypochlorite, therefore, are alkaline, the pH of the solution depending on the concentration of hypochlorite ions present. Solutions of NaOCl are relatively unstable decomposing to give sodium chlorate and hydrochloric acid:

$$NaOCl + 2HOCl \leftrightharpoons NaClO_3 + 2HCl$$

Stability decreases with decrease in pH, which leads to formation of HOCl and, once initiated, acceleration of decomposition occurs. As shown in Table 2, dilution of hypochlorites produces a drop in pH as a result of displacement of the equilibrium and the formation of HOCl. Alternatively, if NaOCl solution is buffered to pH 9.0, the solution is stable for 7–8 weeks or more. The stability of hypochlorite formulations has been reviewed [5,6]. Exposure to light causes breakdown of hypochlorite, producing chlorite (ClO_2) and oxygen; consequently solutions should be stored in dark containers. Heavy metal ions (e.g., copper, cobalt, and manganese) cause catalytic decomposition of hypochlorites, producing chloride and oxygen. High temperatures also reduce the stability of hypochlorite solutions, as the reactions described in the foregoing are accelerated at high temperatures.

TABLE 2 pH of NaOCl, NaDCC and Chloramine-T Solutions at Various Concentrations

Concentration	pH of solution		
(ppm $AvCl_2$)	NaOCl	NaDCC	Chloramine-T
5000	11.2	7.0	7.5
2500	11.0	6.8	7.4
1000	10.6	6.8	7.0
500	9.9	6.8	6.8
100	8.3	6.8	6.5
50		6.7	6.5

3. Sodium Dichloroisocyanurate

Sodium dichloroisocyanurate (NaDCC) is used as a disinfectant either by itself or formulated into various products that are used as general surface disinfectants or as laundry dry bleaches, scouring powders, and industrial sanitizing compounds. These organic chlorine-releasing agents are increasingly used in disinfection as an alternative to NaOCl because they are solid, more stable, and less inactivated by organic material. NaDCC is available commercially as effervescent tablets and granules that dissolve in water, liberating chlorine. Sodium dichloroisocyanurate as shown in Figure 1 dissociates in water to give HOCl, monochloroisocyanurate, and isocyanurate. Complete dissociation of NaDCC yields two molecules of HOCl.

By means of dissociation constants determined by Brady et al. [7], the proportion of each cyanurate species and the total hypochlorite (HOCl + OCl^-) was calculated for NaDCC solution at pHs 6.0 and 9.6 [8]. For NaDCC, at both pH values, only 50% of total available chlorine is actually present as free available chlorine (Table 3). The remaining 50% is combined available chlorine (either mono- or dichloroisocyanurate, according to pH).

Although NaDCC is stable in powder form, its solutions are relatively unstable compared with NaOCl, having a shelf life of about 7 days. This is probably related to their lower pH range (6.5–7.0) [6]. The pH of NaDCC solutions over the range of 10–10,000 ppm is 6.4–7.0, regardless of the concentration (see Table 2).

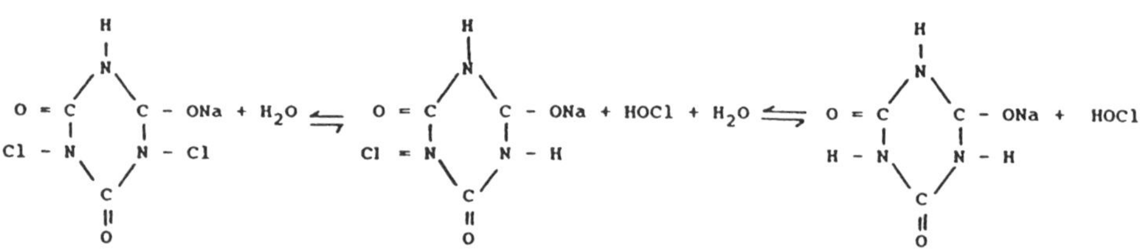

sodium dichloroisocyanurate

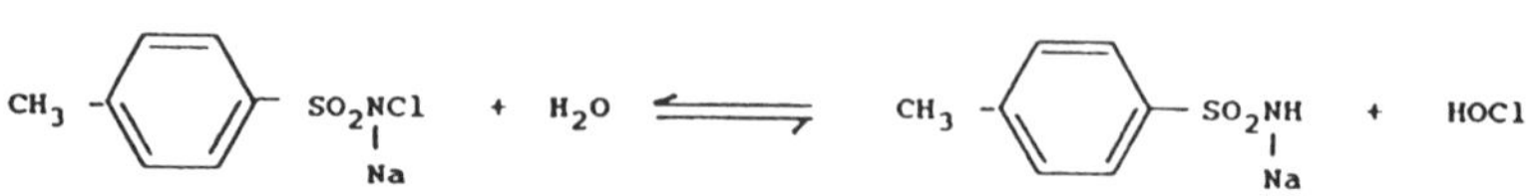

chloramine - T

FIGURE 1 Chemical structures of sodium dichloroisocyanurate and chloramine-T.

Table 3 Calculated Values of Proportions of Dissociated Species Present in NaDCC Solutions at pH 6.0 and 9.6

	Combined available chlorine as proportion of total available chlorine			Free available chlorine as proportion of total available chlorine
pH	Isocyanurate	Monochloroiso-cyanurate	Dichloroiso-cyanurate	$HOCl + OCl^-$
6.0	0.35	0.31	0.35	0.5
9.6	0.14	0.71	0.14	0.49

4. *N*-Chloro-*p*-toluene Sulfonamide (Chloramine-T)

Chloramine-T is formed by the reaction of *p*-toluene sulfonamide with sodium hypochlorite. The chlorine in this compound, as shown in Figure 1, has the linkage N-Cl, which hydrolyzes to varying degrees in water to form the =NH group and HOCl. Chloramine-T is a white, crystalline powder, having a slight chlorous odor. It is freely soluble in water, a saturated solution containing about 15% w/v. Solutions of chloramine-T are stable; neither moderate exposure to heat nor to light causes appreciable decomposition. The pH of solutions of chloramine-T containing 10–5000 ppm $AvCl_2$ ranges from 6.5 to 7.5 (see Table 2). For solutions of chloramine-T, calculations with dissociation constants suggest that it is about 25% dissociated at pH 7.4.

5. Available Chlorine

Although chlorine compounds are bactericidal by virtue of their chlorine content, analysis of available chlorine per se does not give a true measure of bactericidal activity, since it measures only total chlorine available to enter into a reaction. Bactericidal activity also depends on the rate of release of chlorine which, in turn, is determined by a number of factors. The *total available chlorine* may be defined as a measure of the oxidizing capacity and is expressed in terms of equivalent amount of elemental chlorine.

Total available chlorine content of solutions of CRAs can be measured by titration of liberated iodine from an acidic solution of potassium iodine with standard sodium thiosulfate solution, as indicated in the following equation:

$$2I^- + 2OCl^- + 2H^+ \leftrightharpoons 2Cl^- + 2OH^- + I_2$$

The quantity of iodine liberated by this oxidation process is a direct measure of the available chlorine of the original compounds. From this equation it can be seen that 1 mol of NaOCl is equivalent to 1 mol of chlorine of molecular weight (MW) 70.9 (since 1 mol of Cl_2 = 1 mol of HOCl); therefore,

$$\% \text{ available chlorine} = \frac{70.9}{74.5} \times 100$$
$$= 95.2\ \%\ \text{available chlorine}$$

Since NaDCC (MW 220) and chloramine-T (MW 281.69) is equivalent to 2 mol of chlorine, the available chlorine content of these molecules is 64.5% and 25.2%, respectively.

The total available chlorine associated with a given quantity of a particular CRA may be in the form of free or combined chlorine. The term free chlorine is usually applied to chlorine that exists in water as Cl_2, HOCl, or OCl^-. The term combined available chlorine is applied to *N*-chloro compounds that are formed when compounds, such as ammonia, cyanurates, proteins, or other, are added to solutions of free chlorine. The reaction is freely reversible, and the equilibrium concentrations of free and combined chlorine are determined by any number of factors, such as pH, total $AvCl_2$ concentration, and the N/Cl ratio.

C. Mechanism of Action of Chlorine-Releasing Agents

Experimental evidence [9–13] indicates that the bactericidal action of CRAs results from oxidative interaction with sulfydryl groups of key enzymes within the cell membrane or cell protoplast. Chlorine-releasing antibacterials are characterized by their extreme reactivity as oxidizing agents, and the rapid germicidal power of chlorine is undoubtedly due to its high oxidizing reactivity, by which the activity of cellular proteins is destroyed. In solutions of CRAs, these two forms are in an equilibrium that is dependent on pH:

$$Cl_2 + H_2O \leftrightarrows \underset{\displaystyle HCl}{\overset{}{HOCl}} \underset{H^+}{\overset{OH^-}{\leftrightarrows}} \underset{\displaystyle Cl^-}{OCl^-} + H_2O$$
$$\qquad\quad + \qquad\qquad +$$

Early workers suggested that in solutions of CRAs hypochlorous acid is the active species that is responsible for the destruction of microorganisms [14]. This hypothesis was based on observations showing that the activity of CRAs is increased at an acid pH at which hypochlorous acid is almost exclusively in its undissociated form. It may well be that OCl^- has some activity or that HOCl is formed as it is used up. Fair et al. [15] and Morris [16] calculated a theoretical curve for relative disinfecting efficiency of HOCl and OCl^- to produce 99% kill of *Escherichia coli* at various pH levels within 30 min and found that the OCl^- ion possesses about 1/80 of the germicidal potency of HOCl under these conditions. These basic mechanisms also apply to organic *N*-chloro compounds, such as NaDCC and chloramine-T, that hydrolyze in varying degrees to HOCl.

Investigations by various workers suggest that the interaction of CRAs with cell protein and organic material is complex and may involve not only oxidation of sulfydryl and other groups, but also reversible *N*-chlorination and, to a certain extent, irreversible decarboxylation and *C*-chlorination reactions. Langheld [17] reacted 14 α-amino acids with NaOCl and showed that there was rapid breakdown of the α-amino acids to aldehydes, ketones, ammonia, and CO_2. Ingols et al. [18] reported the oxidation of sulfydryl groups by chlorine in its interaction with cysteine.

Dakin [19] investigated the reversible formation of *N*-chloro compounds and showed that aromatic sulfochlor compounds act as chlorinating agents only when the opportunity for chlorine to leave the chloramine and attach to a second nitrogen (one in which the nitrogen is united to less acidic groups) is less than in the original compound.

Formation of *N*-chloro compounds within the bacterial cell does not appear to contribute to bactericidal action, but consumes chlorine, making it unavailable as an oxidizing agent to contribute to bacterial action. Therefore, in any particular situation, as summarized in Figure 2, the net result of adding CRAs to organic material will depend on the nature of the CRAs, the pH, the nature of cell or protein, and the N/Cl ratio. The reaction equilibrium achieved by adding a particular CRA to a bacterial cells will depend on all these factors and will determine the antimicrobial action that results. From this, it is expected that the extent of this action would be significantly different for different CRAs containing the same total $AvCl_2$ concentration, as is found to be the case.

The interaction of CRAs with intact bacterial cells is extremely complex, since the interaction is generally nonselective, such that the total interaction with a given cell mass will be made up not only of relative amounts of chlorine responsible for primary and secondary lesions that cause cell death, but also the amounts used up in other nonbactericidal reactions that consume available chlorine. Equally important is that the extent of interaction with intracellular structures will also be determined by the penetration of outer cell layers to sites of action on or in the cell. As stated earlier, that CRAs are more effective at an acidic pH is thought to be due to potentiation of oxidation, but it may also be due to the fact that

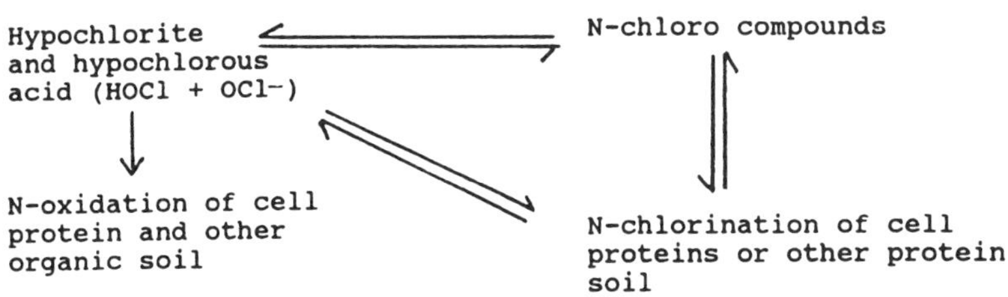

FIGURE 2 A summary of the net result of adding CRAs to organic material.

CRAs penetrate cell outer layers more readily in the nonionized form. Jacobs [20] reported that cellular uptake of most undissociated organic weak acids, particularly those with nonpolar hydrocarbon portions, is rapid, which could explain the higher residual antibacterial activity of NaDCC.

Spores are generally more resistant to all types of biocides, including CRAs. Many chemical agents are known to be effective against vegetative bacteria, but only a few agents, such as CRAs, are sporicidal, a necessary requirement for a chemical sterilizer. Experimental evidence suggests that development of resistance to antimicrobial agents during sporulation results mainly from the production of the spore outer layers (coat and cortex) that are relatively impermeable, thereby preventing access of the agent to its site of action on the underlying protoplast. This is illustrated by the studies [21–24] showing that treatment of spores with agents that achieve progressive removal of a spore's outer coat protein and inner cortex peptidoglycan layers is associated with successive increases in sensitivity to agents, such as CRAs, iodine, and glutaraldehyde.

Further investigations [24–28] indicate that sporicidal agents, such as CRAs, may themselves disrupt outer spore layers and that this contributes to their sporicidal action. Investigations showed that CRAs produced extraction of both spore coat protein and spore cortex peptidoglycan, but that the extent of the extraction varied in a way that correlated with sporicidal activity [24]. Results suggest that degradation of the spore cortex causes rehydration of the spore protoplast, facilitating diffusion of the agent to its site of action in the protoplast. Whereas NaOCl and NaDCC, in the presence of 0.4% NaOH, produced significant degradation of spore coat and cortex material and were rapidly sporicidal under these conditions, by contrast, NaDCC in buffer alone, and chloramine-T in buffer or in 0.4% NaOH produced relatively little coat and cortex degradation and were not effective.

D. Antimicrobial Properties of Chlorine-Releasing Antibacterials

1. Antimicrobial Activity

All solutions of CRAs containing free chlorine have a wide antimicrobial spectrum that includes vegetative bacteria, mycobacteria, bacterial spores, viruses, algae, and protozoa. A review of various studies of the antibacterial action of free available chlorine [4] indicates that concentrations from 0.05 to 5.0 ppm $AvCl_2$ will produce kill of a range of vegetative bacteria within 15 s to 5 min, whereas for *Mycobacterium tuberculosis*, a concentration of 50 ppm (30 s contact) is required. The activity of NaDCC is comparable with that of NaOCl [29], although solutions of NaDCC show significantly higher bactericidal capacity against a range of bacterial species when compared with NaOCl [8,30].

Studies with *Aspergillus niger* and *Rhodotorulla flava* [4] indicate that concentrations of 100 ppm $AvCl_2$ are fungicidal, producing 90% kill of these organisms within 30 min.

Recent studies have confirmed the potential for using organic *N*-chloramines as broad-spectrum disinfectants [32–34]. Early workers generally disagreed on the merits of chloramine-T when compared with hypochlorites, suggesting that the activity of chloramine-T was considerably "slower" than that of hypochlorites [35], although recent evidence suggests that this applies mainly to sporicidal action.

The most resistant forms of microbial life are bacterial spores, and recent studies have shown that chlorine compounds are among the most potent sporicidal agents. Various studies [36,37] have shown sodium hypochlorite, at 100–200 ppm and pH 7.0, will produce a 4- to 5-log reduction in *Bacillus subtilis* spores within 5 min. However, whereas buffered solutions of NaOCl, NaDCC, and chloramine-T show similar activity against vegetative cells of *Bacillus subtilis*, *B. subtilis* spores were not only more resistant to their action, but there was considerable variation in activity. For NaDCC and chloramine-T, 200-ppm solutions had little or no sporicidal action. For NaDCC, 1000 ppm was required to produce 99% kill in 5 min, whereas for chloramine-T 5000 ppm was required to produce 90% kill within 30 min.

Investigations of antiviral activity suggest that hypochlorites are active against a broad spectrum of both lipophilic and hydrophilic viruses [38,39]. The virucidal action of NaDCC has been confirmed [40,41]. The activity of CRAs against hepatitis B virus and human immunodeficiency virus (HIV) has also been evaluated [42–44].

One of the significant advantages of CRAs is their very rapid antimicrobial action that, in the absence of organic soil against vegetative bacteria, may be complete within 30 s to 1 min. Where CRAs are used to achieve sporicidal or cysticidal action or, in some cases, in the presence of organic soil, the antimicrobial effect is increased by prolonged contact.

2. Factors Affecting the Biocidal Activity

The conditions under which disinfectants are effective should be qualified. The conditions of use and appropriate use concentrations are decided on the basis of in vitro tests that, as well as defining the useful biological activity, determine the influence of such factors, as pH, temperature, organic matter, and dilution, on the activity and stability of compounds. Published information concerning CRAs is mostly derived from studies with hypochlorites. Although the biocidal activity of a particular chlorine compound will largely depend on the concentrations of undissociated hypochlorous acid in solution, experimental evidence suggests that for *N*-chloro compounds it will also depend on the rate and extent of hydrolysis producing release of the hypochlorite ion. Because of this, solutions of different chlorine-releasing compounds, even though having the same total available chlorine, can differ markedly in their properties. Formulations of NaDCC and NaOCl containing the same total $AvCl_2$ at the same pH did not show equal activity [8], implying that the relations among activity, pH, and $AvCl_2$, were different for each formulation.

Other factors that influence the effectiveness of CRAs as disinfectants and antiseptics are discussed in the following sections.

3. Effect of pH on Activity

Several studies with bacteria, spores, and viruses have shown that the antimicrobial activity of CRAs is pH-dependent, being greater at lower pH values. [These various studies are reviewed in more detail in Refs. 2–4.] The results of these studies suggest that the effects of pH vary considerably, according to the conditions of use, and are relatively much less with vegetative bacteria than with spores. Various studies indicated that hypochlorite solutions, buffered to pH 10.0 or more, have little or no activity against *B. subtilis* spores at concentrations up to 500 ppm [21,30,35]. Whereas NaOCl at pH 10.6 showed satisfactory disinfectant activity against vegetative bacteria at 200 ppm $AvCl_2$ under clean conditions, formulations at this pH, even at 5000 ppm, were not effective if sporicidal action was required or if substantial organic soiling was present [13].

Study results [45] suggested that NaDCC and chloramine-T are also sensitive to pH changes, but whereas the activity of the chloroisocyanurates was only slightly affected over the pH range 6.0–10.0, chloramines were very sensitive with up to 1000-fold change in activity over the same pH range.

4. Effect of Organic Material on Activity

Since chlorine is a highly reactive chemical, it readily reacts with organic matter, causing loss of antimicrobial activity. The CRAs react readily with all types of organic matter, including milk, blood, feces, and tissues, and are inactivated by oxidation of organic material and formation of *N*-chloro compounds. This leads to a reduction in free $AvCl_2$, although the formation of *N*-chloro compounds is reversible. A margin of safety should thus be included when using CRAs to disinfect areas with organic soil.

Results of various investigations have shown that *N*-chloro compounds are less affected than hypochlorites; the addition of milk or plasma caused a greater loss of $AvCl_2$ from NaOCl solutions, compared with NaDCC solutions [46,47]. This was associated with a greater reduction in bactericidal activity of the NaOCl solution.

Investigations [4] have indicated that, whereas concentrations of 5–10 ppm $AvCl_2$ were sufficient to produce 99.99% kill of vegetative bacteria in the absence of soil, concentrations of 250 and 500 ppm were required to produce the same effect in the presence of 1% w/v yeast and 10% v/v serum, respectively. Results of other studies [30,48] suggest that NaDCC at 2500–3000 ppm $AvCl_2$ was effective against vegetative bacteria in the presence of 20% plasma, compared with only 10% plasma for sodium hypochlorite. In the presence of 10% v/v blood, a concentration of 10,000 ppm $AvCl_2$ or more was required.

The intense chemical reactivity of chlorine compounds does not cover all types of organic compounds. The CRAs do not react with most carbohydrates, and they have been found to be inert against methyl and ethyl alcohols, glycerol, starch, sodium oleate, palmitate, and acetate [49].

5. Other Factors Affecting Activity

In formulating and using CRAs as disinfectants in the health care field, several other factors may need to be considered. The calcium and magnesium ions in hard water do not inactivate chlorine disinfectants, but ferrous or manganous cations, and nitrate and sulfide anions reduce active hypochlorous acid to inactive chloride, possibly by oxidation or chlorination of the hydrocarbon molecules. Inactivation owing to incompatibility may also occur if chlorine compounds are used in conjunction with cationic detergents. The activity of CRAs against vegetative bacteria can be enhanced by the addition of bromine [50,51].

The sporicidal action of sodium hypochlorite (200 ppm available chlorine) can be potentiated by 1.5–4% w/v NaOH [21], low concentrations of ammonia [52], and in the presence of bromine [53]. Potentiation of action also occurs with buffered methanol/sodium hypochlorite mixture, which kills spores more rapidly than buffered hypochlorite alone [36,54].

E. Practical Application of Chlorine Disinfectants in the Health Care Situation

Disinfectants formulations containing CRAs are employed in hospitals to reduce and control the spread of infection and may be applied:

1. As general environmental disinfectants for decontamination in high-risk situations
2. For disinfection of clean items and medical equipment
3. As skin and wound disinfectant

1. Chlorine as a General Environmental Disinfectant

Disinfection policies in most hospitals now recommend less disinfectant usage for general hospital cleaning and more emphasis on good hygiene. Formulations of CRAs are still recommended, however, and are widely used in circumstances during which significant risk of infection transfer may be identified (e.g., transport of bedpans for disposal, treatment of spillage of contaminated exudates from infected patients, or for general disinfection in high-risk areas, such as operating theaters and intensive care units).

For maximum cost-effectiveness in using CRAs, surfaces should be cleaned before disinfection with solution containing 200–1000 ppm $AvCl_2$. In situations during which soiling is unavoidable, concentrations up to 2500 ppm are usually recommended. In achieving safe disposal of blood and body spillages, decontamination of spilled material may be desirable, particularly for the protection of health care workers. In this situation, concentrations of 10,000 ppm are usually recommended. In environmental situations in which relatively large quantities of disinfectant may be required, phenolic and chlorine-releasing disinfectants are generally the agents of choice because they are relatively economical as well as effective. In the past, phenolic disinfectants, rather than CRAs, were preferred for these situations, owing to their greater ability to withstand inactivation by organic matter. However, because of their superior virucidal action, CRAs are increasingly recommended for the disinfection of blood spillage from hepatitis carriers or patients with hepatitis B and those with acquired immunodeficiency syndrome (AIDS), and in all situations for which viral contamination is suspected.

In a series of experiments, suspension test methods were used to compare phenolic and CRA disinfectants [47]. With plasma, to simulate soiled conditions, phenolic disinfectants at recommended-use dilution produced satisfactory disinfection (5-log reduction in 5 min) against vegetative bacteria in the presence of up to 50% v/v plasma. Although NaDCC at 2500 ppm was effective in the presence of up to 20% v/v plasma, compared with only 10% plasma for sodium hypochlorite 2500 ppm, sensitivity of NaDCC to inactivation was still greater than for phenolics. In the presence of blood both CRAs and phenolics were extensively inactivated, although, in this situation, the activity of NaDCC at 10,000 ppm was equivalent to the phenolics. The results indicated that CRAs at 10,000 ppm may be ineffective for treatment of blood spills unless applied at a v/v ratio of 9 parts disinfectant to 1 part blood. In this situation, chlorine-releasing powder formulations, which produce higher $AvCl_2$ concentrations and contain the spilled material, offer an effective alternative. Further evaluation [44] has shown that addition of NaDCC and NaOCl solution (10,000 ppm) to equal volumes of blood contaminated with HIV was sufficient to produce inactivation of the virus within 2 min.

2. Disinfection of Medical Equipment

In the disinfection of critical items of medical equipment, such as endoscopes, incubators, or other, that cannot be heat treated, CRAs, as currently formulated, are not usually recommended because of their corrosive action on metals, rubbers, and fabric.

For disinfection of endoscopes, 2% glutaraldehyde is probably the most widely used agent, but has a number of disadvantages associated with its use, most particularly its irritant and, in some persons, sensitizing effects. Babb et al. [55] also demonstrated that when sporicidal as well as bactericidal activity was required, with glutaraldehyde, a contact time of 3 h was required, compared with hypochlorite solution ppm at pH 7.5, which achieved an equivalent effect within 10–30 min. The effectiveness of two hypochlorite formulations for chemosterilization of medical equipment was evaluated [36,54,55]. These workers showed that NaOCl solution at 2000 ppm (pH 7.0) containing a corrosion inhibitor, and

NaOCl solution at low concentration (200 ppm) combined with methanol, were effective as sporicidal agents at contact periods of less than 1 h.

3. Sanitization in the General Hospital Environment

Chlorine-releasing agents (CRAs), either as NaOCl or organic chloramines, are widely used in the dairy, food production, and all areas of the catering industries because they do not leave toxic residues. Wherever possible, controlled cycles of hot water (at a temperature of not less than 80°C) with rinsing are used for disinfection of food-processing equipment and eating utensils, but for food preparation surfaces, which cannot be treated by this method and which are in constant use for a succession of different processes, indications are that chemical disinfection between operations is desirable to prevent cross-contamination and that hypochlorite formulations are effective for this purpose [56,57]. In hospitals, as in other public eating places, food serving operations are concentrated within a few hours, during which time serving utensils may be reused several times. The consequences of this is that food, eating, and drinking utensils may not be properly washed and, most often, it is possible to give only a short rinse with disinfectant. Chlorine, in various forms, but hypochlorites in particular (50 ppm), are considered suitable for this quick rinse because of their fast-disinfecting property.

It must be assumed that cleaning agents, such as scouring powders that contain solid bleach compound, such as NaDCC, must exert some bactericidal action, but there is no published data available.

Chlorine-releasing disinfectants are used for disinfection of feeding utensils in the hospital ward or clinic situations, for which hot water disinfectants are not available. A study by Anderson and Gatherer [58] showed that, although, under strictly controlled laboratory conditions, boiling was consistently effective for disinfection of infant-feeding utensils, where disinfection of utensils was carried out by the user in the domestic situation, disinfection failures occurred more frequently when boiling, as opposed to hypochlorite (250 ppm $AvCl_2$), treatment was used.

4. Disinfection of Water

Chlorination is one of the most widely employed methods of disinfection of public water supplies, swimming pools, spa baths, and hydrotherapy pools [59]. The concentration of chlorine employed varies according to the pH and chemical purity of the water. In good quality waters, less than 0.5 mg/L is adequate, whereas as much as 20 mg/L may be necessary in heavily contaminated water. For drinking water, a residual chlorine content, after satisfying the chlorine demand of water, of 0.2–0.4 mg/L is considered sufficient to give satisfactory action. In swimming pool water, where a greater variety of organisms are constantly introduced into the pool water by various swimmers and where rapid kill is necessary, a chlorine residual of 1–3 mg/L is recommended for pools treated with sodium or calcium hypochlorite [60]. For pools treated with chloroisocyanurates, 3–4 mg/L is recommended, 5 mg/L when hand dosing is employed [61].

5. Skin and Wound Disinfection

Dakin [1] introduced a solution of sodium hypochlorite, sodium carbonate, and sodium bicarbonate, containing 0.5–0.55% w/v $AcCl_2$, pH 9.5 (Dakin's solution) for treatment of open and infected wounds. This was later replaced by Eusol BPC (solution of calcium hypochlorite and boric acid containing not less than 0.25% w/v available chlorine, pH 8–8.4) because hypochlorites of less alkaline pH were found to be less irritating and, hence, more suitable for wound disinfection.

In more recent years, sodium, rather than calcium, hypochlorite has been used for treatment of leg ulcers, pressure sores, and so on. NaDCC, which current evidence suggests is less adversely affected by organic matter, has also been recommended as an alternative to Eusol [7]. In many situations, it would seem that the effectiveness of alkaline hypochlorite formulations arises from their effectiveness in achieving dissolution of contaminated slough from infected wounds and pressure sores [62].

Dilute solutions of hypochlorite in water or saline have been used as irrigation solutions for disinfection of the bladder, vagina, and urethra. Hypochlorites at 0.1–1.2% w/v concentrations have been used in the control of athlete's foot [63] and oral bleeding following tooth extraction [64]. It is also an accepted method of management of burns; irrigation with hypochlorites being used in the prevention of infection.

Although chlorinated solutions are still widely used in the treatment of wounds, there is increasing evidence to suggest that chlorine formulations applied to wounds can delay healing and may be associated with toxic effects [65–67]. Barton and Barton [67] reported cases of renal failure associated with topical application of chlorinated solutions to pressure sores, which they attributed to the release of toxic lipid A material from bacteria, causing bacteremia or endotoxic shock.

Although chlorine-release disinfectants are no longer used for surgical or hygienic hand disinfection in hospitals, the investigations of Semmelweis in 1847 in which chlorinated lime solutions were introduced for decontamination of the hands must still remain as one of the most convincing demonstrations of the importance of hand hygiene in preventing the transfer of infection in the clinical situation.

F. Toxicological and Environmental Considerations

Chlorine is an effective and reliable disinfectant that is used in the treatment of drinking water to ensure water of consistent and high microbiological quality. The fact that chlorine reacts with many of the organic substances in raw water or sewage with the formation of chlorinated hydrocarbons has led to concern and has stimulated a series of investigations [68–72]. The concentration of the most common by-products, chloroform and trichloroacetic acid, in drinking water seldom exceeds 50 μg/L, but typically levels are 1 μg/L or less. Also, the addition of 6 mg/L NaOCl to primary settled sewage resulted in less than 1% being incorporated into organic materials, giving mainly monochlorinated compounds, the main reaction being the formation of NaCl. Since many mono- and dichlorinated hydrocarbons are biodegradable, concentrations will be even lower in treated sewage. Among the by-products of water chlorination are a number that have been shown to be mutagenic and some that are animal carcinogens. However, the levels of these compounds required to produce toxic effects are orders of magnitude greater than those found in water, and at the present time it is considered that there is no hazard to health or the environment from use of CRAs for disinfection purposes.

II. IODINE AND IODINE COMPOUNDS

A. Introduction

Iodine was first officially recognized as an antimicrobial agent in 1830, when preparations of iodine used for skin disinfection were introduced into the *United States Pharmacopeia*, but it was not until 1874, that experiments demonstrating the antimicrobial properties of iodine were first reported. Since then, iodine preparations have been widely used, most particularly for disinfection of inanimate surfaces, water treatment, and various other uses. The use of iodine preparations for skin disinfection and antisepsis is described in Chapter 4.

B. Structural, Chemical, and Physical Properties

Elemental iodine is only slightly soluble in water (0.3 g/L at 25°C), but is soluble in organic solvents, such as alcohol and acetone, and in potassium iodide solution. The solubility of iodine in 10% aqueous solution of cetomacrogol is about 1% w/v at 20°C and increases linearly with cetomacrogol concentration and temperature [73].

Although solutions of iodine in alcohol and iodine in potassium iodide (e.g., Lugols solution) have been used for many years, these formulations have now largely been replaced by the solubilized preparation of iodine known as iodophors, most particularly solutions of iodine in polyvinylpyrrolidone (PVP). These iodophor formulations have the advantage of being less corrosive, less irritating, and are nonstaining, compared with aqueous and alcoholic solutions.

Iodine undergoes a number of reactions in aqueous solution. The chemistry of aqueous iodine solution is described by the following equations [74]:

$I_2 + H_2O \rightleftharpoons [H_2OI]^+ + I^-$	(hydrolytic ionization)
$[H_2OI]^+ \rightleftharpoons HOI + H^+$	(dissociation of $[H_2OI]^+$)
$HOI \rightleftharpoons OI^- + H^+$	(dissociation of HOI)
$3HOI \rightleftharpoons IO_3^- + 2I^- + 3H^+$	(disproportion of HOI)
$I_2 + I^- \rightleftharpoons I_3^-$	(formation of triodide)

From this it can be seen that iodine in an aqueous solution may exist in several forms, depending on pH. The position of the equilibria at acidic, neutral, and alkaline pH is indicated by the length of the arrows in the following equations:

Acidic pH: $I^- + HOI + 2H^+ \rightleftharpoons I_2 + HOH + H^+$

Neutral pH: $2H^+ + OI^- \rightleftharpoons HOI + I^- + H^+ \rightleftharpoons I_2 + HOH$

Alkaline pH: $3I_2 + 6OH^- \rightleftharpoons 3I^- + 3OI^- + 3H_2O \rightleftharpoons 2I^- + IO_3^-$

Below pH 7.0, diatomic iodine exists as the main species, with HOI and I_3 molecules occurring in minute quantities. Above pH 7.0, there is an increase in iodate formation, and the solution becomes unstable [74]. In an aqueous iodine solution, molecular iodine, hypoiodous acid (HOI), and the iodine cation (H_2OI^+) have strong germicidal properties.

Investigations by Wyss and Strandskov [75] suggest that I_3 is not an effective antimicrobial agent, but where solutions of I_2 in KI are used, free iodine is immediately released, as it is used up in the antimicrobial action. Gottardi [74] recommended addition of KI to iodine formulations to decrease formation of unstable IO_3^-. The term available iodine is applied to iodine that exists in solution as I_2, HOI, and I_3. It is referred to as the total amount of iodine that can be determined iodometrially.

C. Mode of Action of Iodine Formulations

The antibacterial activity of iodine in aqueous solutions has been shown by various workers to be due to mainly molecular iodine, rather than one or more of the ionic species that may occur [75]. The exact manner by which iodine exerts its disinfecting action is yet unknown, although many suggestions have been put forward.

Indications are that the disinfecting action of iodine, like that of chlorine, results from direct interaction of free iodine molecules with essential cell enzymes or proteins [76–78]. Dunn [76] studied the actions of halogens on enzymes and stated that they may oxidize −SH groups to −S−S− groups in the following manner:

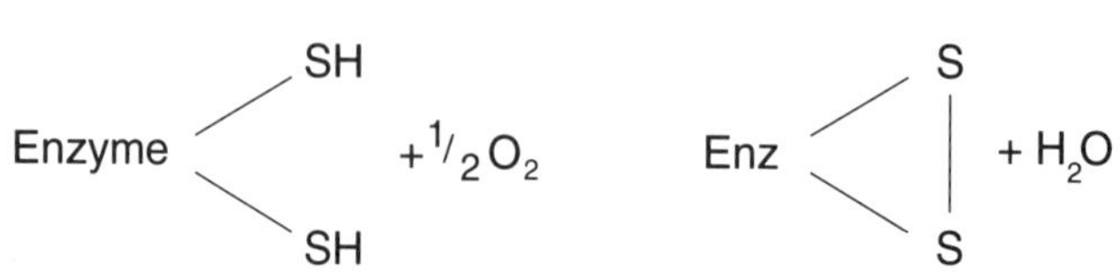

The reaction represents a mild type of oxidation that can be reversed by reducing agents. Work by Dunn was discussed in more detail by Gottardi [77], who stated that iodine has the ability to substitute covalent hydrogen and that it reacts in the following way:

1. Iodine reacts with the N−H group of amino acids, forming *N*-iodo compounds, thereby causing lethal changes to protein structure.
2. The S−H group of the amino acid cysteine is oxidized and loses its ability to make disulfide bonds, thereby disrupting protein synthesis.
3. Iodine reacts with the phenol group of the amino acid tyrosine. The size of the iodine atoms in the ortho position can sterically prevent the formation of hydrogen bonds with the OH group.
4. Iodine reacts by addition to unsaturated fatty acids and olefinic double bonds, producing changes in the physical properties of the lipid membrane, and thereby decreasing its fluidity.

A comparative study of iodine and chlorine-releasing agents [79,80] indicates that against vegetative bacteria, iodine is more effective as a bactericidal agent, on a molecular weight basis, even though it is less chemically reactive than chlorine. This could be because I_2 penetrates cell outer layers more readily than HOCl, or because iodine is less active in the formation of *N*-halo compounds with cell proteins, thereby reducing the level of free halogen within the cell. That iodine forms *N*-halo compound less readily than chlorine probably also accounts for the fact that, under use conditions, iodine is less affected by organic soiling.

Against bacterial spores, on the other hand, indications are that the activity of iodine is equal to (on a molar basis) or less than (on a weight basis) that of NaOCl [24,79–81]. Studies indicate that, as with CRAs, the resistance of spores to iodine, when compared with vegetative cells, results from the spore outer layers that limit penetration to its site of action on the underlying protoplast [24,80,84].

Investigations have suggested that, although iodine like NaOCl produces disruption of outer spore coats, it does not cause degradation of the spore cortex [24,80], as is required for rehydration of the spore protoplast, facilitating diffusion to its site of action on the protoplast. It is suggested that this is responsible for the fact that iodine is effective only against spores at relatively high concentrations and with long contact times, as required to achieve adequate penetration through the spore outer layers.

D. Antimicrobial Properties of Iodine Formulations and Antimicrobial Activity

1. Spectrum of Activity

Iodine formulations, similar to CRAs, show a broad-spectrum activity that includes fungi, mycobacteria, algae, protozoa, and viruses, as well as vegetative bacteria.

Investigations of the activity of iodine against vegetative bacteria have been reported [83–85]. A parallel series of experiments [79,80] showed that the activity of solutions of I_2 in KI (pH 7.0) against vegetative bacteria was superior to that of NaOCl (pH 7.4). These investigations showed that, whereas solutions of 20–80 ppm AvI_2 produced >4-log kill within 5 min, concentrations of >100 ppm $AvCl_2$ were required to give an equivalent effect [79,80]. The exception to this was *B. subtilis* vegetative cells, which were sensitive to CRAs, but relatively more resistant to iodine.

Gershenfeld et al. [86] studied, the effect of iodine on *M. tuberculosis* var. *hominis*. Under the conditions of the experiments, concentrations of free iodine of 0.0625% or more were tuberculocidal after a 5-min exposure at room temperature. In vivo experiments showed that symptoms of tuberculosis were entirely absent in guinea pigs injected with the tubercle bacilli 10 mg/ml and exposed to 0.5 and 0.05% free iodine. A concentration of 0.005% free iodine also markedly reduced the virulence of the strain of *M. tuberculosis* used in the study. Studies on the lethal effect of iodine on tubercle bacilli were also carried out by Lawrence et al. [87].

The sporicidal activity of iodine has been investigated [75,80,85,88,89]. These studies suggest that, with bacterial spores, the activity of iodine is significantly less than that of sodium hypochlorite. Parallel studies [79,80] showed that, whereas NaOCl pH 7.2 produced a 5-log reduction of *B. subtilis* spores within 5 min, with iodine solution in KI, a concentration of 2000–5000 ppm AvI_2 was required to produce an equivalent effect within 15–30 min.

Early workers also found that iodine solution was less sporicidal than hypochlorite: 1000 ppm of AvI_2 from an iodophor destroyed *B. subtilis* in 6–10 h, but 800 ppm took up to 24 h [87]. Cousins and Allan [21] were in general agreement and found that 800 ppm of AvI_2 at pH 2.0 was ineffective against those spores after 4 h, whereas 100 ppm $AvCl_2$ at pH 8.0 killed them after 1 h. Povidone–iodine (Betadine) was reported to be effective against spores of *B. subtilis* within 2–5 h at 20°C [88]. More recently, dry spores of *B. subtilis* were shown to be resistant to PVP-I at 10,000 pm [85]. In other studies [88,90,91], iodine solution was effective in destroying spores of *Bacillus anthracis*, *B. subtilis*, *B. megaterium*, *B. mesentericus*, *Clostridium tetani*, and *C. welchii*.

Iodine has good fungicidal and fungistatic activity. Opinions on the actual lethal concentrations of iodine against fungi vary somewhat. Iodine has been reported to be effective against *Trichophyton gypsum* and *Monilia albicans* [91]. It was also effective against *Epidermophyton inguinale* [92]. Gomez-Vega [93] reported on the effectiveness of iodine against *Monilia*, *Torula*, *Epidermophyton*, and *Tricophyton* species. Other studies of fungicidal activities are also reported [94].

Studies with different species indicate that iodine has good virucidal activity: 8 ppm AvI_2 inactivate the poliomyelitis virus in 5–10 min [95], although, as much as 125–375 ppm is needed to accomplish the same inactivation in 1 min [96].

Against Newcastle disease virus suspended in allantoic fluid, 25 ppm was claimed to be effective, and 50 ppm inactivated poliomyelitis virus [90]. Further studies of the inactivation of poliovirus are reported by Tyler et al. [39].

Iodine has been used prophylactically against herpes virus [97], and therapeutically in cases of smallpox, chickenpox, and herpes [98–100]. The inactivation by iodine of poliomyelitis, influenza, herpes, vaccinia, rabies, tobacco mosaic, and other viruses has been reviewed [96]. Berg et al. [101] discussed the dynamics of inactivation of enteroviruses by iodine and presented inactivation rates for iodine against three enteroviruses.

2. The Effect of pH on Antimicrobial Activity

Although iodine exerts its antimicrobial effect over a wide pH range, published data have consistently shown that the disinfecting efficiency of iodine, like chlorine, increases with decrease in pH, and vice versa. From these observations it is generally assumed that I_2 is the active species [75,77,83,85,102].

3. The Effect of Organic Matter on Antimicrobial Activity

The interaction of iodine with organic matter was studied by Gershenfeld and Witlin [103] who reported decrease in activity of iodine against bacterial cells with an increase in plasma concentration. Kojima [104] reported that numerous organic agents were capable of neutralizing the effect of iodine. Among the organic compounds used were serum, glycerin, syrup, feces, ascitic fluid, egg, milk, sputum, and urine. A reduction in activity of iodine against *Staphylococcus aureus* in the presence of 10% serum has also been observed [105].

Iodine, like chlorine and bromine, is nonselective in its interaction with organic matter, bacteria, dead cells, or dissolved proteins. Such organic matter can be oxidized or iodinated, thereby using up available iodine and reducing its antimicrobial activity. As stated previously, the inactivation of iodine by organic soil is generally less than that of sodium hypochlorite because of the lower chemical reactivity of iodine.

4. The Effect of Temperature on Antimicrobial Activity

The temperature coefficient of iodine is reported to be relatively low, compared with that of most other bactericides, with the results that iodine is more effective than other agents at lower temperatures. Hugo and Newton [105] stated that increasing the temperature from 20 to 37°C did not increase the bactericidal action of iodine against *Staph. aureus*, but Chambers et al. [106] reported that, at a given exposure time and pH, the concentration of iodine required to kill at 2°–5°C was higher than that at 20°–26°C. It has also been shown [107] that the sporicidal activity of iodophors (aqueous Betadine and alcoholic Videne) is markedly temperature-dependent. Betadine, compared with Videne, had negligible activity over a 60-min period at 20°C. Increasing the temperature to 37°C resulted in a radical improvement, with the rate of kill approximating that obtained with Videne at 20°C.

E. Practical Application of Iodine Formulations as a Disinfectant in the Health Care Situation

Iodine has been recommended for the emergency sterilization of various types of surgical equipment, especially when heat sterilization must be avoided. This includes the disinfection of thermometers, rectal catheters, rubber goods, plastic apparatus, knife blades, and brushes [108–110]. Iodine is also used for the disinfection of bronchcoscopes [111–113], contact lenses [114,115], inhalation equipment [116,117], root canal cones [118], and hydrotherapy equipment [119,120]. Although suitable for disinfection of general inanimate surfaces, iodine formulations are used relatively seldom in this situation compared with chlorine-releasing and phenolic disinfectants. When iodine formulations are used for disinfection of surfaces, the possibility of damage to the surface must always be considered.

Iodine formulations, however, are widely used in the disinfection of dairy equipment and as a general sanitizer in cleaning of food-processing plants, such as in poultry processing [121–124].

REFERENCES

1. H. D. Dakin, the antiseptic action of hypochlorites, *Br. Med. J. 2*:809 (1915).
2. G. Sykes, The halogens, *Disinfection and Sterilization*, 2nd ed., E. and F. Spon, London (1965).
3. J. R. Trueman, The halogens, *Inhibition and Destruction of the Microbial Cell* (W. B. Hugo, ed.), Academic Press, London, 1971, p. 137.
4. G. R. Dychala, Chlorine and chlorine compounds, *Disinfection, Sterilization and Preservation*, 3rd ed. (S. S. Block, ed.), Lea & Febiger, Philadelphia, 1983, p. 157 (1985).
5. P. N. Hoffman, J. E. Death, and D. Coates, The stability of sodium hypochlorite solutions, *Disinfectants. Their Use and Evaluation of Effectiveness* (C. H. Collins, M. C. Allwood, S. F. Bloomfield, and A. Fox, eds.), Society of Applied Bacteriology, Technical Series. No. 16, Academic Press, London, 1981, pp. 77–83.
6. S. F. Bloomfield, and T. J. Sizer, Eusol BPC and other hypochlorite formulations used in hospitals, *Pharm. J. 235*:153–157.
7. A. P. Brady, K. M. Sancier, and G. Sirine, Equilibria in solutions of cyanuric acid and its chlorinated derivatives, *J. Am. Chem. Soc. 85*:3101–3104 (1963).
8. S. F. Bloomfield, and G. A. Miles, The antibacterial properties of sodium dichloroisocyanurate and sodium hypochlorite formulations, *J. Appl. Bacteriol. 46*:65–73 (1979).
9. C. Venkobachar, L. Iyengar, and A. G. S. Prabhakara Roa, Mechanism of disinfection, *Water Res. 9*:119–124 (1975).
10. C. Venkobachar, L. Iyengar, and A. V. S. Prahbakara Rao, Mechanism of dis-

infection: Effect of chlorine on cell membrane functions, *Water Res. 11*:727–729 (1977).
11. A. K. Camper, and G. A. Mcfeters, Chlorine injury and enumeration of waterborne coliform bacteria, *Appl. Environ. Microbiol. 37*:633–641 (1979).
12. E. E. Uso, Antibacterial properties of sodium hypochlorite and sodium dichloroisocyanurate, M. Phil. Thesis, University of London, 1983.
13. W. E. Knox, P. K. Stumpf, D. E. Green, and V. H. Auerbach, The inhibition of sulphydryl enzymes as the basis of the bactericidal action of chlorine, *J. Bacteriol. 55*:451–458 (1948).
14. R. J. Baker, Types and significance of chlorine residuals, *J. Am. Water Works Assoc. 51*:1185–1190 (1959).
15. G. M. Fair, J. C. Morris, S. L. Chang, I. Weil, and R. P. Burden, The behaviour of chlorine as a water disinfectant, *J. Am. Water. Works Assoc. 40*:1051–1061 (1948).
16. J. C. Morris, The mechanism of the hydrolysis of chlorine, *J. Am. Chem. Soc. 68*: 1692–1698 (1946).
17. K. Langheld, Uber das verhalter von-Aminosauren gegen Natrium-Hypochlorite, *Ber. 42*:2360–2365 (1909).
18. R. S. Ingols, H. A. Wyckoff, T. W. Kethley, H. W. Hodgen, E. L. Fincher, J. C. Hildebrand, and J. E. Mandel, Bactericidal studies of chlorine, *Ind. Eng. Chem. 45*:996–1000 (1953).
19. H. D. Dakin, J. B. Cohen, M. Daufresene, and J. Kenyon, The antiseptic action of substances of the chloramine group, *Proc. R. Soc. (B) 89*:232–237 (1916).
20. M. H. Jacobs, *Cold Spring Harbor Symp. Quant. Biol. 8*:30–32 (1940).
21. C. M. Cousins, and C. D. Allan, Sporicidal properties of some halogens, *J. Appl. Bacteriol. 30*:168–174 (1967).
22. W. M. Waites, and C. E. Bayliss, The effect of changes in spore coat on the destruction of *Bacillus cereus* spores by heat and chemical treatment, *J. Appl. Biochem. 1*:71–76 (1979).
23. S. P. Gorman, E. M. Scott, and E. P. Hutchinson, The effect of sodium hypochlorite–methanol combinations on spores and spore forms of *Bacillus subtilis*, *Int. J. Pharm. 17*:291–298 (1983).
24. S. F. Bloomfield, and M. Arthur, Mechanisms of inactivation and resistance of spores to chemical biocides. *Fundamental and Applied Aspects of Bacterial Spores* (G. W. Gould and A. D. Russell, eds.). Society for Applied Bacteriology Symposium Series No. 23, Academic Press, London, 1994, 16, 91s–104s.
25. L. J. Rode, and M. G. Williams, Utility of sodium hypochlorite for ultrastructure study of bacterial spore integuments, *J. Bacteriol. 92*:1772–1778 (1966).
26. A. Kulikovsky, H. S. Pankratz, and H. L. Sadoff, Ultrastructural and chemical changes in spores of *Bacillus cereus* after action of disinfectants, *J. Appl. Bacteriol. 38*:39–46 (1975).
27. C. E. Bayliss, W. M. Waites, and N. R. King, Resistance and structure of spores of *Bacillus subtilis*. *J. Appl. Bacteriol. 50*:379–390 (1981).
28. P. M. Foegeding, and F. F. Busta, Proposed mechanism for sensitization by hypochlorite treatment of *Clostridium botulinium* spores, *Appl. Environ. Microbiol. 45*:1374–1379 (1983).

29. L. F. Ortenzio, and L. S. Stuart, A standard test for efficacy of germicides and acceptability of residual disinfecting activity in swimming pool water, *J. Off. Agric. Chem. 47*:540–547 (1964).
30. S. F. Bloomfield, and E. E. Uso, The antibacterial properties of sodium hypochlorite and sodium dichloroisocyanurate as hospital disinfectants, *J. Hosp. Infect. 13*:231–239 (1984).
31. D. Worley, and H. D. Burkett, The stability in water of a new chloramine disinfectant as a function of pH, temperature and water quality, *Water Res. 20*:365–368 (1984).
32. H. D. Burkett, J. H. Faison, H. H. Kohl, W. B. Wheatley, S. D. Worley, and N. Bodor, A novel chloramine compound for water disinfection, *Water Res. 17*:874–879 (1981).
33. S. D. Worley, D. E. Williams, U. Barnela, and L. J. Swango, A comparison of two novel organic halamines with free chlorine as water disinfectants [Abstract]. *J. Am. Chem. Soc. 189*:23 (1985).
34. D. E. Williams, S. D. Worley, W. B. Wheatley, and L. J. Swango, Bactericidal properties of a new water disinfectant, *Appl. Environ. Microbiol. 49*:637–643 (1985).
35. D. B. Charlton, and M. Levine, Germicidal properties of chlorine compounds, *Bull. Iowa Eng. Exp. Stat. 132*: (1937).
36. J. E. Death, and D. Coates, Effect of pH on sporicidal and microbicidal activity of buffered mixtures of alcohol and sodium hypochlorite, *J. Clin. Pathol. 32*:148–153 (1979).
37. S. F. Bloomfield, and M. Arthur, Effect of chlorine-releasing agents on *Bacillus subtilis* vegetative cells and spores, *Lett. Appl. Microbiol. 8*:101–104 (1989).
38. M. Klein, and A. Deforest, The inactivation of viruses by germicides, Proceedings 49th Annual Meeting of Chemical Specialty Association, 1983, pp. 116–118.
39. R. Tyler, and G. A. J. Ayliffe, A surface test for virucidal activity of disinfectants: Preliminary study with herpes virus, *J. Hosp. Infect. 9*:22–29 (1987).
40. V. P. Shirobokov, A. G. Nikitenko, and A. I. Yakimenko, Viricidal activity of the sodium salt of dichloroisocyanuric acid and detergent compositions based on it in relation to enteroviruses, *Mikrobiol. Zh. (Kiev) 37*:2:205–211 (1975).
41. R. Hopkins, Chemical disinfectants in the hospital and laboratory, *A Comprehensive Review of Hospital and Health Service Supplies*, 1981, pp. 143–146.
42. O. Thraenhart, E. K. Kuwert, R. Dermietzel, N. Scheiermann, and F. Wendt, Influence of different disinfection conditions on the structure of the hepatitis B virus as evaluated in the morphological alteration and disintegration test (MADT), *Zentralbl. Bakteriol. Hyg. (B) A242*:299–314 (1978).
43. W. W. Bond, M. S. Favero, N. J. Petersen, and J. W. Ebert, Inactivation of hepatitis B virus by intermediate-to-high level disinfectant chemicals, *J. Clin. Microbiol. 12*:535–538 (1983).
44. S. F. Bloomfield, C. A. Smith-Burchnell, and A. G. Dalgleish, Evaluation of hypochlorite-releasing disinfectants against the human immunodeficiency virus (HIV), *J. Hosp. Infect. 15*:273–278 (1990).
45. L. F. Ortenzio, and L. S. Stuart, The behaviour of chlorine bearing organic compounds in the AOAC available chlorine germicidal equivalent concentration test, *J. Assoc. Off. Agric. Chem. 42*:630–633 (1959).

46. S. F. Bloomfield, and G. A. Miles, The relationship between residual chlorine and disinfection capacity of sodium hypochlorite and sodium dichloroisocyanurate solutions in the presence of *Escherichia coli* and of milk, *Microbios Lett. 10*:33–43 (1979).
47. D. Coates, A comparison of sodium hypochlorite and sodium dichloroisocyanurate products, *J. Hosp. Infect. 6*:31–40 (1985).
48. S. F. Bloomfield, and E. A. Miller, A comparison of hypochlorite and phenolic disinfectants for disinfection of clean and soiled surfaces and blood spillages, *J. Hosp. Infect. 13*:231–239 (1989).
49. L. A. Allen, and E. Brooks, The interaction of chlorine-releasing agents with organic materials, *Proc. Soc. Appl. Bacteriol. 15*:155–159 (1952).
50. J. R. Brown, D. M. Mclean, and M. C. Nixon, Bromine disinfection of swimming pools, *Can. J. Public Health 54*:267–270 (1963).
51. L. Shere, M. J. Kelly, and J. H. Richardson, Effect of bromide–hypochlorite bactericides on micro-organisms, *Appl. Microbiol. 10*:538–541 (1962).
52. G. R. Weber, and M. Levine, Factors affecting the germicidal efficiency of chlorine and chloramine, *Am. J. Public Health 32*:719–728 (1944).
53. H. Farkas-Himsley, Killing of chlorine resistant bacteria by chlorine-bromide solutions, *Appl. Microbiol. 12*:1–6 (1964).
54. D. Coates, and J. E. Death, Sporicidal activity of mixtures of alcohol and hypochlorite, *J. Clin. Pathol. 31*:148–152 (1978).
55. J. R. Babb, C. R. Bradley, and G. A. J. Ayliffe, Sporicidal activity of glutaraldehyde and hypochlorites and other factors influencing their selection for the treatment of medical equipment, *J. Hosp. Infect. 1*:63–75 (1980).
56. E. Scott, S. F. Bloomfield, and C. G. Barlow, Evaluation of disinfectants in the domestic environment under 'in use' conditions, *J. Hyg. (Cambr.) 92*:193–203 (1984).
57. E. Scott, and S. F. Bloomfield, A comparison of the effectiveness of detergent washing and chemical disinfection on contamination of cleaning cloths, *J. Appl. Bacteriol. 66*:279–285 (1990).
58. J. A. D. Anderson, and A. Gatherer, Hygiene of infant-feeding utensils, practices and standards in the home, *Br. Med. J. 1*:20–23 (1970).
59. G. C. White, *Handbook of Chlorination*, Van Nostrand Reinhold, New York, 1972, p. 596.
60. Swimming pool disinfection systems using sodium hypochlorite and calcium hypochlorite—a survey of the efficacy of disinfection, Her Majesty's Stationery Office, London ISBN 011751606/6.
61. Swimming pool disinfection using chloroisocyanurates—a survey of the efficacy of disinfection, Her Majesty's Stationery Office, London, ISBN 001171550/7.
62. W. E. Finch, *Disinfectants, Their Values and Uses*, Chapman & Hall, London, 1958, p. 120.
63. S. M. Costigan, Fungicidal studies of some products commonly used in foot baths. Presented at the 59th Annual Convention of Proprietary Association of America, New York, 1941.
64. J. Bunyan, The use of hypochlorite for the control of bleeding, *Oral Surg. 13*:1026 (1960).

65. D. A. Morgan, Chlorinated solutions: (e) useful or (e) useless? *Pharm. J.* Aug 19: 219–220 (1989).
66. S. Cameron, and D. Leaper, Antiseptic toxicity in open wounds, *Care Sci. Pract.* June:19–20 (1987).
67. A. Barton, and M. Barton, *The Management and Prevention of Pressure Sores*, Faber & Faber, London, 1981.
68. R. G. Tarcliff, Health effects of organics: Risk and hazard assessment of ingested chloroform, *J. Am. Water Works Assoc. 41*:658–664 (1977).
69. E. K. Weisburger, Carcinogenicity studies on halogenated hydrocarbons, *Environ. Health Perspect. 21*:7–16 (1977).
70. H. E. Stokinger, Toxicology and drinking water contaminants, *J. Am. Water Works Assoc. 41*:399–402 (1971).
71. H. F. Kraybill, Evaluation of public health aspects of carcinogenic/mutagenic biorefractories in drinking water, *Prev. Med. 9*:212–218 (1980).
72. C. A. Ravacha, Determination of oxidants formed upon te disinfection of drinking water with chlorine, *J. Environ. 19*:943–957 (1984).
73. W. B. Hugo, and J. M. Newton, The antibacterial activity of complex of iodine and a non-ionic surface active agent, *J. Pharm. Pharmacol. 16*:189–197 (1964).
74. W. Gottardi, Iodine and iodine compounds. Iodine, *Disinfection Sterilization and Preservation*, 3rd ed. (S. S. Block, ed.), Lea & Febiger, Philadelphia, 1983, pp. 183–196.
75. O. Wyss, and F. B. Strandokov, The germicidal action of iodine, *Arch. Biochem. 6*:261–267 (1945).
76. C. G. Dunn, Mechanisms of action of some germicidal and inhibitor agents, *Am. Brewer 83*:25–30 (1952).
77. W. Gottardi, The influence of chemical behaviour of iodine on the germicidal action of disinfectant solutions containing iodine, *J. Hosp. Infect. 6*(Suppl.):1–11 (1985).
78. B. Carroll, The relative germicidal activity of triiodide and diatomic iodine, *J. Bacteriol. 69*:413–417 (1955).
79. M. Arthur, Studies of the mode of action of chlorine-releasing antibacterial agents, PhD thesis, University of London, 1991.
80. R. Megid, Studies of the mode of antibacterial action of iodine, PhD thesis, University of London, 1991.
81. N. D. Williams, and A. D. Russell, The effects of some halogen containing compounds on *Bacillus subtilis* endospores, *J. Appl. Bacteriol. 70*:427–436 (1991).
82. E. P. McErlean, S. P. Gorman, and E. M. Scott, Physical and chemical resistance of ion-exchange and coat defective spores of *Bacillus subtilis*, *J. Pharm. Pharmacol. 32*:32 (1980).
83. J. M. Newton, A pharmaceutical study of formulated iodine preparations, PhD thesis, University of Nottingham, 1962.
84. R. W. Lacey, Antibacterial activity of povidone–iodine towards non-sporing bacteria, *J. Appl. Bacteriol. 40*:113–119 (1979).
85. W. W. Bond, N. J. Paterson, and L. A. Carson, Observations on the sporicidal, bactericidal and virucidal activity of iodophors (G. A. Degenis, and J. Ansell, eds.),

Proceedings of the International Symposium on Povidone, University of Kentucky Press, Lexington, 1983, pp. 167–177.
86. L. Gershenfeld, W. Flagg, and B. Witlin, Iodine as a tuberculocidal agent, *Milit. Surg. 114*:172–183 (1954).
87. C. A. Lawrence, C. H. Carpenter, and A. W. C. Naylor-Foote, Iodophors as disinfectants, *J. Am. Pharm. Assoc. (Sci. Ed.) 46*:500–505 (1957).
88. L. Gershenfeld, and B. Witlin, Iodine solution as a sporicidal agent, *J. Am. Pharm. Assoc. (Sci. Ed.) 41*:451–452 (1952).
89. G. Sykes, The sporicidal properties of chemical disinfectants, *J. Appl. Bacteriol. 33*:147–156 (1970).
90. P. G. Bartlett, and W. Schmidt, Disinfectant iodine complexes as germicides, *J. Appl. Microbiol. 5*:355–359 (1957).
91. C. W. Emmons, Fungicidal action of some common disinfectants on 2 dermatophytes, *Arch. Dermatol. Syphilol. 28*:15–21 (1933).
92. J. G. Schamberg, H. Brown, and M. J. Harkins, The chemotherapy of ring worm infection, *Arch. Dermatol. Syphilol. 24*:1033–1052 (1931).
93. P. Gomez-Vega, Mycostatic studies on certain moniliae and related fungi, *Arch. Dermatol. Syphilol. 32*:49–58 (1935).
94. F. G. Reddish, *Antiseptics, Disinfectants, Fungicides and Chemical and Physical Sterilization*, 2nd ed., Lea & Febiger, Philadelphia, 1957, pp. 234–235; 238–249.
95. S. L. Chang, and J. C. Morris, Elemental iodine as a disinfectant for drinking water, *Ind. Eng. Chem. 45*:1009–1001 (1953).
96. L. Gershenfeld, Iodine as a virucidal agent, *J. Am. Pharm. Assoc. (Sci. Ed.) 44*:177–182 (1955).
97. H. Hruszek, Weitere Resistenzprüfungsversuche am Herpesvirus durch in vivo-experimente, *Dermatol. Wochenschr. 98*:485–494 (1934).
98. L. Boudreau, Iodine infections, *J. Méd. Bordeaux 87*:39–41 (1916).
99. L. M. Hale, The treatment of corneal ulcer with povidone–iodine (Betadine), *N. C. Med. J. 30*:50–56 (1969).
100. L. M. Hale, Treatment of experimental herpes simplex keratilis with povidone iodine, *Am. J. Opthalmol. 68*:520–522 (1969).
101. G. Berg, S. L. Chong, and E. K. Harris, Devitalization of microorganisms by iodine, *Virology 22*:469–481 (1964).
102. D. Horn, and W. Ditter, Physical–chemical fundamentals of the microbicidal action of povidone–iodine, International Symposium on Povidone, 1983, pp. 120–139.
103. L. Gershenfeld, and B. Witlin, Antibacterial activity of concentrations of iodine and human blood fractions, *J. Am. Pharm. Assoc. (Sci. Ed.) 37*:136–138 (1948).
104. K. Kojima, Theory and practice of the bactericidal action of various kinds of disinfectants, *Zikkenigaku Zassi 24*:410–419 (1940).
105. W. B. Hugo, and J. M. Newton, the Adsorption of iodine from solution by microorganisms and by serum, *J. Pharm. Pharmacol. 16*:49–55 (1964).
106. C. W. Chambers, P. W. Kabler, G. Malancy, and A. Bryant, Iodine as a bactericide, *Soap. Sanit. Chem. 28(10)*:149–151, 153, 163, 164 (1952).
107. S. P. Gorman, E. M. Scott, and E. P. Hutchinson, Effects of aqueous and alcoholic povidone–iodine on spores of *Bacillus subtilis*, *J. Appl. Bacteriol. 59*:99–105 (1985).

108. V. V. Thompson, Thermometer disinfection, *Am. J. Nurs. 63*:113–115 (1963).
109. N. A. Allawalla, and S. Riegelman, The properties of iodine in solutions of surface active agents, *J. Am. Pharm. Assoc. (Sci. Ed.) 44*:396–401 (1953).
110. V. O. Andzhelov, A. L. Belyaev, and L. Z. Skala, Chemical disinfection of some objects in epidemic keratoconjunctivitis, *Vest. Ofthalmol. 1*:40–42 (1970).
111. J. M. Gwaltney, and M. Sande, A rapid method of disinfecting the broncofiberscope, *Am. Rev. Respir. Dis. 114*:1198–1200 (1976).
112. J. C. Donaldson, D. R. Stoop, and S. A. Williams, Control of high density bacterial contamination of fibroptic bronchoscope, *Chest 72*:10–12 (1977).
113. H. G. Weinstein, R. C. Bone, and W. E. Ruth, Contamination of a fibreoptic bronchoscope with a *Proteus* species, *Am. Rev. Respir. Dis. 116*:541–543 (1977).
114. H. I. Silverman, R. Pazzano, Kluza, and D. Korb, An investigation of a method of cold sterilization of the hydrophilic contact lens polyhydroxy ethylmethacylate by the use of a polyvinyl pyrrolidone–I_2 complex, *Am. J. Optom. Arch. Am. Acad. Optom. 49*:896 (1972).
115. J. K. Andrews, G. F. Ruder, and M. F. Turner, Kinetics of a transient iodine disinfecting system for hydrophilic contact lenses, *J. Pharm. Pharmacol. 29*: (Suppl.) (1977).
116. R. Castaing, J. Bentegeat, and B. Favarel-Garrigues, The use of polyvinyl pyrrolidone for disinfection in a respiratory intensive care unit, *Bordeaux Med. 4*:3759–3766 (1971).
117. P. Berche, J. C. Chassia, and M. Veron, In vitro study and clinical application of an antiseptic solution containing iodinated polyvinylpyrrolidone, *Pathol. Biol. 21*:883–893 (1973).
118. S. Montgomery, Chemical decontamination of guttapercha cones with polyvinylpyrrolidone iodine, *Oral Surg. 31*:258–266 (1971).
119. A. Simonetti, R. Miller, and J. Ciristina, Efficiency of povidone iodine in the disinfection of whirl pool baths and hulbard tanks, *Phys. Ther. 52*:1277–1282 (1972).
120. A. M. Spencer, and T. E. Jaworek, Povidone iodine as an adjunct to hydrotherapy, *J. Am. Podiat. Assoc. 62*:226–227 (1972).
121. W. S. Mueller, Bactericidal effectiveness of iodophor detergent, *J. Milk Food Technol. 18*:144–146; 140 (1955).
122. J. L. Wilson, W. G. Mizuno, and C. S. Bloomberg, *Soap Chem. Spec. 37*:100–143 (1961).
123. J. G. Davis, Iodophors as detergent-sterilizers, *J. Appl. Bacteriol. 25*:195–201 (1962).
124. J. G. Davis, and G. A. Howe, An investigation of iodophors in dairy practice, *J. Appl. Bacteriol. 25*:11–19 (1962).

9

Antimicrobial Effects of Hydrogen Peroxide as an Antiseptic and Disinfectant

Andrea M. Lever
Bausch & Lomb, Inc.
Rochester, New York

Scott V. W. Sutton
Alcon Laboratories, Inc.
Fort Worth, Texas

I. INTRODUCTION

Hydrogen peroxide was first discovered by the French chemist Louis Jacques Thénard, in 1818. Although other researchers, such as Gay-Lussac and Davy, had also done work on what were called "oxygenated acids," it was Thénard who was credited as being the father of hydrogen peroxide. Thénard, in his extensive work on what he called "oxygenated water" observed that it was an extremely reactive compound that could be rapidly decomposed by certain other compounds, without adversely affecting them. By the mid-1800s several papers had been published on the preparation and catalytic decomposition of hydrogen peroxide. In initial production efforts the resulting hydrogen peroxide was very unstable, and yields were about 33%. Thénard determined that acidifying hydrogen peroxide has a stabilizing effect on the compound, and he was also able to increase manufacturing yield to produce an almost pure anhydrous form.

Although the importance of hydrogen peroxide was recognized in the early discovery and development years, the initial uses for it were limited to that of an external irritant in medicine and for the restoration of paintings obscured by hydrogen sulfide. The English physician B. W. Richardson observed, in 1858, that hydrogen peroxide was able to eliminate foul odors and, therefore, he proposed it for use as a disinfectant. This observation and subsequent proposal by Richardson led directly to the early commercial use of hydrogen peroxide as a disinfectant marketed under the trade names of Sanitas. The manufacturing process proposed by Thénard persisted until 1950, when an electrochemical process was developed that produced pure preparations in high concentrations that were stable at increased temperatures and had extended shelf lives [48].

Early in its evolution hydrogen peroxide was used in low concentrations as a preservative for milk and water [18] and for the sterilization of cocoa milk beverages [65]. Hydrogen peroxide has been investigated for use as a sterilant for drinking water [69] and milk [40].

Mentel and Schmidt [38] determined that hydrogen peroxide exhibits virucidal activity against rhinovirus, both on carriers and in suspension at concentrations as low as 0.75×10^4 ppm. Toledo et al. [59] discovered that hydrogen peroxide, at concentrations of 10–25%, had adequate sporocidal properties. In experiments, D-values of 0.8–7.3 min at 24°C were obtained for various spore-forming bacteria (*Bacillus* spp. and *Clostridium* spp.). Leaper [29] examined the effect that increased temperatures and increased concentrations of hydrogen peroxide had on the spores of *B. subtilis* SA 22. It was concluded that increasing the temperature from 20° to 45°C decreased the time to kill spores 10–20 times. Increasing the concentration of hydrogen peroxide from 17.7 to 35.4% decreased the time required to kill spores by three to four times.

Hydrogen peroxide is also a potent biocidal agent against non–spore-forming, vegetative bacteria. Experiments by Kunzmann [28] show that within 1 h, 1000 ppm of hydrogen peroxide is an effective biocide against *Staphylococcus aureus*, *Escherichia coli*, and *Salmonella typhi*. Nambudripad et al. [41] showed that 500 ppm of hydrogen peroxide was an effective biocide against *E. coli*, *Aerobacter aerogenes*, *Sarcina* spp., *Streptococcus lactis*, and *Enterococcus faecalis* when exposed to these organisms for 10–30 min (*E. coli* and *A. aerogenes*), 150 min (*Sarcina* spp. and *S. lactis*) or 240 min (*E. faecalis*). Wardle and Renninger [64] determined that 30 ppm of hydrogen peroxide was an effective biocidal agent against *Micrococcus* spp. and *Staphylococcus epidermidis* within 10 min.

The reaction of the superoxide ion with peroxide to form the hydroxyl radical is the mechanism by which hydrogen peroxide decomposes [12; see Eq.

(1)]. The hydroxyl radical is the an extremely potent oxidant [13] and is the agent responsible for the destruction of bacterial cells. The hydroxyl radical has the ability to react with essential cell components, such as membrane lipids and DNA.

$$O_2^- + H_2O_2 \rightarrow OH^{\cdot} + OH^- + O_2 \qquad (1)$$

Several investigators have determined that the rate of formation of the hydroxyl radical and the biocidal efficacy of hydrogen peroxide is enhanced in the presence of transition metals [8,10,69].

II. HYDROGEN PEROXIDE AS AN ANTISEPTIC

A germicide that is used on living tissues, such as skin, for the purpose of inhibiting or destroying microorganisms is, by definition, an *antiseptic*. In contrast, a germicide that is used primarily for destroying microorganisms on nonliving objects or surfaces is a *disinfectant* [11].

A. Hydrogen Peroxide and Wounds

The antiseptic uses of hydrogen peroxide have been recognized for many years. Hydrogen peroxide has been commonly used as an antiseptic for all types of wounds, as an adjunct to surgery, and as a treatment for chronic conditions. The effect of hydrogen peroxide on living tissue and wound healing was investigated by Lineaweaver et al. [31], who determined that 3% hydrogen peroxide was cytotoxic when directly applied to human fibroblast cultures. In studies using Sprague–Dawley rats, these investigators found that 3% hydrogen peroxide did not adversely affect wound healing in this animal model. However, it was noted that a 3% concentration of hydrogen peroxide was required to yield adequate antiseptic activity, although this level of peroxide was cytotoxic to the human fibroblast cultures.

B. Hydrogen Peroxide and the Oral Cavity

Hydrogen peroxide, alone and with other compounds, is used as an antiseptic in the treatment of periodontal diseases and for root canal debridement during endodontic therapy [56]. The principal etiologic agents of periodontal disease are microorganisms found in bacterial plaque [52]. Chemotherapeutic reduction of the bacteria associated with periodontally diseased sites and with bacterial plaque by agents, such as hydrogen peroxide, has been the subject of contrasting papers by Rosling et al. [47] and Wolff et al. [68].

In the first study, Rosling et al. [47] determined the efficacy of periodontal scaling and root planing, with and without adjunctive topical chemotherapy, in patients with adult periodontitis. Since the main etiological factor for the onset and continuation of periodontal disease is bacterial plaque, therapy should include agents for the eradication of the periodontopathic microflora. Test subjects were treated identically (half of the dentition of all subjects was supragingivally and subgingivally scaled) to the control group with the following exceptions: (a) during subgingival scaling, pockets were irrigated with povidone–iodine (Betadine), and a solution of hydrogen peroxide, sodium chloride, and sodium bicarbonate was applied subgingivally; (b) during the study the test subjects applied the peroxide mixture twice daily and had it subgingivally applied every other week. It appeared that the combination of the topical chemotherapy, periodontal scaling, and root planing was a successful treatment for the arrest of disease-caused periodontal tissue destruction in the test subjects. Additionally, it was suggested that repeated use of the peroxide mixture may enhance periodontal healing following scaling and root planing.

Subsequently, Wolff et al. [68] compared a salt and hydrogen peroxide oral hygiene regimen with a conventional oral hygiene regimen (regular brushing and the use of dental floss) to determine which treatment was most effective in shifting the oral microbial flora in the direction associated with periodontal health. All test subjects had an initial tooth polishing, scaling, and root planing. The patients in the test group were instructed to floss and brush with sodium bicarbonate mixed to a paste with equal parts of 3% hydrogen peroxide and water. Following tooth brushing and flossing, the test subjects were instructed to irrigate their gingival areas with a premixed saturated solution of sodium chloride. Subgingival plaque samples were collected from both groups of subjects at 8, 16, and 24 months and compared with the baseline results. Cellular morphology of the bacteria recovered from the plaque samples was examined, and clinical readings, such as probing depth and attachment level, were collected as indices of periodontal health. For both test groups the number of cocci had increased and the number of motile rods had decreased at the 8-month visit, but had returned to the baseline levels by the 16-month visit. The occurrence of spirochetes decreased and remained at a low level throughout the duration of the study in both test groups. No change in the levels of bacterial rods, filaments, fusiforms, or curved rods was observed in either group. The researchers concluded, in contrast to the Roseling et al. [47] investigation that given these microbiological and clinical results, the use of the salt and hydrogen peroxide regimen was not superior to the conventional oral hygiene regimen. When combined with professional care, both oral hygiene regimens appeared to facilitate a favorable and significant change in the oral microbial flora, leading to increased periodontal health.

The two previous studies discussed have contrasting conclusions concerning the efficacy of hydrogen peroxide regimens, along with professional treatment, in the arrest of periodontal disease. However, because of the rigorous nature of these regimens and the delicate and possibly compromised condition of the gingiva, it is reasonable to assume that these regimens may have an adverse effect on tissues of the oral cavity.

The toxic effects of salt and hydrogen peroxide regimens on the oral cavity have been the subject of few studies. Herrin et al. [19] demonstrated that erosive gingival lesions, sometimes found in patients using the salt and hydrogen peroxide regimen, are a result of the saturated sodium chloride solution used to irrigate the gingiva, and not a result of the use of salt and hydrogen peroxide paste. The saturated sodium chloride solution appears to dehydrate the gingival epithelium, with subsequent separation of keratinized epithelial cells, thus inducing the visible lesion.

An in vitro study by Ramp et al. [45] examined the apparent inhibition of glucose metabolism and collagen synthesis in bone by hydrogen peroxide. The effects of hydrogen peroxide on bone were evaluated in an organ culture system (tibiae from chick embryos). The tibiae were incubated for up to 3 days in a medium containing 0.07–20 m*M* of hydrogen peroxide. The results suggest that low concentrations of hydrogen peroxide inhibit bone energy metabolism and growth in vitro. This inhibition was measured by glucose metabolism and collagen synthesis. These findings may indicate the possibility exists for damage to the oral cavity from exposure to hydrogen peroxide at therapeutic levels. Patients with deep periodontal pockets, or those who have had periodontal surgery, may be at an increased risk of exposing bone to hydrogen peroxide when used as a therapeutic mouth rinse. Bone is also very likely to be exposed to high concentrations of hydrogen peroxide during root canal debridement. Because of the potential for hydrogen peroxide exposure of bone during periodontal treatment and root canal procedures and the demonstration that hydrogen peroxide inhibits chick embryo bone in vitro, it has been suggested that clinical studies should be performed to reevaluate the use of hydrogen peroxide in therapy.

There is clearly controversy surrounding both the beneficial and harmful effects of the use of hydrogen peroxide on sensitive tissues such as the gingiva, wounds, and bone. Therefore, caution should be used by both the patient and practitioner when hydrogen peroxide is employed as an antiseptic for wound cleansing and for dental regimens and therapies.

III. HYDROGEN PEROXIDE AS A DISINFECTANT

Disinfection is the process of eliminating all recognized pathogenic microorganisms present on an inanimate object. In contrast with sterilization, which is the complete eradication of all microbial life, the scope of disinfection does not extend to forms such as bacterial endospores and, accordingly, is considered less lethal than sterilization. The effectiveness of a disinfectant is influenced by various factors, including the number and type of microorganisms present, the amount of organic matter present, the type and concentration of disinfectant used, the nature of the object being disinfected, and the time and temperature of exposure. Hydrogen peroxide, at various concentrations, provides good disinfection of materials, ranging from septage to surfaces and medical equipment [11].

A. Hydrogen Peroxide Disinfection of Septage

A study performed by Stramer and Cliver [54] examined the effectiveness of hydrogen peroxide as a disinfectant of septage. Septage is a highly concentrated form of waste, both solid and liquid, that is periodically pumped out of septic tanks. Because of the origin of septage material, the possibility exists for it to contain various pathogens that may bacterial, viral, or protozoan. To avoid the possibility of pathogen contact with humans, either by recreational waters or contaminated food or drinking water, it has been suggested that septage be disinfected before disposal. Stramer and Cliver [54] examined various treatments, including hydrogen peroxide, to reduce the concentration of bacteria and polioviruses in septage. The disinfection treatments tested included technical- and analytical-grade glutaraldehyde, hydrogen peroxide plus trichloroacetic acid (added to retard catalase-mediated peroxide degradation), hydrogen peroxide plus trichloroacetic acid at 55°C, and exposure to heat at 55°C. Septage was inoculated with septage-derived fecal coliform, a fecal streptococcal isolate, and commercially available poliovirus type 1. Survival of the bacteria after disinfectant treatment was determined by pour plates (fecal coliforms) and the most probable number (MPN) technique (fecal streptococci). Survival of poliovirus was assessed by the plaque method. The results indicate that the combination of hydrogen peroxide and heat (5 mg of hydrogen peroxide plus 0.33 mg of trichloroacetic acid at 55°C) was the most efficacious disinfection regimen and killed all bacteria in the septage within 1 h. The heat treatment (55°C) alone inactivated the inoculated polioviruses within 30 min. The researchers concluded that application of this disinfection regimen to septage before disposal into the environment (e.g., onto agricultural land) would decrease the potential health threat that may exist in contaminated septage. However, the authors noted that, although the hydrogen peroxide and heat treatment is the most effective procedure for septage disinfection, it may not be practical for on-site use. It was recommended that further studies on septage disinfection be conducted.

B. Hydrogen Peroxide Disinfection of Medical Supplies and Equipment

The disinfection of delicate, or in-use, medical supplies or equipment, that is not amenable to autoclaving, is generally achieved by the use of chemical disinfectants such as hydrogen peroxide. The effectiveness of hydrogen peroxide as a disinfectant has been the subject of many studies focusing on the disinfection of urinary drainage bags, endoscopes, and renal dialysis equipment.

Urethral catheterization accounts for more than 80% of all hospital-acquired infections of the urinary tract [61]. Microbial contamination of the urine drainage system is generally found before the development of bacteriuria, indicating that the drainage bag acts as a reservoir for contaminants. The periodic instillation of various disinfectants, including hydrogen peroxide, into urinary drainage systems has been investigated as a measure for the prevention of bacteriuria in the catheterized patient.

Holliman et al. [21] performed a study to determine if drainage bag disinfection would affect the rate of hospital-acquired, catheter-associated urinary tract infection. Two groups of catheterized patients were observed: the control and the test group. The test group comprised catheterized patients whose drainage bags were disinfected with 3% hydrogen peroxide. The control group patients had no disinfectant instillation into their drainage bags. Results indicate that there was a significant reduction in the number of catheter-associated urinary tract infections in patients whose drainage bags were disinfected, when compared with the control group. It was concluded that the disinfection of urinary drainage systems by hydrogen peroxide is an effective treatment for reduction of catheter-associated urinary tract infection.

In contrast to the finding of Holliman et al. [21], studies by Sweet et al. [57] and Thompson et al. [58] found that hydrogen peroxide disinfection of urinary drainage systems had no effect on the reduction of catheter-associated bacteriuria. In both the Thompson et al. [58] and Sweet et al. [57] studies, one group of catheterized patients had their drainage systems disinfected every 8 h by instillation of 30 ml of 3% hydrogen peroxide, and the control group of catheterized patients had no added disinfectant. Urine samples were aseptically taken from both groups of patients on a daily basis until the catheter was removed or the patient developed bacteriuria. The urine samples were examined for bacterial contamination. Any contaminants isolated from the urine samples were cultured and identified. Results from both studies indicate that there was no difference between the control and test groups relative to prevention of catheter-associated urinary tract infections.

The disinfection of dialysis equipment by hydrogen peroxide is the subject of papers by Klein [25] and Gordon et al. [15]. Dialysis equipment must be disinfected daily and sterilized periodically to reduce microbial contamination and, thereby, reduce the risk of patient exposure to potentially harmful organisms. However, great care must be taken to ensure that the patient is not inadvertently exposed to residual levels of disinfectants, such as hydrogen peroxide. Klein [25] indicated that the effect of residual hydrogen peroxide on hemodialysis patients, when they are directly exposed during the dialysis process, is direct protein oxidation and hemolysis. The paper by Gordon et al. [15] describes an incident of exposure of hemodialysis patients to hydrogen peroxide. The source of hydrogen peroxide contamination was thought to be the water treatment system, which was inadequately rinsed after disinfection. The peroxide (10–20 ppm)-contaminated dialysis fluid induced a significant decline in the blood hemoglobin level of the exposed patients, who required packed red blood cell transfusions. The authors concluded that stringent testing methods should be employed to detect residual concentrations of disinfectants and sterilants.

Bilotta et al. [3] describes the use of a 3% hydrogen peroxide rinse as a final step in the sterilization of endoscopes. The paper details an epidemic of hydrogen perioxide enteritis found among patients in a gastrointestinal endoscopy unit over a 2-week period. The causative agent of the enteritis was peroxide left in the endoscope during the final rinse stage of the sterilization procedure. During endoscope use, peroxide was released into the intestine during lens-cleaning procedures. When an endoscope lens is cleaned, water is flushed through the instrument; this procedure released trapped hydrogen peroxide remaining from the sterilization procedure. It was recommended that if hydrogen peroxide is going to be used as part of the sterilization regimen for endoscopes, a 12-h period of aeration or a final flushing of the channels with water be employed to effectively eliminate peroxide residue.

C. Hydrogen Peroxide Disinfection of Contact Lenses

Hydrogen peroxide (3%) was the first "nonthermal" disinfection solution investigated for contact lenses. Although hydrogen peroxide systems for contact lens care were developed in the 1970s [23,62], it was not until 1983 that the Food and Drug Administration (FDA) approved hydrogen peroxide for use as a contact lens disinfectant in the United States. Since 1983, hydrogen peroxide contact lens disinfection has gained popularity as an alternative to heat and other cold chemical disinfection regimens. Hydrogen peroxide systems, in regimen, provide effective antimicrobial activity and, with proper neutralization, avoid postdisinfection ocular side effects. Additionally, during hydrogen peroxide disinfection, the dimensional and optical parameters of most soft contact lenses are maintained [20].

A variety of hydrogen peroxide contact lens disinfection systems have been, or are presently available to consumers. All of the systems employed require the following regimen steps that may be integrated or separate: (a) clean, (b) rinse, (c) disinfect, (d) rinse, and (e) neutralize and store. Neutralization is critical to ensure that the eye is protected from oxidative damage. The neutralization method used varies with the regimen. Some systems use multiple saline rinses to remove hydrogen peroxide by simple dilution [62]. It is generally recognized that the degree to which a patient complies with a recommended regimen may vary [7]. Therefore, chemical and catalytic neutralizers have been developed to minimize the risk of ocular complications that may result from exposure to incompletely neutralized hydrogen peroxide [17].

Currently used catalytic neutralizers are either enzymatic or metallic. Enzymatic neutralization is achieved through the action of the naturally occurring enzyme catalase. This process is catalytic, rather than stoichiometric; therefore, the catalase enzyme is required to be present in only very low concentrations to provide rapid neutralization of hydrogen peroxide [Eq. (2)]. Catalase thus would appear to be an ideal neutralizer for peroxide systems. However, catalase is a protein material from natural sources and, as such, has the potential to induce an ocular allergic response in susceptible individuals [42,46].

$$2\ H_2O_2 + \text{catalase or platinum} \rightarrow 2\ H_2O + O_2 \quad (2)$$

Metallic neutralization in a commercial lens care product is accomplished with platinum, in the form of a metallic coating on a plastic disk. Through contact with the metal surface of the disk, hydrogen peroxide is reduced to oxygen and water [see Eq. (2)]. This neutralization process occurs at a slower rate than the rapid enzymatic catalase reduction. During repeated normal use the neutralization effectiveness of the platinum disk declines, which necessitates replacement of the disk at appropriate time intervals.

Neutralization may be achieved under appropriate conditions with a variety of chemical reducing agents. As reviewed in Krezanoski et al. [27], neutralization of hydrogen peroxide by pyruvate proceeds by oxidative cleavage of pyruvate to acetate and carbon dioxide. This cleavage occurs with the reduction of peroxide to water. The process in contact lens care systems occurs completely within 10 min. [Eq. (3)] [27].

$$H_2O_2 + CH_3COCO_2^- \rightarrow CH_3CO_2^- + CO_2 + H_2O \quad (3)$$

Sorbic acid-preserved saline, containing sodium sulfite, can also be used to neutralize hydrogen peroxide. In this process, sodium sulfite is oxidized to sodium sulfate, and the hydrogen peroxide is converted to water. In contact lens care systems this reaction occurs rapidly [10 min; Eq. (4); 27].

$$H_2O_2 + Na_2SO_3 \rightarrow Na_2SO_4 + H_2O \quad (4)$$

Chemical neutralization of hydrogen peroxide by thiosulfate proceeds to form tetrathionate and hydroxide [see Eq. (5)]. This reaction requires buffering to maintain mildly basic conditions to prevent precipitation of elemental sulfur and its subsequent deposition onto contact lenses. This reaction is complete within 30 min after addition of the thiosulfate-neutralizing reagent [42].

$$H_2O_2 + 2\ S_2O_3^{2-} \rightarrow S_4O_6^{2-} + 2\ OH^- \quad (5)$$

Peroxide neutralization is essential, since exposure of unneutralized hydrogen peroxide (3%) to the eye can induce several conditions. These range from stinging and tearing, to hyperemia, edema, blepharospasm, and possible corneal damage [16,24,26]. Various studies have been performed, both in vivo and in vitro, to assess the response of ocular tissues to hydrogen peroxide.

Paugh et al. [43] conducted experiments on human subjects to determine the threshold level of hydrogen peroxide toxicity. The introduced peroxide into the ocular environment by high-water-content hydrogel contact lenses. The study lenses were presoaked in saline solutions containing 0, 25, 50, 100, 200, 400, and 800 ppm of hydrogen peroxide. These investigators measured subjective comfort, conjunctival hyperemia, corneal and conjunctival epithelial staining, and corneal oxygen uptake after treated lenses had been worn for 5 min. For patient comfort, the results indicated that higher levels of hydrogen peroxide were associated with greater discomfort and increased conjunctival hyperemia. Interestingly, no significant corneal or conjunctival epithelial staining or alteration of corneal aerobic response was observed after ocular exposure to 800 ppm of hydrogen peroxide. However, given the subjects' discomfort and hyperemia, it was concluded that residual concentrations of hydrogen peroxide in excess of 100 ppm should be avoided in contact lens care systems.

Chalmers and McNally [5] measured the discomfort threshold for a range of hydrogen peroxide concentrations administered to the eye directly by drops and hydrogen peroxide-soaked contact lenses. They found that the threshold for hydrogen peroxide administered by drops is in excess of 800 ppm and for soaked lenses the threshold is less than 300 ppm. The researchers noted that the detection threshold was higher than that found in previous studies. These differences in threshold levels of hydrogen peroxide were attributed to their experimental methods, which were modified to reduce the occurrence of false-positive subject responses. Additionally, this detection threshold does not define the maximum concentration of hydrogen peroxide that could be left in contact lens care solutions, but simply defines what could be present in a solution before detection by stinging. It was concluded that, since hydrogen peroxide disinfection systems are designed to have residual hydrogen peroxide concentrations in the eye of no more than 50–60 ppm, an adequate safety margin exists relative to patient comfort.

Human clinical studies were performed by McKenny et al. [37] on corneal permeability to fluorescein after exposure to 50 and 500 ppm of hydrogen peroxide. It was determined that topical application (no contact lenses were worn by the subjects) of solutions containing 50–500 ppm of hydrogen peroxide did not cause significant changes in corneal permeability, when compared with a saline control.

Tripathi and Tripathi [60] investigated the cytotoxicity of hydrogen peroxide by the exposure of primary human corneal epithelial cell cultures to a single dose of hydrogen peroxide in concentrations ranging from 30 to 100 ppm. In these experiments, cell death was observed within a few minutes following exposure to hydrogen peroxide in concentrations ranging from 70 to 100 ppm. Hydrogen peroxide at 50 ppm resulted in the immediate cessation of cellular activity. Exposure of cells to concentrations of hydrogen peroxide as low as 30 ppm induced cell retraction, cessation of cell movement, and loss of mitotic activity; cell death occurred within 7–8 h after peroxide exposure. It was concluded that further study is advisable to determine the potential long-term effects of residual hydrogen peroxide on the cornea [60]. However, although these results may be significant, overinterpretation of in vitro cell culture results should be avoided, as they are indicative, but not necessary predictive.

Chalmers et al. [6] measured the rate of in vivo neutralization of hydrogen peroxide delivered to the eye at 50 ppm on hydrogel contact lenses. Hydrogen peroxide was removed surprisingly fast, particularly when the eyelids were closed. In the closed eyelid experiment all detectable peroxide was removed from the lens within 30 s. In contrast, in the open-eye trials, 50 ppm of peroxide was removed from the lenses after 60 s. The amount of residual hydrogen peroxide measured on the lenses was significantly different between the opened- and closed-eye conditions at 15 and 30 s. It was suggested that these differences could be attributed to endogenous catalase and other reducing enzymes present in the conjunctival epithelium and ciliary body [1,2]. Although hydrogen peroxide levels that induce patient discomfort have been reasonably well defined, the natural ability of the eye to neutralize residual hydrogen peroxide, as well as the potential long-term effects of chronic ocular exposure to low levels of peroxide are worthy of further study.

Even though neutralization of residual hydrogen peroxide is necessary for the prevention of oxidative damage to the eye, the disinfection step in hydrogen peroxide contact lens care regimens is essential for the destruction of microorganisms that may be responsible for ocular infections. Properties of the conjunctival mucous membrane, such as temperature, pH, nutrients, and humidity, make this an attractive environment for a number of bacteria. Most of these bacteria are constituents of the normal conjunctival flora and, as such, are rarely etiologic agents for ocular infection. However, contact lens wearers may be exposed to virulent transient bacteria, seldom encountered in the conjunctival flora, by virtue of contact lens handling and wear [14,36]. The hydrophilic nature of the lens can allow the proliferation of microorganisms between the lens and cornea, where the nutritional environment is conducive to microbial growth. The lens may interfere with normal tear flow and natural removal of contaminants [4]. Contact lenses may also reduce oxygen delivery to the cornea, which may increase the risk of hypoxic damage and infection [14].

The disinfectant efficacy of hydrogen peroxide in contact lens care systems has been extensively studied and is well documented [22,35,44,49,53]. Stokes et al. [53] examined the effect of various concentrations of hydrogen peroxide on *S. aureus*, *E. coli*, *Pseudomonas aeruginosa*, *Clostridium sporogenes*, and *Candida albicans*. Their results show complete kill of all organisms tested, except *C. albicans*, within 2 min when exposed to 1.5% hydrogen peroxide solution; *C. albicans* was completely killed within 20 min. It was suggested that a 1.5% hydrogen peroxide solution could, together with normal precleaning, be a satisfactory system for disinfection of soft contact lenses. Additionally, Sibley et al. [49] examined the microbiological effectiveness of a 5-min, 3% hydrogen peroxide disinfection lens care regimen, including a cleaning step. Organisms examined in this study were *P. aeruginosa*, *Serratia marcescens*, *Staphylococcus epidermidis*, and *C. albicans*. All organisms tested were successfully removed or disinfected with the use of this regimen (5-min peroxide soak time). However, neither the Stokes study nor the Sibley study showed data for the effect of hydrogen peroxide on molds such as *Aspergillus fumigatus*. The FDA requires that all contact lens disinfecting solutions exhibit sufficient activity against mold, specifically *A. fumigatus*, as well as against a yeast, *C. albicans*, and three bacteria: *S. marcescens*, *P. aeruginosa*, and *S. epidermidis* [63].

Houlsby et al. [22] evaluated several nonthermal disinfecting solutions, including hydrogen peroxide (3%), by the D-value method. The D-value (expressed in minutes) is the time required to kill 90% (1 log) of the cells present in the initial inoculum. D-values are used as a guide to determine the minimum time required for adequate disinfection. All five FDA-required organisms were examined. The D-values obtained for *S. epidermidis* and *P. aeruginosa* were 1.7 min, with r^2 being 0.95 and 0.84, respectively. The D-value obtained for *S. marcescens* was 3.1 min, with $r^2 = 0.94$. *Aspergillus fumigatus* and *C. albicans* had substantially longer D-values than those of the bacteria: *A. fumigatus* D-value = 25.1 min, $r^2 = 0.90$; *C. albicans* D-value = 27.9 min, $r^2 = 0.92$. It was suggested that the rate of disinfection, based on the experiments, appears to be linear. Despite the apparent linearity in this particular study, it is clear that chemical disinfecting systems, including hydrogen peroxide, frequently do not exhibit first-order kinetics (i.e., show nonlinear kill) against many organisms [55].

Penley et al. [44] reported experiments in which D-values for hydrogen peroxide (3%) were determined for a variety of microorganisms including *S. marcescens*, *C. albicans*, and *A. fumigatus*. The result for *S. marcescens* (D-value = 2.8 min, r^2 not reported) is not substantially different from that reported by other investigators [22,62]. In contrast with the work of Houlsby et al. [22], the D-values for *A. fumigatus* and *C. albicans* (11 min and 48 min, respectively, r^2 values not reported) obtained by Penley et al. [62] were different both in magnitude and direction. Given the D-value results, the 10-min disinfection regimens in hydrogen peroxide (3%) appear to be long enough for adequate disinfection of bacteria. However, owing to the high D-value obtained for *C. albicans*, it was recommended that the soaking period for disinfection of contact lenses in 3% hydrogen peroxide be no less than 45–60 min.

Although D-value determination is a popular means of reporting rates of kill, it is essential that the kill rates be linear for the D-value to be a valid measure. Dramatically different D-values can be derived from the same nonlinear kill curve, depending on the specific method of numeric D-value generation. This nonlinearity of chemical kill curves (and the variation in the derived D-values) is inconsistent with the first-order kinetics observed for thermal disinfection, and has led to the suggestion that D-values are not an appropriate measure of disinfecting efficacy for systems other than heat [55].

Marques et al. [35] examined the effect of commercial hydrogen peroxide contact lens disinfecting solutions against ocular bacterial strain growth. The organisms tested were *P. aeruginosa*, *E. coli*, *S. epidermidis*, and *S. marcescens*. The bacteria were inoculated into the hydrogen peroxide solutions and allowed to soak for the manufacturers recommended minimal disinfection times. Samples were taken from the hydrogen peroxide disinfecting solutions, and the survivors were enumerated. Results indicate that all of the 3% hydrogen peroxide-disinfecting solutions tested completely killed *P. aeruginosa*, *E. coli*, and *S. epidermidis*. However, some growth of *S. marcescens* was noted in two of the four solutions tested. It was suggested that the catalasic activity (the inherent ability of the organism to decompose hydrogen peroxide) of *S. marcescens* was superior to that of the other organisms tested and that this enables *S. marcescens* to survive longer than the other organisms.

In addition to bacteria and fungi, free-living protozoans of the genus *Acanthamoeba* pose a serious threat to ocular health. The two forms of this protozoan are the motile, trophozoite form, and the sessile, cystic form. The cystic form is double-walled, which makes it more resistant to freezing, desiccation, antimicrobials, and most contact lens cleaning and disinfection regimens [33]. Several studies have been performed attempting to assess the effect various contact lens disinfecting solutions have on both the trophozoite and cystic forms of this organism. Silvany et al. [50] examined the efficacy of contact lens disinfectants against *Acanthamoeba* spp. The results indicate that hydrogen peroxide (3%) is an effective disinfectant against *A. castellanii* and *A. polyphaga* (cysts and trophozoites), as measured by the growth of trophozoites within 14 days of incubation. They suggested that solutions containing 3% hydrogen peroxide could be used as a disinfectant against *Acanthamoeba* spp., if the agent was not prematurely neutralized and an extended soak time of 2–3 h (before neutralization) was used. In contrast to the foregoing studies, in-regimen experiments by Lindquist et al. [32], Ludwig et al. [34], and Connor et al. [9], all indicate that hydrogen peroxide in a 3% solution is not an effective disinfecting agent, when used according to the manufactures instructions, against *Acanthamoeba* spp. and that heat disinfection is the most reliable system for eradication of these organisms [39]. These varied efficacy results against *Acanthamoeba* spp. may arise from differing experimental procedures (whether the disinfectant is tested in-regimen or not) and conditions such as different ratios of cysts to trophozoites. It appears that a more standardized protocol needs to be established for the evaluation of disinfectant activity against *Acanthamoeba* spp.

Many studies have assessed the degree to which contact lens cases are contaminated with microorganisms [30,51,66,67]. Some of these studies examined the efficacy of hydrogen peroxide systems in curtailing such contamination. Wilson et al. [67] compared microbial contamination of contact lens storage cases from asymptomatic patients who used a variety of lens care regimens. One study group had the manufacturers lens care instructions periodically reinforced and the other did not. No differences in the rate of case contamination were observed between the reinforced group and nonreinforced group of hydrogen peroxide system users, although the peroxide group showed a significantly lower incidence of lens case contamination compared with individuals who used other chemical disinfection systems. It was suggested that storage containers (including those for hydrogen peroxide systems) for lenses be regularly cleaned, disinfected, and replaced periodically to avoid the formation of biofilm that may shield bacteria from the action of disinfecting solutions [30].

In a further study, Wilson et al. [66] examined the effect of different disinfecting solutions, including 3% hydrogen peroxide, against artificially generated biofilms of *P. aeruginosa*, *S. marcescens*, *S. epidermidis*, *Streptococcus pyogenes*, and *C. albicans* deposited in the wells of polyethylene lens cases. The "biofilms" were exposed to the manufacturers minimum disinfection time for each solution tested and were evaluated for viable organisms. It was suggested that 3% hydrogen peroxide was the most effective disinfectant against these artificially generated biofilms. The relative order of resistance of monoclonal, in vitro bacterial biofilm to peroxide was as follows: *S. marcescens* > *P. aeruginosa* > *S. epidermidis* > *S. pyogenes*.

Although the results of the Wilson studies [66,67] suggest that hydrogen peroxide is a more effective disinfectant for both the reduction of lens case contamination in patient-used lens cases and reduction of viable organisms in artificially generated biofilms than other disinfecting solutions, Simmons et al. [51] found no difference in effects of lens care systems on naturally occurring biofilms. These investigators studied the incidence of bacterial contamination in patient-used contact lens cases that had been used in conjunction with different chemical care systems, including those formulated with 3% hydrogen peroxide, polyquaternium 1, and polyaminopropyl biguanide. In contrast with the Wilson et al. [66] study on artificial biofilms, this real-world study indicates that there is no difference in contamination rate between hydrogen peroxide and non-hydrogen peroxide disinfecting systems.

It is clear that there are inconsistencies in the literature relative to hydrogen peroxide-induced ocular toxicity, discomfort, and disinfectant efficacy. However, when hydrogen peroxide is used as a disinfectant in a complete contact lens care regimen (including proper neutralization), it is both safe and effective when used by the compliant patient.

ACKNOWLEDGMENT

The authors gratefully acknowledge O. William Lever Jr., Ph.D. for his valuable comments during the preparation of this manuscript.

REFERENCES

1. L. Atalla, M. A. Fernandez, and N. A. Rao, Immunohistochemical localization of catalase in ocular tissue, *Curr. Eye Res.* *6*:1181 (1987).
2. K. C. Bhuyan, and D. K. Bhuyan, Superoxide dismutase of the eye. Relative functions

of superoxide dismutase and catalase in protecting the ocular lens from oxidative damage, *Biochim. Biophys. Acta 542*:28 (1978).
3. J. J. Bilotta, and J. D. Waye, Hydrogen peroxide enteritis: The "snow white" sign, *Gastrointest. Endosc. 35*:428 (1989).
4. P. S. Binder, D. M. Rasmussen, and M. Gordon, Keratoconjunctivitis and soft contact lens solutions, *Arch. Ophthalmol. 99*:87 (1981).
5. R. L. Chalmers, and J. J. McNally, Ocular detection threshold for hydrogen peroxide: Drops vs. lenses, *ICLC 15*:351 (1988).
6. R. L. Chalmers, M. Tsao, G. Scott, and L. Roth, The rate of in vivo neutralization of residual H_2O_2 from hydrogel lenses, *Contact Lens Spectrum 4*:21 (1989).
7. M. J. Collins, and L. G. Carney, Compliance with care and maintenance procedures amongst contact lens wearers, *Clin. Exp. Optom. 69*:174 (1986).
8. L. Colobert, Mechanism of bactericidal activity of hydrogen peroxide and ascorbic acid on *Escherichia coli*, *Rev. Corps. Sante Armees 3*(Suppl.):495 (1962).
9. C. G. Connor, Y. Blocker, and D. G. Pitts, *Acanthamoeba culbertsoni* and contact lens disinfection systems, *Optom. Vis. Sci. 66*:690 (1989).
10. H. R. Dittmer, I. L. Baldwin, and S. P. Miller, The influence of certain inorganic salts on the germicidal activity of hydrogen peroxide, *J. Bacteriol. 17*:203 (1930).
11. M. S. Favero, and W. W. Bond, Sterilization, disinfection, and antisepsis in the hospital. In *Manual of Clinical Microbiology*, 5th ed. (A. Balows, ed.), American Society for Microbiology, Washington, DC, 1991, pp. 183–200.
12. I. Fridovich, The biology of oxygen radicals, *Science 201*:875 (1978).
13. I. Fridovich, Oxygen: boon and bane, *Am. Sci. 63*:54 (1975).
14. P. G. Galentine, E. J. Cohen, P. R. Laibson, C. P. Adams, R. Michaud, and J. J. Arentsen, Corneal ulcers associated with contact lens wear, *Arch. Ophthalmol. 102*: 891 (1984).
15. S. M. Gordon, L. A. Blannd, S. R. Alexander, H. F. Newman, M. J. Arduino, and W. R. Jarvis, Hemolysis associated with hydrogen peroxide at a pediatric dialysis center, *Am. J. Nephrol. 10*:123 (1990).
16. P. Gyulai, A. Dziabo, W. Kelly, R. Kiral, and C. H. Powell, Relative neutralization ability of six hydrogen peroxide disinfection systems, *Contact Lens Spectrum 2*:61 (1987).
17. M. G. Harris, Practical consideration in the use of hydrogen peroxide disinfection systems, *CLAO 16*(Suppl. 1):S53 (1990).
18. P. G. Heinemann, The germicidal efficiency of commercial preparations of hydrogen peroxide, *JAMA 60*:1603 (1913).
19. J. R. Herrin, C. A. Squier, and W. C. Ribright, Development of erosive gingival lesions after use of a home care technique, *J. Periodontol. 58*:785 (1987).
20. B. Holden, A report card on hydrogen peroxide for contact lens disinfection, *CLAO 16*(Suppl. 1):S61 (1990).
21. R. Holliman, D. V. Seal, H. Archer, and S. Doman, Controlled trial of chemical disinfection of urinary drainage bags, *Br. J. Urol. 60*:419 (1987).
22. R. D. Houlsby, M. Ghajar, and G. Chavez, Microbiological evaluation of soft contact lens disinfecting solutions, *J. Am. Optom. Assoc. 55*:205 (1984).
23. L. E. Janoff, The Septicon system: A review of pertinent scientific data, *ICLC 11*:274 (1984).

24. J. E. Josephson, and B. E. Caffery, Exploring the sting, *J. Am. Optom. Assoc. 58*:288 (1987).
25. E. Klein, Effects of disinfectants in renal dialysis patients, *Environ. Health Perspect. 69*:45 (1986).
26. H. L. S. Knopf, Reaction to hydrogen peroxide in a contact-lens wearer, *Am. J. Ophthalmol. 97*:796 (1984).
27. J. Z. Krezanoski, and R. D. Houlsby, A comparison of new hydrogen peroxide disinfection systems, *J. Am. Optom. Assoc. 59*:193 (1988).
28. T. Kunzmann, Investigation on the disinfecting action of hydrogen peroxides, *Fortschr. Med. 52*:357 (1934).
29. S. Leaper, Comparison of the resistance to hydrogen peroxide of wet and dry spores of *Bacillus subtilis* SA22, *J. Food Technol. 19*:695 (1984).
30. M. W. LeChevallier, C. D. Cawthon, and R. G. Lee, Inactivation of biofilm bacteria, *Appl. Environ. Microbiol. 54*:2491 (1988).
31. W. Lineaweaver, S. McMorris, D. Soucy, and R. Howard, Cellular and bacterial toxicities of topical antimicrobials, *Plastic Reconst. Surg. 75*:394 (1985).
32. T. D. Lindquist, D. J. Doughman, J. B. Rubenstein, J. W. Moore, and R. C. Campbell, Acanthamoeba-contaminated hydrogel contact lenses, *Cornea 7*:300 (1988).
33. G. E. Lowther, Contact lens solutions in clinical practice, *Clinical Ocular Pharmacology*, 2nd ed. (J. D. Bartlett and S. D. Jaanus, eds.), Butterworths, Boston, 1989, pp. 357–382.
34. I. H. Ludwig, D. M. Meisler, I. Rutherford, F. E. Bican, R. H. S. Langston, and G. S. Visvesvara, Susceptibility of *Acanthamoeba* to soft contact lens disinfection systems, *Invest. Ophthalmol. Vis. Sci. 27*:626 (1986).
35. M. S. Marqués, S. Lluch, M. D. Merindano, M. Gonzalez, and C. Saona, Effect of different disinfecting contact lens solutions against ocular bacterial strain growth, *Contact Lens J. 19*:9 (1991).
36. M. S. Mayo, W. L. Cook, R. L. Schlitzer, M. A. Ward, L. A. Wilson, and D. G. Ahearn, Antibiograms, serotypes and plasmid profiles of *Pseudomonas aeruginosa* associated with corneal ulcers and contact lens wear, *J. Clin. Microbiol. 24*:372 (1986).
37. C. M. McKenney, R. L. Chalmers, and N. Burstein, Human corneal epithelial permeability after instillation of hydrogen peroxide [Abstract]. *Optom. Vis. Sci. 66*:239 (1989). Presented at American Academy of Optometry Annual Meeting, New Orleans, December 7–11, 1989.
38. R. Mentel, and J. Schmidt, Investigations on rhinovirus inactivation by hydrogen peroxide, *Acta Virol. 17*:351 (1973).
39. M. B. Moore, *Acanthamoeba* keratitis and contact lens wear: The patient is at fault, *Cornea 9*(Suppl. 1):S33 (1990).
40. K. Naguib, and L. Hussein, The effect of hydrogen peroxide on the bacteriological quality and nutritive value of milk, *Milchwessenschaft 27*:758 (1972).
41. V. K. N. Nambudripad, I. I. Laxminarayana, and K. K. Lya, Bactericidal efficiency of hydrogen peroxide. 1. Influence of different concentration on the rate and extent of destruction of bacteria of dairy importance, *Indian J. Dairy Sci. 4*:38 (1949).
42. L. Ogunbiyi, The use of sodium thiosulfate for inactivating residual hydrogen peroxide on contact lenses after disinfection, *Clin. Exp. Optom. 69*:16 (1986).

43. J. R. Paugh, N. A. Brennan, and N. Efron, Ocular response to hydrogen peroxide, *Am. J. Optom. Physiol. Opt. 65*:91 (1988).
44. C. A. Penley, C. Liabrés, L. A. Wilson, and D. G. Ahearn, Efficacy of hydrogen peroxide disinfection systems for soft contact lenses contaminated with fungi, *CLAO 11*:65 (1985).
45. W. K. Ramp, R. R. Arnold, J. E. Russell, and J. M. Yancey, Hydrogen peroxide inhibits glucose metabolism and collagen synthesis in bone, *J. Periodontol 58*:340 (1987).
46. M. Rogan, Systems for hydrogen peroxide disinfection of soft contact lenses, *Trans BCLA Conf.* p. 40 (1985).
47. B. G. Rosling, J. Slots, L. A. Christersson, and R. J. Genco, Topical chemical antimicrobial therapy in the management of the subgingival microflora and periodontal disease, *J. Periodont. Res. 17*:541 (1982).
48. W. C. Schumb, C. N. Satterfield, and R. L. Wentworth, *Hydrogen Peroxide*, Van Nostrand Rheinhold, New York, 1955, pp. 813–816.
49. M. J. Sibley, K. L. Shih, and J. Hu, Evaluation of a new thimerosal-free 5-minute hydrogen peroxide disinfection lens care regimen, *Can. J. Opt. 44*:5 (1984).
50. R. E. Silvany, J. M. Dougherty, J. P. McCulley, T. S. Wood, R. W. Bowman, and M. B. Moore, The effect of currently available contact lens disinfection systems on *Acanthamoeba castellanii* and *Acanthamoeba polyphaga*, *Ophthalmology 97*:287 (1990).
51. P. A. Simmons, T. B. Edrington, L. Hsieh, and L. Wang, Bacterial contamination rate of soft contact lens cases, *ICLC 18*:188 (1991).
52. S. S. Socransky, Microbiology of periodontal disease—present status and future considerations, *J. Periodontol. 48*:497 (1977).
53. D. J. Stokes, and D. J. Morton, Antimicrobial activity of hydrogen peroxide, *ICLC 14*:146 (1987).
54. S. L. Stramer, and D. O. Cliver, Septage treatments to reduce the numbers of bacteria and polioviruses, *Appl. Environ. Microbiol. 48*:566 (1984).
55. S. V. W. Sutton, R. J. Franco, D. A. Porter, M. F. Mowrey-McKee, S. C. Busschaert, J. F. Hamberger, and D. W. Proud, D-value determinations are an inappropriate measure of disinfecting activity of common contact lens disinfecting solutions, *Appl. Environ. Microbiol. 57*:2021 (1991).
56. T. A. Svec, and J. W. Harrison, Chemomechanical removal of pulpal and dentinal debris with sodium hypochlorite and hydrogen peroxide vs. normal saline solution, *J. Endodontal. 3*:49 (1977).
57. D. E. Sweet, H. C. Goodpasture, K. Holl, S. Smart, H. Alexander, and A. Hedari, Evaluation of H_2O_2 prophylaxis of bacteriuria in patients with long-term indwelling Foley catheters: A randomized controlled study, *Infect. Control 6*:263 (1985).
58. R. L. Thompson, C. E. Haley, M. A. Searcy, S. M. Guenthner, D. L. Kaiser, D. H. M. Gröschel, J. Y. Gillenwater, and R. P. Wenzel, Catheter-associated bacteriuria, *JAMA 251*:747 (1984).
59. R. T. Toledo, F. E. Escher, and J. C. Ayers, Sporocidal properties of hydrogen peroxide against food spoilage organisms, *Appl. Microbiol. 26*:592 (1973).
60. B. J. Tripathi, and R. C. Tripathi, Hydrogen peroxide damage to human corneal epithelial cells in vitro, *Arch. Ophthalmol. 107*:1516 (1989).

61. M. Turck, and W. E. Stamm, Nosocomial infection of the urinary tract, *Am. J. Med. 70*:651 (1981).
62. F. J. Turners, Hydrogen peroxide and other oxidant disinfectants. *Disinfection, Sterilization, and Preservation*, 3rd ed. (S. S. Block, ed.), Lea & Febiger, Philadelphia, 1983, pp. 240–250.
63. U. S. Food and Drug Administration, Testing guidelines for class III soft (hydrophilic) contact lens solutions (draft), U.S. Food and Drug Administration, Washington, DC (1985).
64. M. D. Wardle, and G. M. Renniger, Biocidal effect of hydrogen peroxide on spacecraft bacterial isolates, *Appl. Microbiol. 30*:710 (1975).
65. J. B. Wilson, W. R. Turner, and J. W. Sale, Studies on bottled cocoa beverages. II. Efficiency of hydrogen peroxide in preserving cocoa-milk beverages, *Am. Food J. 22*:347 (1927).
66. L. A. Wilson, A. D. Sawant, and D. G. Ahearn, Comparative efficacies of soft contact lens disinfectant solutions against microbial films in lens cases, *Arch. Ophthalmol. 109*:1155 (1991).
67. L. A. Wilson, A. D. Sawant, R. B. Simmons, and D. G. Ahearn, Microbial contamination of contact lens storage cases and solutions, *Am. J. Ophthalmol. 110*:193 (1990).
68. L. F. Wolff, B. L. Pihistrom, M. B. Bakdash, E. M. Schaffer, J. R. Jensen, D. M. Aeppli, and C. L. Bandt, Salt and peroxide compared with conventional oral hygiene II. Microbial results, *J. Periodontol. 58*:301 (1987).
69. Y. Yoshpe-Purer, and E. Eylan, Disinfection of water by hydrogen peroxide, *Health Lab. Sci. 5*:233 (1968).

10

Alcohols for Antisepsis of Hands and Skin

MANFRED L. ROTTER
Hygiene Institute, Allgemeines Krankenahus
Vienna, Austria

I. DEFINITION

Alcohols are organic compounds, of which the general chemical formula is R−OH. (R stands for an alkyl or substituted alkyl radical.)

Alcohols, therefore, can be described as hydroxyl derivatives of paraffins (= hydrocarbons). Whereas the functional hydroxyl group determines the kind of possible chemical or physiological reaction, the speed and extent of these reactions are mainly controlled by properties of the alkyl radical.

II. NOMENCLATURE AND CHEMISTRY OF ALCOHOLS USED IN SKIN ANTISEPSIS

Alcohols (or "alcanols") are named by terminating the name of the hydrocarbon with the ending "-ol," for example methan*ol*.

The *number* of hydroxyl groups in a paraffin molecule determines the valency of an alcohol. Methanol, therefore, is a monovalent alcohol; 1,2-ethandiol ("glycol") is a divalent alcohol, and glycerol, which contains three OH-groups, is a trivalent alcohol.

According to the *position* of the hydroxyl groups, an alcohol is classified as primary, secondary, or tertiary:

Primary $R{-}CH_2{-}OH$

Secondary $R{-}CH{-}OH$ (with a second R attached to the CH carbon)

Tertiary $R{-}C{-}OH$ (with two further R groups attached to the central C)

Aliphatic (chain) alcohols are distinguished from aromatic (ring) alcohols.

Only the short chain, aliphatic alcohols are used in antisepsis, mainly for practical reasons. The only aromatic alcohol of some importance is benzyl alcohol; it is used virtually exclusively as an additive to other alcoholic antiseptics to support the antimicrobial effect.

III. PHYSICOCHEMICAL PROPERTIES

An important feature for their usability in antisepsis is the miscibility of alcohols with water. As can be seen from Table 1, only short-chained alcohols, such as methanol, ethanol, and the propanols, are completely miscible. Higher alcohols, such as hexanol, heptanol, and so on, or benzyl alcohol, have little or no solubility in water. Alcanols with chain lengths above the undecyl alcohols are not liquid at 20°C, thereby rendering their use impractical.

The longer the alkyl chain, the more lipophilic (and less hydrophilic) alcohols become. This is somehow counteracted by branching of the chain. The extent of the lipophilic characteristics is expressed by the partition coefficient, P, of an alcohol in systems such as octanol/water or diethyl ether/water [1]. As an example, some values are given for the system diethyl ether/water in Table 1.

If concentrations are given in fractions of 100 ("percent"), they must be clearly defined as percentage by weight (g/g or w/w) or by volume (ml/ml or v/v). The transformation of percentage by volume into percentage by weight is not possible by merely multiplying volume percent by the specific weight of the alcohol, since, in mixtures with water, a volume contraction of the alcohol occurs that is greatest when mixing equal parts of alcohol and water. Thus, for example, adding 50 ml of water to 50 ml of absolute ethanol results in a total volume of 96.3 ml, rather than 100 ml. Therefore, to prepare 100 ml, one would have to top up 50 ml of absolute alcohol with (distilled) water right to the 100-ml mark of a measuring cylinder. Table 2 lists the corresponding percentages by volume and weight for ethanol. As the corresponding values for propan-2-ol and propan-1-ol were lacking, they were generated especially for this chapter in laboratory experiments [H. Felsch, personal communication]. According to these measurements the corresponding curves are similar to, although not identical with, that of ethanol. In the area of greatest deviation, 60% by volume of propan-2-ol corresponds to 53.7% by weight, and of propan-1-ol to 54%. The best way to avoid difficulties in preparation of alcoholic solutions, however, is to work with concentrations by weight, rather than by volume, as early recommended by Beyer [2] in 1911 and, later, by Price [3]. The short-chain, aliphatic alcohols are inflammable liquids. Table 3 lists various safety data for these.

Because of their polar structure, low-weight alcohols possess considerably less surface tension than water. This is the reason for their better wetting characteristics, an important feature for antiseptics used on the skin surface. In this context, an old observation by Neufeld and Schütz [4] showed that ethanol absorbs six times more air than water, thus, it more easily enters crevices and blind-ending glandular ducts of the skin.

IV. HISTORY OF ALCOHOLS AS ANTISEPTICS

The name *alcohol* is taken from the Arabic *Al Kohl*, which was used for refined antimony sulfide used by ancient Egyptians as a remedy against eye infections and, in combination with a blue dye, as a cosmetic. In the early 16th century, the famous physician, scientists, and philosopher Paracelsus Theophrastus von Hohenheim gave the name "Al-Kool" to the distillate of wine for its antiseptic properties [5]. In antiquity and medieval times, alcohol was used in the form of wine or its distillate probably for wound antisepsis only. According to Beck [6] it was recommended for this purpose by Galen (131–201 AD) and Guy de Chauliac (1363), and Dittrich [7] reports that the surgeon Augustin Belloste (17th century) favored its use as did I.-F. Bathailé.

L. Buchholtz [8] was probably the first to investigate the antimicrobial activities of ethanol on a scientific basis in 1875 and to relate these to the possible application for antiseptic purposes. Robert Koch, however, concluding from the results of his experiments with anthrax spores, which he had dried onto silk threads, that ethanol was completely unsuitable for antiseptic and disinfection purposes [9]. Despite this erroneous belief, he was right in his observation that alcohols are not (or hardly) sporicidal. This was later often demonstrated [10–18].

Preoperative skin disinfection with alcohol was first carried out in the latter half of the last century [6] by Nelaton using alcohol soaks at the site of operation.

P. Fürbringer [19] was the first to recommend alcohol in 1888 for hand disinfection of surgeons and medical staff.

However, it was a long time until the controversies over the necessary concentrations and times of action were resolved by the discovery that the pres-

TABLE 1 Synopsis of Some Physical Properties of Short-Chain Alcohols

Alcohol	Mol wt	Specific gravity at 20°C	Boiling point (°C)	Melting point (°C)	Solubility in water at 20°C (g/100 g water)	Partition coefficient (log P)[a]	Viscosity (mPa·s)	at (°C)[b]
Methanol	32.04	0.791	64.7	−98	Unlimited		0.52	20
Ethanol	46.07	0.789	78.4	−117	Unlimited	−0.58	1.22	20
Propan-1-ol[c]	60.10	0.805	97.2	−127	Unlimited	0.28	2.75	20
Propan-2-ol[d]		0.785	82.0	−89	Unlimited	−0.19	2.27	20
Butan-1-ol[e]	74.12	0.810	117.0	−89	7.7	0.89	2.95	20
Butan-2-ol[f]		0.808	99.8	−114	12.5	0.84	4.21	20
2-Methylpropan-1-ol[g]		0.802	108	−108	9.5	0.65	6.68	20
2-Methylpropan-2-ol[h]		0.779	82.9	24	Unlimited	0.34	3.30	30
Pentan-1-ol[i]	88.15	0.815	138	−79	2.7		3.68	20
Pentan-2-ol[j]		0.808	119	−50	13.5			
3-Methylbutan-1-ol[k]		0.816	131	−117	2.5			
2-Methylbutan-1-ol[l]		0.816	128	70	3.6		5.10	20

2-Methylbutan-2-ol[m]		0.800	102	−12	17.7 (30°C)	3.70	25
Hexan-1-ol	102.20	0.814	157	−45	0.6	0.45	24
Heptan-1-ol	116.21	0.824	175	−34	Slight		
Octan-1-ol	130.23	0.827	195	−16	Slight	8.40	20
Benzyl alcohol	108.13	1.042	204.7	−15.3	3.5		

[a]System: Diethylether/water [54]
[b]Centipoise
[c]*n*-Propyl alcohol
[d]Isopropyl alcohol
[e]*n*-Butyl alcohol
[f]*sec*-Butyl alcohol
[g]Isobutyl alcohol
[h]*tert*-Butyl alcohol
[i]*n*-Amyl alcohol
[j]*sec*-Amyl alcohol
[k]Isoamyl alcohol
[l]*act*-Amyl alcohol
[m]*tert*-Amyl alcohol
Source: Courtesy of Merck Corp., Darmstadt; Refs. 250, 251.

TABLE 2 Corresponding Concentrations as Percentages by Weight and Volume for Ethanol, Propan-2-ol and Propan-1-ol

Concentrations by			
Weight (% g/g) Alcohol	Volume at 20°C (% ml/ml) Ethanol	Propan-2-ol	Propan-1-ol
40	47.4	47.0	46.5
50	57.8	57.0	56.5
60	67.7	67.2	66.2
70	76.9	76.5	76.1
80	85.4	85.0	84.5
90	93.2	93.8	93.5
95	96.7	97.5	97.2
100	100.0	100.0	100.0
Volume at 20°C (% ml/ml)	Weight (% g/g)		
40	33.4	33.5	34.0
50	42.6	43.0	43.4
60	52.1	53.7	53.9
70	62.6	62.8	63.5
80	73.4	73.5	74.5
90	85.8	86.0	86.0
95	92.5	92.8	93.0
100	100.0	100.0	100.0

Source: Courtesy H. Felsch, 1992.

ence of water plays an important role in the antimicrobial activity of alcohols [20–24], the necessary amount of water depending mainly on the status—dry or moist—of the bacterial cell. The hypothesis is that a high percentage, or pure, alcohols deprive bacterial cell walls of water, thereby inducing the formation of impermeable protein layers which, in turn, prevent the alcohol from penetrating into the cell [25]. Hence, as a prerequisite for antimicrobial activity of alcohols on *dry* cells, the absorption of water by the cell wall is necessary [2,26,27]. But this was not true for moist cells, as Morton [16] has demonstrated in his experiments. Therefore, the statement, found in many textbooks, that high-percentage alcohols, such as 95% ethanol, are practically worthless as germicides needs to be qualified.

In contrast with European countries, in many of which the performance of

TABLE 3 Various Safety Data on Alcohols Used for Antiseptic Purposes

Alcohol	Vapor pressure at 20°C hPa	Vapor density at 20°C	Flash point[a] (°C)	Explosion limits in air (vol %)	Autoignition temp. (°C)	Odor threshhold (ppm)
Methanol	128	1.1	11	6.0–36.0	385	5
Ethanol	59	1.6	13	3.3–19.0	363	10
Propan-1-ol	19	2.1	15	2.2–13.7	410	30
Propan-2-ol	43	2.1	12	2.0–12.7	399	40
Butan-1-ol	7	2.5	30	1.4–11.3	343	10
Butan-2-ol	17	2.6	23	1.7–9.8	405	0.6
Isobutanol	12	2.6	29	1.7–10.9	430	40
tert-Butanol	40	2.6	11	2.3–8.0	478	73
Pentan-1-ol	3	3.0	48	1.3–10.5	300	1
Pentan-2-ol	3	3.0	41	1.2–9.0	347	
Isoamyl alcohol	3	3.0	42	1.2–8.0	350	70
akt-Amyl alcohol	51[b]	?	50	1.9–9.3	340	
tert-Amyl alcohol	13	3.0	20	1.3–9.6	425	2.3
Hexan-1-ol	1	3.5	63	2.1–7.7	292	5.2
Heptan-1-ol	0.5	4.0	70		350	0.5
Octan-1-ol	0.3	4.5	81	0.8	270	
Benzyl alcohol	0.1	3.7	101		436	

[a]For concentrated solutions
[b]At 50°C
Source: Courtesy of Merck Corp., Darmstadt; and Refs. 250, 251.

alcohols has remained the standard of hand and skin antisepsis since the beginning of this century, in the United States the attitude toward the use of ethanol or propan-2-ol has changed completely during the last five decades. In the 1930s, ethanol was the favorite hand and skin antiseptic. In 1938, the surgeon P. B. Price added further evidence for the excellent performance of ethanol in hand and skin antisepsis, as demonstrated by the results of his intelligently designed tests [28–30]. Even in his later papers reporting on his findings with the then more modern preparations that were based on quaternary ammonium compounds, hexachlorophene, and iodophors [31,32], he mentions that "seventy per cent alcohol by weight is still believed to be the solution of choice for disinfection of the skin." In his survey of 1948 [33], 64% of the hospitals in the United States were using ethanol (although nearly 70% also used organic mercurials, and 48% used a quaternary ammonium compound). In 1965, however, only 18% of North American hospitals had alcohol on their list of hand and skin disinfectants [34] and, today (1992), alcohols are seldom used for this purpose, presumably for reasons of flammability and their allegedly drying effect on skin [35]. During the period discussed, alcohols were replaced on the North American continent mainly by hexachlorophene and iodophors, the former of which was withdrawn from the market in the 1970s because of its potential toxicity. To a lesser extent, preparations containing chlorhexidine gluconate, triclosan, or phenolic compounds such as *p*-chloro-*m*-xylenol (PCMX) are used.

V. ANTIMICROBIAL PROPERTIES

The antimicrobial activity of alcohols is highly dependent on their concentration and on the status—dry or moist—of the microorganism. The short-chain, aliphatic monovalent alcohols, ethanol and the propanols, seem to be the most suitable for antisepsis because they are miscible with water to an unlimited degree, hardly toxic, and are nonallergenic, fast-acting, and are microbicidal, rather than microbiostatic, if used in the correct range of active concentrations. The bi- or trivalent alcohols do not play a role in the field of antisepsis, nor do aromatic alcanols. Some of them are used as solvents and for preservation because of their bacteriostatic and fungistatic characteristics. They will not be discussed in this chapter.

A. Mechanisms of Antimicrobial Action

The mechanism for the disinfection effect of alcohols is generally believed to originate from the coagulation of proteins [36–39]. Kamm [36] explained the biological effects of alcohols more precisely with both their capability to coagulate proteins and their solubility in lipids. Both features become more prominent with increasing molecular weight. For example, the relative efficiencies of molar solutions to inactivate enzymes for ethanol, propan-1-ol, and butan-1-ol are 1, 2.6, and 8.0, respectively.

The coagulation of proteins occurs within certain concentration limits encompassing the concentration optimum. Coagulation of enzymatic protein results in the loss of specific cellular functions. For bacterial dehydrogenases, Sykes [40] has demonstrated an increasing effect in the following range of order:

ethanol < propan-2-ol < butan-1-ol < pentan-1-ol

An alcohol-induced prolongation of the growth lag phase of *Enterobacter aerogenes* could be counteracted by the addition of amino acids [41]. Ethanol- and, to a lesser extent, propan-1-ol–induced effects, however, were "curable" by only certain amino acids, such as D,L-methionine, L-leucine, L-glutamic acid, or L-histidine and D,L-tryptophan; the addition of some other amino acids such as D,L-alanine, D,L-serine, or D,L-aspartic acid caused the inverse effect (i.e., a further increase of the lag phase). Therefore, this bacteriostatic action may be caused by an alcohol-induced inability of the cell metabolism to produce some metabolites necessary for growth. Other workers [42] succeeded in resuscitating cells of *E. coli*, which were apparently killed by a 10-min exposure to 20% ethanol, by "treating" them with several metabolites of the citrate cycle, such as *cis*-aconitic acid,α-ketoglutarate, or a mixture of several metabolites, including both of these. This could also be demonstrated under conditions of hand disinfection [43]. The activity of cholinesterase was repressed, within certain concentration limits, by the normal C_3 to C_6 alcohols, such that, with increasing chain length, the activity decreased. Beyond certain concentration, the enzyme was completely inactivated. This critical concentration for each alcohol also decreased with increasing chain length [44].

Bacterial enzymes localized in the cell wall are easier to inactivate than those inside the cell [45]. The coagulation of plasma proteins is visually demonstrable in electron micrographs, in which mixing of cell plasma and nucleoid can be observed [46]. However, coagulation of nucleoproteins does not occur [27]. Another target in the microbial cell is the cytoplasmic membrane, the destruction of which leads to an outflow of cytoplasm, amino acids, and nucleic acid constituents [47–49] at relatively low concentrations. As with other disinfectants, bacterial lysis may occur at relatively low concentrations (about twice the minimum inhibitory concentration; MIC) [50]. Also, lysis of human *Mycoplasma* strains has been reported [51].

There are associations between the structure and the biological effects of alcohols. Those of monovalent alcohols were summarized by Friedemann and Treuhoff [52]. An extensive account on the dependence of biological effects on the chemical and physical properties of alcohols was given by Franke and Kramer [53]. An important review on alcohols as disinfectants has been written by Morton [38]. A more recent survey was presented by Heeg et al. [54].

The main facts are the following: The antibacterial effect of monovalent, aliphatic alcohols increases with the boiling point [55]. Elongation of the unbranched alkyl radical up to six carbon atoms improves the bactericidal effect, further elongation impairs it. Eggensperger [56], for example, demonstrated that *Staphylococcus aureus* was killed by a 2-min exposure in suspensions of alcohols with the following critical concentrations (by weight): methanol 60%, ethanol 50%, propan-1-ol 20%, butan-1-ol 10%, hexan-1-ol 5%, but dodecan-1-ol at 15% (see also Table 4).

The branching of the alkyl radical increases water solubility, but diminishes the disinfection effect. In the foregoing experiments, for instance, a concentration of 40% was necessary for propan-2-ol and tertiary butanol to achieve an effect comparable with that of the corresponding normal alcohols.

Among the isomers, the following range order of the antibacterial effect has long been known: *n*-primary > *iso*-primary > secondary > tertiary alcohols [22,57]. Of the alcohols that are completely miscible with water, propan-1-ol, therefore, is the most active [11,58,59].

As with all disinfectants, there is a clear association of temperature and disinfection efficacy [11,27,60,61].

Controversial reports exist on the impairment of disinfection efficacy by the presence of organic material, especially proteins. Whereas Friedemann [62] showed that bacteria suspended in serum or wound secretions are protected by coagulated protein from the killing effect of alcohols, and Schürmann and Eggers [61] found a concentration-dependent inhibition of the antiviral activity of an ethanol/propan-2-ol mixture, other authors [63,64] have shown that, under practical conditions, doubling the exposure time easily compensates for this factor. Rotter and colleagues [65] found, in practical experiments, that the mean log reductions of the release of test bacteria (*E. coli*) from the fingertips of artificially contaminated hands by rubbing in 60% (v/v) propan-2-ol for a total of 60 s were only slightly (and not at all significantly) different if the test organisms were suspended in broth, horse serum, or skim milk, and then dried on the hands for 3 min before disinfection. The respective log reductions were: 4.74, 4.69, and 4.48.

With some combinations of alcohols, synergistic effects have been observed. For example, with the combination of methanol/butan-1-ol, the killing times were reduced by 30–80%, compared with those of either alcohol alone [66].

Development of resistance against alcohols does not play a role, at least in concentration ranges suitable for disinfection. In 50 passages of *Staphylococcus aureus* and four gram-negative bacterial species on media containing 7% and 3.5% propan-2-ol, respectively, Wille [67] could not detect any indications of an increase in alcohol resistance. In experiments by Wigert et al. [68] using concentrations of various alcohols below the bacteriostatic concentrations, the sensitivity of *S. aureus* was decreased by 50% (i.e., one dilution step), and for gram-negative bacteria, the change was even less [69]. In addition, this behavior of the test strains was not stable, but disappeared immediately after termination of the exposure to alcohol. Therefore, any genetic basis for it can be excluded [70].

B. Antimicrobial Spectrum

In this section, the current knowledge of the antimicrobial activity of alcohols as used in disinfection and antisepsis (against bacterial spores, vegetative forms of bacteria, fungi, and viruses) will be presented in the order that this knowledge was accumulated during the last century.

1. Bacterial Spores

As pointed out in Section IV, alcohols are generally considered to be completely inactive against bacterial spores. It would be imprecise, however, to state that they are never sporicidal. Even if it is true that spores of *Bacillus subtilis* stay alive in 95% ethanol for years [39], such that a high ethanol concentration was used as a suspension liquid to preserve them as standard inoculum for antibiotic assays, this is not true for all bacterial spores and not under all conditions. Regamay [71], for instance, has demonstrated that spores of anaerobic bacteria indeed survived in 10% ethanol for more than 10 months, but in concentrations of 40% and 80% they survived for only 7½ and 4 weeks, respectively. Morton [38] produced evidence that, in contrast with Koch's and other authors findings, spores of *B. anthracis* may well be sensitive to ethanol concentrations between 42 and 100%. Under his experimental conditions they were killed within 48 h. With propan-2-ol this was possible even with concentrations of 30–40% within 2–5 min [72], but methanol was reported to be completely inactive against spores of this species in concentrations of 0.004–95% [17].

Spores of other species, namely *B. subtilis* and *Clostridium novyi*, remained unaffected by exposure to 20–91% propan-2-ol over 1 h [73]. Although one cannot advocate the use of alcohols for indications that require sporicidal actions, one would often wish that the differences in activity were better defined in terms of an exact description of the bacterial strains, the condition of the spores—dry or moist—and accompanying factors. Some of these may well explain controversial findings, since sporicidal activity of alcohols can be augmented or induced by the addition of agents such as alkali, mineral acids [74], H_2O_2 (0.5%), or some tensides [75,76] and formaldehyde [66].

2. Vegetative Bacteria

When ethanol was recommended in 1888 by Fürbringer [19] for disinfection of hands, the idea was to use it as a cleansing agent [see also Ref. 77] for its alleged fat-dissolving properties, rather than as a bactericide. The actual disinfection with a mixture of 0.2% sublimate in 3% carbolic acid was the next step in Fürbringer's recommendation. In those days, others [78,79] explained the reduction of bacterial release from the hands, as observed after the application of alcohol, by the fixation of microbial cells to the skin or by trapping them in the dephth of the alcohol-indurated skin surface. This hypothesis, however, was soon refuted (a) by the results of laboratory experiments proving that bacteria are readily killed by ethanol [22,55,80], (b) by the elegant experiment of Neufeld and Schiemann [12] demonstrating that an alcohol rub reduced the release of *E. coli*, but not that of *B. subtilis* spores from hands previously contaminated with a mixture of both organisms, and (c) by Bernstein [81], who demonstrated that a 2-min alcohol treatment of hands did not reduce the release of epithelial cells from the skin.

Table 4 summarizes the antimicrobial activity of various alcohols in terms of the concentrations necessary either to inhibit growth or to kill *S. aureus* and *E. coli*. As pointed out in Section V.A, it is evident that the antimicrobial activity increases with the chain length of the alcohols and that tertiary alcohols are less

TABLE 4 Comparison of Minimum Bacteriostatic and Microbicidal Concentrations of Various Alcohols in Suspension Tests

	Minimum effective concentrations (% v/v)							
Effect	Growth inhib.	Killing effect in suspension test						
Test Organism	*S. aureus*	*S. aureus*				*E. coli*		
Exposure (min): Alcohol		1	2	2	10	1	2	12
Methanol	9	65		67[c]		60–65		
Ethanol	7	50		58[c]		45–50		
Propan-1-ol	4	20		23[c]		17		
Propan-2-ol	4	45		46[c]		26		
Butan-1-ol	3	9		11[c]		5		
Isobutanol	3	16				6.5		
sec-Butanol		15				11		
tert-Butanol	5		26	46[a]			14	
Pentan-1-ol	1				3	2		
Isopentanol			4			2.75		
sec-Pentanol		7.5			4			
tert-Pentanol	3	15				9		
Hexan-1-ol		0.6[a]	5[a]		0.7			
Heptan-1-ol		0.12[b]						0.12
Octan-1-ol		0.06[b]				0.06		
Ref.	22	55	55	56	55	55	55	55

[a]In 30 min.
[b]In >30 min.
[c]Converted from % w/w.

active than primary or secondary ones. This is also true for exposure of the test organisms on carriers such as batiste patches or glass slides (Table 5).

From these tables and from reports in the literature, the following conclusions can be drawn: Methanol exerts a definite, although comparatively low, activity [22,55,56,58,66]. Concentrations of above 50% (v/v) are necessary for bactericidal effects. However, it was noted by some authors [83] that, in contrast with other alcohols, it is also active on dried bacteria (not spores) without the presence of water. The combination of ethanol (96%) with methanol (10%) for denaturation of the former does not impair its disinfection efficacy [86].

Ethanol exerts bacteriostatic effects at approximately 10% (v/v). At this concentration germination of bacterial spores is also inhibited [39]. Bactericidal concentrations exist from about 30% (v/v) upward, depending on the bacterial species, its water content, and the exposure time [16]. Table 6 gives examples of the times necessary to kill various bacterial species in ethanol at the optimum concentrations of 60–80% (v/v). Except for *Streptococcus pyogenes*, all species listed are killed within 1 min. This is also true for *Mycobacterium tuberculosis* in this experiment. A very high percentage (90% up to absolute ethanol) is less active than the less-concentrated solutions.

An informative picture of the influence of test organism status—dry or moist—and of the weak action of highly concentrated ethanol is given by the results of experiments carried out by Harrington and Walker [20] with *S. aureus* on moist and dried threads. Table 7 demonstrates that, in moist conditions, the test

TABLE 5 Comparison of Minimum Bactericidal Concentrations of Various Alcohols in Carrier Tests

	Minimum effective concentrations (% v/v)								
Test Organism	*Staph. aureus*			*E. coli*	*M. tuberculosis*				
Exposure (min): Alcohol	2	30	30	0.25–15s	10[b]	10[b]	15	30	120
Methanol		57[a]	70	95					90
Ethanol	80	43[a]	60	80	96	80		80	
Propan-2-ol	50		50	50	60			20	
Propan-1-ol		23[a]	30	20	50	50		20	
Butan-1-ol								5	
Isobutanol									10
Propen(1)-ol(3)								20	
Benzyl alcohol							~4		
Refs.	82	11	83	83	84	82	85	85	85

[a]Converted from %w/w.
[b]Dried in sputum on slide.

TABLE 6 Bactericidal Efficacy of Ethanol at Various Concentrations (% [v/v]) in Suspension Tests

Bacterial species	Killing time (s)		
	60%	70%	80%
Staphylococcus aureus	15	15	10
S. epidermidis	30	30	
Streptococcus pyogenes			90
Escherichia coli	60	30	30
Serratia marcescens		10	
Salmonella typhi		10	
Pseudomonas aeruginosa		10	
Mycobacterium tuberculosis	60	30	30

Source: Ref. 39.

organism is killed in 5 min by the highest ethanol concentration of 99%, but on dried threads, this concentration is still inactive after 24 h. Later, Morton [16] produced results comparable with these, but with more bacterial species and with anthrax spores. He points out that the results of alcohol on dried organisms are unjustly taken as typical of the germicidal action of ethanol.

The effect of ethanol against *Mycobacterium tuberculosis* depends on the suspending medium—sputum or water—and on its moisture content—liquid or dried. Table 8 summarizes results of Smith [87] demonstrating the influence of these variables on the tuberculocidal effect of ethanol. From these results the author concluded that 95% ethanol was the most effective on wet surfaces, 50% on dry, and 70% on either wet or dry surfaces. The reports of other authors [82,84,85], as summarized in Table 8, show the same concentration effects. Ethanol is also active against other mycobacterial species, such as *M. phlei* [88].

A bactericidal effect of ethanol was also shown for *Mycoplasma* strains by Razin and Argaman [51], who found a concentration range of 4.3–5.4 mol effective, and for *Spiroplasma* strains by Stanek et al. [89], who described effective concentration/time relationships from 20%/5 min to 30%/40 min.

The two propanols, which are homologues with the highest molecular weight, are miscible with water in any proportion. Of the water-soluble alcohols, propan-1-ol (syn: *n*-propanol, *n*-propyl alcohol) was reported to be the most active [11,58,59,90–92]. For the more sensitive species, such as *E. coli*, propan-1-ol is bactericidal at 13% [55], the optimum concentration being 40–60% (v/v). The antibacterial spectrum is comparable with that of ethanol [25,39,54,83; see also Tables 4 and 5].

TABLE 7 Bactericidal Activity of Various Concentrations of Ethanol Against *Staphylococcus aureus* on Moist and Dried Threads

Conc. (%)	Exposure of test organism[a]																	
	Moist threads									Dried threads								
	min					h				min					h			
	5	10	15	30	45	1	2	7	24	5	10	15	30	45	1	2	7	24
15	+	+	+	+	+	+	+	+	+	+	+	+	+	+	+	+	+	+
20	+	+	+	+	+	+	+	+	−	+	+	+	+	+	+	+	+	+
25	+	+	+	+	+	+	+	−	−	+	+	+	+	+	+	+	+	−
30	+	+	+	+	+	+	−	−	−	+	+	+	+	+	+	+	−	−
40	−	−	−	−	−	−	−	−	−	+	+	−	−	−	−	−	−	−
50	−	−	−	−	−	−	−	−	−	+	−	−	−	−	−	−	−	−
60	−	−	−	−	−	−	−	−	−	−	−	−	−	−	−	−	−	−
70	−	−	−	−	−	−	−	−	−	−	−	−	−	−	−	−	−	−
80	−	−	−	−	−	−	−	−	−	+	−	−	−	−	−	−	−	−
85	−	−	−	−	−	−	−	−	−	+	+	−	−	−	−	−	−	−
90	−	−	−	−	−	−	−	−	−	+	+	+	−	−	−	−	−	−
94	−	−	−	−	−	−	−	−	−	+	+	+	+	+	+	+	+	−
99	−	−	−	−	−	−	−	−	−	+	+	+	+	+	+	+	+	+

[a]+, growth; −, no growth.
Source: Ref. 20.

TABLE 8 Tuberculocidal Effect of Ethanol Under Various Conditions

	Tuberculocidal exposure	
Condition	Conc. (% v)	Time
In sputum (wet)	100	30 s
	95	15 s
	70	30 s
	50	60 s
In water	100	30 s
	95	15 s
	70	60 s
In sputum (dried)		
Thin layer	70	60 s
	50	60 s
Thick layer	100	>60 min
	95	30 min
	70	>5 min

Source: Ref. 87.

For its outstanding antimicrobial properties propan-1-ol is a common constituent of commercial preparations for skin and hand disinfectants in Europe. At a concentration of 60% (v/v), it is now the standard, the activity of which acts as a reference for every new surgical hand disinfectant to be approved in German-speaking countries [93,94]. Another reason for its frequent employment is its relatively high flash point of 23°C. In many European countries inflammable liquids with a flash point below 21°C are not permitted for general use in hospitals. Therefore, many alcoholic antiseptic solutions contain propan-1-ol to raise the flash point.

Propan-2-ol (syn: isopropanol, isopropyl alcohol), the antimicrobial activity of which ranges between those of ethanol and propan-1-ol (see Tables 4 and 5) also requires the presence of water around the microorganism or in mixture with the alcohol to attain a microbicidal effect [95]. The optimum bactericidal concentration ranges between 50 and 70% (v/v) [39,73,74,82,83,96,97]. Tuberculocidal activity is achieved in 10 min at 40% [85]. In sputum, however, 50–70% is necessary [82,84,98].

Butane- and pentane-derived alcohols are of little importance in this context. Except for the tertiary alcohols, their antimicrobial activity seems to be higher than that of the propanols (see Tables 4 and 5). Their low solubility in water and their offensive smell are obstacles for practical use. Nevertheless, some commercial preparations contain one or the other as an additive in low concentration.

Allylalcohol and benzyl alcohol are active against *M. tuberculosis* at 20% and in saturated solution (ca. 3.5%), respectively, in 15 min [85].

3. Fungi

The fungicidal activity of short-chain, monovalent alcohols, mainly ethanol, is well documented. As with their activity against bacteria, with methanol higher concentrations (50%) are necessary to kill *Candida* or *Aspergillus* species within 1 min than with ethanol (35%) or with the propanols, which are active in the range of 10–30% [66].

Table 9 summarizes the fungicidal activity of ethanol as reported in the literature. Except for the results of Szerémi [66], which demonstrate an unusually high sensitivity of *Candida* and *Aspergillus* spp. (killing by 35% ethanol in 1 min), the most effective concentrations range is between 70 and 90%, acting in 5–60 min.

4. Viruses

Controversial opinions appear in the literature on the antiviral activity of alcohols. There is, however, general agreement that enveloped, lipophilic viruses are easier to inactivate by disinfectants than are "naked" viruses [104]. This is also true for alcohols.

Among the enveloped viruses the vaccinia virus has been studied extensively for sensitivity to alcohols. As shown in Table 10, methanol is reported to inactivate the virus at concentrations of 80% in 15 min [105], of 50% in 1 h, and of 20% in 24 h [106]: the latter was not possible with ethanol. However, at 50% [106] to 60% [105] ethanol inactivated the virus within 1 h, even under dried conditions [105]. In a suspension of chick embryo material it seems to be much easier to attain inactivation. Only 2, 3, or 10 min were necessary at the respective concentrations of 70% [107,108], 80% [109], 70 or 40% [110,111]. Compared with ethanol, the propanols seem to be more active against this virus, since lower concentrations are needed for the same effect within the same times or even shorter [105,108–111].

Togaviruses such as yellow fever or Eastern equine encephalitis are readily inactivated by otherwise suboptimal concentrations of ethanol, 17% and 50%, respectively, in 30–60 min [113,114]. The effect of propan-2-ol was comparable with that of ethanol.

At an optimum concentration of 70%, Kewitsch and Weuffen [108] found ethanol effective against influenza A in 2 min; within the same time, only 40% of propan-1-ol caused the same effect. With propan-2-ol the virus was inactivated within 10 min at concentrations of 30 [111] and 48.5% [110].

Against the mumps virus, 36% ethanol was effective in 30 min [115], and another paramyxovirus, Newcastle disease virus, was inactivated by 70% ethanol

TABLE 9 Fungicidal Effect of Ethanol Against Various Fungal Species

	Minimum effective conc. (%)					
	Exposure time (min)					
Fungus	1	5	30	60	Not stated	Ref.
Yeasts		75				99
Candida albicans	35					66
C. krusei	35					66
Cryptococcus neoformans					70	100
Histoplasma capsulatum[a]					70	100
Blastomyces dermatitidis[a]					70	100
Coccidioides immitis[a]					70	100
Dermatophytes		50				99
Microsporum gypseum			80			39
(spores)			85			101
M. audouinii				70		102
(on naturally infected hairs)					70	109
Aspergillus niger	35					66
Penicillium tardum					90	103
(conidia)					(70–96)	

[a]Tissue and culture phase.

in 3 min [116] or by 25% propan-2-ol in 1 h [117]. Herpes simplex lost its infectivity after contact for 10 min with either 30% ethanol or 20% propan-2-ol [111].

These reports demonstrate that enveloped viruses are readily inactivated by short-chain aliphatic alcohols in due time, if these are used at their optimum concentrations. At least, for the viruses discussed, the same alcohol order range of virucidal capacity applies as that for effect against bacteria: methanol < ethanol < propan-2-ol < propan-1-ol.

An unusual behavior among the lipophilic viruses, however, was reported by Jaeger et al. [118] for the rabies virus which could not be inactivated by ethanol within 30 min. Although adenoviruses do not number among the enveloped viruses, they behave like them in their tenacity. Klein and Deforest [111] found them sensitive to ethanol and propan-2-ol at 50% in 10 min in a suspension test model. Propan-2-ol was reported active against the respiratory syncytial virus at 35% in 1 min [119].

In contrast to the foregoing viruses, naked viruses are much more difficult to inactivate with higher alcohols than with ethanol. Especially the picornaviruses pose considerable disinfection problems (Table 11). Against poliovirus type 1, 70% ethanol was found effective in 2 min [108]. Within the disinfection time of 10

TABLE 10 Inactivation of Enveloped Viruses by Short-Chain, Aliphatic Alcohols in Vitro

		Minimum effective conc. (%)								
		Exposure times (min)						(h)		
Virus	Alcohol[a]	0.5	2	3	10	15	30	1	24	Ref.
Vaccina	M					80				105
								50	20	106
Dried	E						70	60		105
								50		106
Suspension			70							107,108
				80						109,112
					70					110
					40					111
	P-1					30	15			105
		40								108
				60						109
	P-2				48.5					110
					30					111
Yellow fever	E						17			113
Eastern equine	E							50		114
Encephalitis	P-2							49.5		114
Influenza	E		70							108
					70					110
					30					111
	P-1		40							108
	P-2				48.5					110
					30					111
Mumps	E						36			115
Newcastle disease	E		70							116
	P-2							25		117
Herpes simplex	E				30					111
	P-2				20					111

[a]M, methanol; P-1, propan-1-ol; P-2, propan-2-ol.

min, which Klein and Deforest [111] have chosen for their suspension test, this concentration was also effective; with only 25%, 240 min was needed for the same effect [120]. With a concentration of 80% (v/v) Steinmann et al. [196] reported increasing reductions of the poliovirus 1 (Sabin) of 2.5, 3.7, and 4.8 logs after 1, 2, and 5 min, respectively. In contrast, Tyler et al. [200] found, in their suspension tests, that the titer of poliovirus (Sabin 1 an) was reduced with ethanol 70% (v/v) by only 0.4 logs in 1 min. Propan-2-ol was inactive even at 95% [111]. Schürmann

Table 11 In Vitro Inactivation of Picornaviruses by Short-Chain, Aliphatic Alcohols

Virus	Alcohol	Minimum effective conc. (%) Exposure times (min) 1	2	4	6	8	10	240	(d) 6	Ref.
Poliovirus type 1	E		70							108
							70			111
								25		120
	P-2						(95[a])			111
	CP[b]				100					121
Coxsackie B	E						60			111
	P-2						(95[a])			111
	CP[b]		100							121
Echo 6	E						50			111
	P-2						90			111
Echo 9 (Hill)	CP[b]					100				121
Echo 11	M	76								122
	E	(76[a])								122
Hoof-and-mouth disease	E								25	123
Hepatitis A	E	(70[a])								124
	P-2	(45[a])								124

[a]*Not* effective at this concentration.
[b]Commercial preparation of ~90% ethanol, 10% propan-2-ol, and 0.1% tetrabromo-6-methylphenol.

and Eggers [121], who have extensively tested the antiviral properties of a commercial preparation containing 78.2% (w/w) of 95.3% (v/v) ethanol (about 90% v/v), 10% (v/v) propan-2-ol and 0.1% of a tetrabromo-6-methylphenol in a standard suspension test, demonstrated a 4-log reduction of polio type 1 in 6 min, but reported only a 2-log reduction of polio type 2 in 10 min. Coxsackievirus B was inactivated by 60% ethanol, but not by 95% propan-2-ol in 10 min [111]. With the aforementioned commercial product containing mainly ethanol, the titer of coxsackievirus B4 was reduced by 4 logs in 2 min, but with coxsackievirus B3, this reduction could not be achieved in 10 min [121]. To inactivate echovirus 6 in 10 min was possible with 50% ethanol, but not with 90% propan-2-ol [111]. A strain of echovirus 9 was reduced by 4 logs in 8 min by the commercial product mentioned earlier [121]. The authors noted, however, considerable differences between the five strains tested. A strain of echovirus 11 was sensitive to methanol, 76% in 1 min, but not to ethanol, propan-1-ol, propan-2-ol, or butan-2-ol, even at 90% [122]. Hoof-and-mouth disease virus was not inactivated in 3, but was in 6 days by 25% ethanol [123].

Hepatitis A, now classified as an enterovirus, is known for its high resistance to chemical and physical influences. In a recent study, 20 disinfectants were tested for their capability of inactivating, within 1 min, the virus contained in a fecal suspension and dried on polished, stainless steel disks. Only 2% glutaraldehyde, sodium hypochlorite (dissolved to give 5000 ppm free chlorine), and a quaternary ammonium compound in 23% HCl were effective; however, 70% ethanol and 45% propan-2-ol were not effective.

Among other enteric viruses, rotaviruses behave as enveloped viruses, being increasingly sensitive to methanol upto butan-2-ol (Table 12) [122]. In contrast, astrovirus was sensitive only to high concentrations (90%) of methanol and ethanol [122].

In the past, the hepatitis B virus (HBV) was considered extremely resistant to the action of chemical disinfectants. However, since the experiments of Bond et al. [125] and Kobayashi et al. [126], who treated dried or liquid human plasma containing high-titer HBV with various chemicals or heat, and inoculated it into susceptible chimpanzees, this belief is no longer justified. Both groups found all treatments effective, including 70% propan-2-ol for 10 min and 80% ethanol for 2 min, respectively.

5. Protozoans

In a suspension test, trophozoites of *Toxoplasma gondii* were killed in 3 min by ethanol 30% and by propan-1-ol 20% [127]. But this information is of importance

TABLE 12 Activity of Short-Chain Alcohols on Enteric Viruses

	Minimum effective conc. (%)[a]		
	Virus tested		
Alcohol	Rota	Astro	Echo 11
Methanol	>50	90	76
Ethanol		90	(76[b])
Propan-1-ol	40	No effect	No effect
Propan-2-ol	40	No effect	No effect
Butan-2-ol	30	No effect	No effect

[a]Minimum concentration reducing virus titer by at least 4 $\log_{10}$ in 1 min in a suspension test system.
[b]Only 3-log reduction.
Source: Ref. 122.

only for laboratory workers handling these stages of *Toxoplasma*, for instance in Sabin–Feldmann tests. For the really relevant question about the activity of alcohols against the persistent stages of this protozoan, as excreted with cat feces, only little is known. The same applies to cysts of *Entamoeba histolytica* or *Cryptosporidium*, all of which can be transmitted with fecal material by hands. For oocysts of *Toxoplasma gondii*, Duby et al. [128,129] have reported an anti-protozoal activity for 95% ethanol in 1 h and of 20% ethanol in 7 days.

C. Antimicrobial Activity in Vivo

Most results of in vitro experiments on the efficacy of disinfectants and antiseptics offer only little information about the real time–concentration relationships of the antimicrobial action in vivo. Therefore, disinfectants and antiseptics must be tested under conditions simulating practical conditions as closely as possible. For disinfectants of hands and skin, for instance, real skin must be used as a carrier for test organisms if the efficacy against transient flora has to be tested, and for surgical hand disinfection, real hands with their natural resident microbial flora must be employed in tests.

1. Microbial Flora of Hands and Skin

All considerations on the microbial flora of hands and skin must take into account two groups of microorganisms: one group that can be found regularly and usually in great numbers because its members reside in and on the skin as autochthonous flora, and the other groups consisting of the flora that happens to be found there by chance. The first group is generally known as "resident flora," whereas the latter is called "transient flora" [28,29]. However, distinguishing between the two may be difficult [130,131].

The term *resident flora*, as specified very cautiously by Price [28,29], covers microorganisms that not only survive on the skin, but also multiply there. By mechanical means—with or without a brush—they are difficult to remove. To significantly reduce their viable counts, suitable disinfectants, antiseptics, or antibiotics are necessary [131–133]. Washing hands with soap and water only halves the release of skin bacteria every 6 min [28].

The composition of the microflora of the healthy and diseased skin has been described extensively in several reviews [134–141]. There are differences between different parts of the body [142–144], between individuals [143,145], between persons of different ages, and between the sexes [146,157]. Possibly, there also exists an association with the seasons [148].

The preponderant part of the normal, resident flora consists of only a few bacterial species, the pathogenic potential of which is usually considered to be low [149], unless they are introduced into body tissue by trauma or by adhering to foreign bodies such as implants [150,151].

Except for very greasy skin areas, most skin bacteria grow aerobically [142,149]. Most of them belong to the family *Micrococcaceae* or are members of the "diphtheroid" group [133,142,152]. The most prevalent species is *Staphylococcus epidermidis*. Fortunately, *S. aureus*, a well-recognized pathogen, only rarely colonizes the healthy skin, mainly in the perineal region [142]. But also hands, face, and the nape are preferential areas of temporary colonization. The skin of the hands of health care personnel, however, is considerably more often colonized with *S. aureus* [28,153,154].

Lipophilic diphtheroids are located mainly in the armpits [145], nonlipophilic ones usually at all visible skin areas. Gram negative rods, mainly *Acinetobacter*, *Enterobacter*, and *Proteus* spp. can be found in moist regions [133,155]. *Klebsiella* has been reported to colonize hands in a few instances [156].

Of the anaerobic bacteria *Propionibacterium acnes* is the most prevalent organism [134]. It multiplies in skin regions with strong sebum production, but can be found all over the body surface as a transient. Other anaerobic bacteria of the skin are peptococci and *Propionibacterium granulosum* and, in moist regions, *P. avidum* [142,157,158].

The population density of resident skin bacteria ranges between "hundreds and thousands," rather than within "billions" per square centimeter, as sometimes assumed [130,142]. One reason for differing reports in the literature may be that different methods of quantifying the skin flora are used. Nevertheless, from studies with comparable sampling techniques, it is known that intertriginous regions are the most densely populated, and forehead, face, nape, and soles are the next most densely colonized skin regions of the human body [143,144]. The individual population density at a site remains remarkably unchanged over a long time [28,143,159]. Long-term changes occur only with skin diseases and as a consequence of forceful interference with the biocenosis by antibiotics, disinfectants, or antiseptics [132,136,145,160]. The strongest short-term (1–2 h) fluctuations can be observed after intensive contact with water, as with bathing or showering [143,161].

As opposed to former assumptions, the deep layers of the epidermis are not colonized by a microbial flora. Except for the anaerobe propionibacteria, which reside in the infundibula of the sebaceous glands, most skin bacteria colonize only the uppermost layers of the stratum corneum [145,162]. There, they are embedded in the pars disjuncta of keratinized epithelial cells, a mass of lipids and cell detritus [143,157,163]. Probably, the fastest microbial multiplication takes place in the upper parts of the hair follicle, which was also demonstrated by electron microscopy [164]. In contrast, the ducts of eccrine and apocrine sweat glands of the healthy skin are usually sterile [142].

Members of the transient flora are all microorganisms that do not multiply in or on the skin, but are found there by chance after contamination. Therefore, pathogens may also be encountered. Usually, transient bacteria do not survive very long on healthy skin [165]. Although the reasons for this are not completely understood, it is generally accepted that possible mechanisms for the colonization resistance of skin could be explained not only by the inhospitable environment resulting from physicochemical conditions, but also by several active self-defense mechanisms [130,166]. The most important factors for colonization control by the skin are (a) free fatty acids liberated from skin lipids by bacterial metabolism—mainly by *P. acnes*; (b) the low water content of the stratum corneum; and (c) the production of bacteriocins and other antibioticlike inhibitors by the resident flora [130,149,167,168]. In contrast, pH and osmotic conditions are of little importance [148]. From the foregoing, the protective role of the autochthonous skin flora against infection becomes evident [169].

The transient flora is easily removable by mechanical means [179]. A simple handwash with soap and water for 1 min reduces the numbers released from the hands by two to three orders of magnitude [59,171,172]. Rubbing hands with water alone is nearly as effective [170,171].

2. Principles of Assessment

Various approaches are possible to assess the effectiveness of alcohols as hand and skin disinfectants. One may, for instance, try to directly assess the influence of using alcohol for hand disinfection on the ratio of nosocomial infections. This seems to be a difficult undertaking, as the frequency of nosocomial infections is influenced by many factors other than merely hand transmission of pathogens. In addition, it would also be a very uneconomical method of assessment: To demonstrate a statistically significant reduction of hand-transmitted infections by 50% from 2 to 1% (at $\alpha = 10\%$, directional, and a testing power of 90%) one would have to observe approximately 2500 patients in each of the treatment and control groups [173].

One could also assess the efficiency of alcoholic handrubs to prevent the transfer of pathogens by measuring their effect on the frequency of indicator bacteria. Again, this is tedious and prone to be influenced by other factors that are difficult to control.

Another in vivo test can be done by either in-use tests or by employing a model that simulates practical conditions in the laboratory. With both, the reduction of the release of organisms from the hands or skin is assessed. With the former, many variables are difficult to control, whereas with the latter, employing volunteers, the antimicrobial efficiency of alcohols can be measured in a standardized and well-controlled manner. Therefore, several models that make use of this possibility have been described [93,94,153,170,173–182].

3. Test Models

In test models of hand disinfection, hands are more realistic test objects than is animal skin. Therefore, laboratory tests with volunteers are preferred. Furthermore, when testing hand disinfectants, the purpose of the antiseptic measure has to be taken into consideration. If the elimination of transient organisms is important, as for example, with health care personnel trying to prevent transmission of pathogens from one patient to the next, a model with artificial contamination should be used to assess the effect of the disinfectant on the release of transient organisms. This type of handcare is known as "postcontamination treatment of hands" and can be carried out either as a "hygienic handrub" (in German-speaking countries—"hygienic hand disinfection") or as a "hygienic handwash" with disinfectant–detergents. An effect against the resident flora is usually not necessary, or even desired, nor is it a sustained effect. If, however, the suitability of a surgical handrub or scrub is to be tested, the reduction in the release of resident skin bacteria must be assessed. Because a sustained action might also be required under a surgeon's glove, this can simultaneously be tested with the split-hand model (one hand gloved for a specified time), as suggested by Michaud et al. [183].

Test models for either type of hand treatment should assess the number of viable bacteria released from the hands, both before and after treatment, to estimate the "degerming" efficacy of the treatment by using the ratio of both values as a measure. In addition, the test model must allow for individual attributes of the volunteers that may influence the test results (e.g., size of the hands and skin quality), as well as the way the treatment is performed. Therefore, it was suggested by Rotter and co-workers [59,174,179,182] that each volunteer acts as his or her own control by performing a standard disinfection procedure in the same test. If, at the same time, the efficacy of the standard disinfection has been chosen such that it may serve as a break-point for the efficacy of hand treatments under test, then the results of evaluation in different laboratories or with different groups of volunteers are rendered homogeneous [179]. At present, for testing hygienic hand disinfection procedures, only the Vienna model [59,173–175,179], which was adopted by the Austrian [93] and German [94] societies for hygiene and microbiology as an official test, and the Birmingham model [172] compare against a parallel standard disinfection procedure. For testing surgical handrubs or scrubs this principle is included in only the official test models of the Austrian [184] and German [94,181] societies and in a test proposed recently by the Birmingham group [181].

VI. ALCOHOLS USED IN SKIN ANTISEPSIS

Together with the information on the principles of assessing the efficacy of hand disinfectants and skin antiseptics, as outlined in the foregoing, the following explanations are intended to summarize the present knowledge on the efficiency of alcohols in hand and skin antisepsis.

A. Hygienic Hand Disinfection

The objective of hygienic hand disinfection is to render hands safe after contact with dangerous pathogens. This may, possibly, be accomplished by a simple handwash with soap or with a disinfectant–detergent (hygienic handwash). To prevent dissemination of the pathogens during and by the wash process itself, inactivation of microorganisms should take place while they are still on the hands. Therefore, procedures that include rubbing a disinfectant on to the hands (hygienic handrub) are to be preferred. Furthermore, as time is usually scarce for postcontamination treatments, only fast-acting chemicals are suitable.

As can be seen from Table 13, alcohols are very fast-acting agents, at least against bacteria. When rubbed onto artificially contaminated hands they reduce the release of test bacteria within 30–60 s by 2.60–5.78 log, depending on the type of alcohol and the concentration used. Longer exposure times, such as 2 to 4 min, do not greatly improve effectiveness.

Although the results obtained by various authors are comparable when similar test methods are employed, the association of effectiveness with the type of alcohol, the concentration used, and the exposure time are better demonstrated by tests with a homogeneous technique. In Table 14, results obtained with the Vienna model are summarized. All data stem from the author's laboratory, ensuring the greatest technical homogeneity possible. In addition, comparisons can be made on the effectiveness of procedures for hygienic handwash. The table demonstrates that

1. There is a clear order range for the effectiveness of the compared alcohols: propan-1-ol > propan-2-ol > ethanol.
2. The efficacy is also associated with the concentration used.
3. A handrub with povidone–iodine aqueous solution is not significantly less effective than with propan-2-ol 60% (v/v), the performance standard for hygienic handrubs in several European countries.
4. Each of the hygienic handwash procedures tested is significantly less effective than the alcohol standard.
5. Of all hygienic handwash procedures tested, only that with povidone–iodine liquid soap meets the requirement of performing significantly better than soft soap, the control treatment [182].

These results clearly demonstrate that, in comparison with other methods of postcontamination hand treatment, alcoholic handrubs are now the most efficient measure to quickly reduce the release of transient flora from the hands. Consequently, it may be inferred that, after contamination, alcohol-treated hands are less likely to transfer bacteria than soap-washed hands. Indeed, this has been shown by Ehrenkranz and Alfonso [193], who demonstrated that, after contact with heavily colonized patients' groins, handwashing with soap failed to prevent the transfer of

Table 13 Efficiency of Short-Chain Alcohols to Reduce the Release of Various Test/Bacteria From Artificially Contaminated Hands at Various Concentrations and Exposure Times in Hygienic Handrubs[a]

Alcohol	Conc. (%v/v)	Test/Bacterium	Mean log reduction Exposure time (min) 0.5	1.0	2.0	4.0	Ref.
Ethanol	60	*Escherichia coli*		3.8			92
	70	*E. coli*	2.6–2.9				186
			3.4	4.1			177
			3.6	3.8	4.5		92
				4.0			179
				4.3	4.9	5.4	171
				4.3	5.1		185
		Staph. saprophyticus	3.5				177
				4.0			187
		Staph. aureus	2.6				188
			3.7				177
	80	*E. coli*		4.5			92
Propan-2-ol	50	*E. coli*	3.4	3.9	4.4		92
	60	*E. coli*		4.0			182
				4.2			189
				4.3			190
				4.4			179
				4.7			185
		Serratia marcescens		4.1			191
	70	*E. coli*	3.4–3.5				186
				4.8			192
				4.9			92
Propan-1-ol	40	*E. coli*		4.3			92
	50	*E. coli*	3.7	4.7	4.9		92
				5.0			92,179
	60	*E. coli*		5.5			92
	100	*E. coli*		5.8			91

[a]Tested with comparable methods.

aerobic gram-negative bacilli by health care workers' hands to urinary catheters in 11 of 12 experiments, but when isopropanol (propan-2-ol) 70% (v/v) was used, transfer occurred in only 2 of 12 experiments. Furthermore, soap failed to prevent subsequent colonization in each of the 12 experiments, alcohol in 5. The risk ratio was calculated with 2.4 (95% confidence interval: 1.2–21.7). From this, the authors concluded that soap was generally ineffective in preventing hand transfer

Table 14 Comparison of the Effectiveness of Various Degerming Procedures to Reduce the Release of *E. coli* from the Artificially Contaminated Hands as Assessed by the Vienna Model

Procedures (1 min each)	Conc. %	Mean ($N = 12–15$) log reduction	Remarks
Hygienic handrub with			
Propan-1-ol	100	5.8	
	60	5.5	
	50	5.0	
	40	4.3	
Propan-2-ol	70	4.9	
	60	4.0–4.3	Standard
	50	3.9	
Ethanol	80	4.5	
	70	4.0	
	60	3.8[a]	
Povidone–iodine sol	1 w/w	4.0	
Hygienic handwash with			
Povidone–iodine liq. soap	0.75 w/w	3.5[a,b]	
Chlorhexidine gluconate liq. soap	4.0 w/w	3.1[a]	
Triclosan liq. soap	0.1 w/w	2.9[a]	
Phenol derivatives liq.soap	2.0 w/w	2.6[a]	
Soft soap Eur. Pharmacop.	20 w/v	2.7[a]	Comparative basis

Con., concentration in % v/v unless otherwise stated.
[a]Significantly less effective than standard for hygienic handrub.
[b]Significantly more effective than comparative basis for hygienic handwash.
Source: Refs. 59, 182.

of gram-negative bacilli to catheters following contact with a heavily contaminated source, whereas alcohol was generally effective.

Except for their tuberculocidal activity, the bactericidal effectiveness of alcohols on the skin does not greatly differ among the bacterial species [64,177]. Ayliffe et al. [177], for instance, found the following log reductions with 70% ethanol: *E. coli* 3.4; *S. epidermidis* 3.5; and *S. aureus* 3.7. Rotter and Koller [65], using propan-2-ol 60% (v/v) observed, in one experiment, log reductions for *S. aureus* of 4.9 and for *E. coli* of 4.8.

As reported previously, alcohols are tuberculocidal. The range of order describing their antibacterial activity is the same as that for other bacteria. Possibly, longer exposure times are required [84]. Unfortunately, exact information on practical conditions is lacking as, to our knowledge, experiments with *Mycobacterium tuberculosis* or *M. bovis* on the hands have not yet been carried out. Models with other mycobacteria, such as *M. aquae* [194] and *M. avium*, *M. terrae*, or *M. smegmatis* [195], have been proposed. Present recommendations for specific methods of hand disinfection, however, are based on results of in vitro experiments. Usually, they include a disinfection time of 1–5 min with 70% ethanol, 60–70% propan-2-ol, or 50–70% propan-1-ol [65].

The antiviral activity of alcohols in vivo has been studied on the model of artificially contaminated hands or fingertips by several groups. One finding of these studies was that the effectiveness of alcohols against some "difficult" viruses, such as entero- and rotaviruses is significantly better than that of handwashing with soap and water [196,199]. Steinmann and co-workers [196] have shown, for instance, that disinfection with absolute ethanol or with propan-1-ol reduces the release of poliovirus from the contaminated hands approximately 100 times more efficiently than a social handwash with soap and water, a postcontamination procedure generally recommended (Table 15). It is notable, however, that the alcoholic handrubs tested take 10 min. Absolute ethanol, which produced a reduction of viral release of 3.2 log, was more effective than 80% (v/v) ethanol (2.2), absolute propan-1-ol (3.0) or absolute propan-2-ol (2.4). In contrast, with an individual handwash of 10–55 s, a reduction of only 1 log was observed.

Schürmann and Eggers [121], who have extensively investigated methods of testing procedures for virus disinfection of hands, found enveloped viruses, such as fowl plague, influenzae A, or the complex vaccinia virus completely inactivated

TABLE 15 Reduction of the Release of Poliovirus Type 1 (Sabin) from Artificially Contaminated Hands by Handrubs[a] with Short-Chain Aliphatic Alcohols and With Social Handwash[b]

Alcohol	Conc. (% v/v)	Mean log reduction[c]
Ethanol	100	3.2
	80	2.2
Propanol-2-ol	100	2.4
Propanol-1-ol	100	3.0
Handwash		1.1

[a]Five consecutive 2-min rubs with 1.4 ml each totaling 10 min with 7 ml.
[b]Free-style handwash with soap and tap water (10–55 s)
[c]log ID_{50} recovered from untreated hands minus log ID_{50} recovered from disinfected hands.
Source: Ref. 196.

on the hands and fingertips when rubbing a commercial preparation of ethanol (90% w/w) plus propan-2-ol (8% w/w) and tetrabrome-6-methylphenol (0.1%) onto the hands (Table 16). Before this, the vaccinia virus had already been shown to be completely inactivated on the skin by 70% propan-2-ol alone, within 5 min [197]. In the foregoing tests, naked viruses were more slowly inactivated, and inactivation was usually incomplete after 10 min. From their results the authors concluded that the high-alcohol–content preparation tested is effective against naked viruses, such as enteroviruses, provided the surrounding conditions (high temperature, large disinfectant/virus volume ratio, low protein load) are favorable [121].

Hendley et al. [198], who investigated the effectiveness of handrubs—liquids or foams—in preventing the transmission of rhinovirus in nasal secretions by direct hand-to-hand contact, found that alcohols were only moderately active when used alone (Table 17). In combination with hexachlorophene or benzalkonium chloride and applied as a foam, ethanol proved more effective, even in a concentration as low as 50% (w/w). The most effective agent, however, was iodine in ethanol or water.

Data about the antiviral activity of alcohols on the skin are also available for the human rotavirus (Table 18) confirming the favorable results of tests in vitro (see Table 12). Propan-2-ol and ethanol, both at 70% (v/v), bring about a reduction

TABLE 16 Reduction of the Release of 11 Viruses from Artificially Contaminated Hands or Fingertips by Handrub[a] or Fingerrub[b] with an Alcoholic Commercial Preparation[c]

	Log reduction[d]			Log reduction[d]	
Virus	Hand	Fingertips	Virus	Hand	Fingertips
Fowl plague	>2.5		Coxsackie B3	1.1	2.9
Influenza A/WSN	>2.5		Coxsackie B4	1.3	3.0
Vaccinia MVA	>1.5		Echo 9 Hill	0.7	2.6
Adeno 5	2.1	>2.3	Echo 9 Barty	1.3	2.2
Polio 1	1.0	2.5	SV 40	0.9	1.8
Polio 2	0.2	0.7			

[a]Five consecutive 2-min handrubs with 1.4 ml each, totaling 10 min with 7 ml.
[b]Five consecutive 2-min fingerrubs with 30 μl/finger, each totaling 10 min with 150 μl/finger.
[c]Commercial preparation of ~90% ethanol, 10% propan-2-ol, and 0.1% tetrabromo-6-methylphenol.
[d]Log virus titer recovered from water-treated hands (fingers) minus log virus titer recovered from disinfected hands (fingers).
Source: Ref. 121.

TABLE 17 Reduction of the Release of Rhinovirus 29 and 39 from Artificially Contaminated Fingertips by Handrubs[a] with Alcoholic or Alcohol-Based Liquids or Foams

	Mean log reduction[b]	
	Rhinovirus	
Preparations and concentrations (%w/w)	29	39
Liquids		
Ethanol, 70%		1.5
Ethanol, 70% in towels		1.8
Ethanol, 79% + propan-2-ol, 10%		2.0
Ethanol, 62% + benzyl alcohol, 3%		2.0
Ethanol, 62% + benzyl alcohol, 6%	0.7	
Ethanol, 10% (placebo)	0.1	1.1
Water		0.8
Foams		
Ethanol, 50% + hexachlorophene, 0.23%		2.7
Ethanol, 50% + benzalc. chlor. 0.2%	1.9	2.5
Ethanol, 10% (placebo)	0.6	

[a]1 ml of liquid or 3-cm–diameter ball vigorously rubbed over the hand and allowed to dry completely.
[b]Geometric mean log ID_{50} recovered from fingertips of untreated hands minus geometric mean log ID_{50} recovered from fingertips of disinfected hands.
Source: Ref. 198.

of viral release from contaminated hands and fingers that is approximately 100 times that of tap water [199].

As opposed to their antiviral activity on skin, alcohols are not sufficiently effective against viruses, such as hepatitis A [124], polio [200], or rhinovirus [201] on hard surfaces, such as steel, glass, or tiles, respectively.

B. Surgical Hand Disinfection

The objective of surgical hand disinfection is the removal of all the readily detachable transient flora and as many of the resident flora as possible to reduce the release of skin bacteria from the surgical team's hands to a safe level for the duration of the surgical operation, should the glove be punctured or broken. Therefore, not only a strong immediate, but also a continuing, antibacterial effect is desirable. Some agents without residual activity, such as alcohols may, however, exert an immediate effect strong enough to cause a long delay in regrowth of the skin flora. Consequently, their activity is also long-lasting [202,203].

TABLE 18 Reduction of Human Rotavirus from Artificially Contaminated Fingertips by Combined Handrub/Wash[a] or Fingertip Treatment[b] with Alcoholic or Alcohol-Based Products and Other Agents [199]

Preparations and concentrations (%)[c]	Mean log reduction[d]	
	Hand	Fingertips
Propan-1-ol, 70%	3.0	2.7
Ethanol, 70%		2.7
Ethanol, 70% + CHG, 0.05% + CETR, 0.5%	2.0	2.1
CHG, 0.008% + CETR, 0.075% in water	0.8	0.7
Liquid soap 10	1.2	0.9
Tap water	0.8	0.8

CHG, chlorhexidine gluconate; CETR, cetrimide.
[a]0.5 ml of preparation vigorously rubbed over the hand for 10 s, then washed with 500 ml of warm tap water, dried with paper towel.
[b]Vial containing 1 ml of preparation placed on fingertip and inverted ten times within 10 s followed by scraping fingertip on inside rim of vial and rinsing with 15 ml tap water for 5 s.
[c]Alcohol concentrations (% v/v), other concentrations (% w/v).
[d]Log virus titer recovered from untreated hands (fingers) minus log virus titer recovered from disinfected hands (fingers).
Source: Ref. 199.

Although evidence that surgical hand disinfection is really necessary is lacking from epidemiological data, it is a logical consequence of the concept of asepsis in surgery. Indirect evidence, however, has been produced by Furuhashi and Miyamae [204] demonstrating that 10^3–10^4 colony-forming units of skin bacteria escaped through the pinholes of experimentally punctured gloves during a period of 5 min when hands had been scrubbed earlier without a disinfectant. However, when the hands and forearms had been brushed with a disinfectant, bacterial counts were no more than 100, even 3–4 h after handwashing. As skin bacteria are known to cause infections, especially in operations involving prosthetic surgery, it is sensible to prevent them from entering the surgical wound.

There are three basic techniques for reducing the skin flora: surgical handwash, surgical handrub, and a combination of both, employing them consecutively. The effects of these techniques are markedly different (Tables 19–21).

Whereas, depending on the preparation and the time taken, surgical handwashes with unmedicated soap or disinfectant–detergents cause immediate 0.2–1.3 log reductions in the bacterial release (see Table 19), with handrubs, applying alcohols or a povidone–iodine aqueous solution, log reductions of 0.6–2.9 can be

TABLE 19 Efficacy of Surgical Handwash on Resident Skin Flora from the Hands

Preparation	Conc. (%)[b]	Time (min)	Mean log reduction after surgical handwash[a] Immediate	3 h	Ref.
Unmedicated soap		5	0.4	−0.1	28,35,192
		4	0.2		204
		2	0.06	0.09	202
Povidone–iodine detergent	0.75	6	0.9		204
		5	0.9–1.1	0.2–0.3	35,190,192
		2	0.5–0.6		204,205
Chlorhexidine gluconate detergent	4.0	6	1.3		204
		5	0.9	0.9	35,192
		3	0.8–1.0	0.8–1.0	35,190,192
		2	0.9	1.5–1.6	202,206
Hexachlorophene detergent	3.0	4	0.3	1.0	183
Triclosan detergent	1.0	5	0.6	0.5	35
Benzethonium detergent	10.0	6	1.3		204
		3	1.0		204
		2	0.9		204
Cetrimide detergent	1.0	2	0.4		205

[a]Log pretreatment counts minus log posttreatment counts.
[b]All alcohol concentrations are % v/v, all other agents are % w/v.

achieved (see Table 20). Thus, the minimum and maximum effectiveness of both techniques differ by factors of 2.5–40, respectively. For the range of achievable sustained effects, reductions observed with surgical handwash techniques were −0.1–1.6 log, or by handrubs 0.6–2.7 log; therefore, handrubs were 5–13 times more effective.

With sequential techniques of surgical hand treatment both strong immediate (1.7–2.9 log) and sustained (1.1–3.0 log) effects were observed by several authors (see Table 21).

A more detailed analysis of Table 19 reveals that, in comparison with the poor effects (log reductions 0.06–0.4) of unmedicated soap [28,35,192,202,204] or cetrimide [295], a surgical handwash with a disinfectant–detergent containing povidone–iodine [35,190,192,204,205], chlorhexidine gluconate [35,190,192, 202,204,206], or benzethonium chloride [204] causes stronger immediate effects of 0.5–1.3 log. With hexachlorophene [183] and triclosan [35], as well as with chlorhexidine gluconate-containing detergents, sustained effects, which differ between 0.5 and 1.6 log from the baseline counts, are clearly demonstrated.

Table 20 Efficacy of Surgical Handrub on Resident Skin Flora from the Hands

Preparation	Conc. (%)[b]	Time (min)	Mean log reduction after surgical handrub[a] Immediate	3 h	Ref.
Propan-1-ol	60	5	2.5–2.9	1.6–2.4	192,207,215
Propan-2-ol	70	5	2.4	2.1	208
		3	1.8	1.2	209
		2	1.2	0.8	202
		<1	0.8		213
	60	5	1.7	1.0	190
	50	2	0.7		204
Ethanol	95	2	2.1		210
	80	5	2.5		211
		2	1.5		212
	70	5	2.0	1.2	35
		2	0.6–1.0	0.6	202, 204
Povidone–iodine sol.	1.0	5	1.9	0.8	189
		2	1.0	0.4	202
Ethanol + chlorhexidine gluconate	70 0.5	2	1.0–1.4	0.7	202,204
Ethanol + chlorhexidine gluconate	90 0.5	2	1.1		204
Propan-2-ol + chlorhexidine gluconate	70.5	5	2.5	2.7	208
		4	1.7	1.1	207
		2	1.5	1.0	202
		<1	0.8	1.2	213

[a]Log treatment counts minus log posttreatment counts.
[a]All alcohol concentrations are %v/v, all other agents are %w/v.

Of the alcohols tested for effectiveness in the surgical handrub, propan-1-ol has long been recognized as the most efficient [11,22,58,90]. This has also been demonstrated by more recent studies [192,207,215]. A 5-min rub with this alcohol at 60% (v/v) can be expected to reduce the release of skin bacteria by 2.5 to 2.9 log. Because of this strong immediate activity, the difference from pretreatment counts is still high (1.6–2.4 log) after 3 h. Propan-2-ol is also a powerful agent for handrubs, but less active than its isomeric alcohol. Ethanol is the least active of the three, but is still more effective than any of the disinfectant–detergents shown in Table 19.

The only nonalcoholic agent with similar efficacy and acceptable skin tolerance is an aqueous solution of povidone–iodine. Its effect is comparable with that of 60% (v/v) propan-2-ol, and similarly it has no residual effect [189,202].

TABLE 21 Efficacy of Sequential Surgical Hand Treatments on Resident Skin Flora

Preparation	Conc. (%)[b]	Time (min)	Mean log reduction after sequential treatment[a] Immediate	3 h	Ref.
—[c]	—	—	2.6	1.1	214
Propan-1-ol[d] (comparison)	60	5			
Liquid soap[c]		2	1.8	1.1	214
Propan-1-ol[d]	60	5			
Liquid soap[c]		3	1.7	1.1	207
CHG in propan-2-ol[d]	0.5 + 70	4			
CHG in detergent[c]	4.0	3	2.5	1.7	207
CHG in propan-2-ol[d]	0.5 + 70	4 (5)	(2.9)	(2.2)	
CHG in detergent[c]	4.0	3	2.5	3.0	204
CHG in ethanol[d]	0.5 + 90	1			
Benzethonium chloride in detergent[c]	10.0	3	2.5	3.0	204
Benzethonium chloride in ethanol[d]	0.5 + 90	1			
Benzethonium chloride in detergent[c]	10.0	3	2.1	3.0	204
CHG in water[d]	0.02	1			

[a]Log pretreatment counts minus log posttreatment counts.
[b]All alcohol: concentrations are %v/v, all other agents are %w/v. CHG, chlorhexidine gluconate.
[c]1st phase: scrubbing with c.
[d]2nd phase: rubbing on d.

Mixtures of alcohols and agents with a real sustained action, such as chlorhexidine gluconate, provide the rapid effect of the former and the residual activity of the latter. Acceptance of some preparations, however, may be a limiting factor because of alleged skin damage.

The principle of combining the effects of both groups of agents can also be applied by using them consecutively in two phases. From the work of Lilly and associates [216], it is known that the alcohol must always follow the antiseptic handwash to obtain best results; but, from Table 21, it is evident that a handwash with unmedicated soap directly preceding the application of alcohol may impair the effectiveness of the latter [214]. Otherwise, two-phase surgical hand treatments ensure strong immediate and long-lasting activity if suitable preparations are chosen [204,207]. Disadvantages may be the longer duration of the total procedure and a higher risk of adverse skin reactions.

An indication for a combined surgical handwash–rub is probably the first hand preparation for the day's surgery, and when a handwash is necessary or desired between and during surgical operations (e.g., when gloves are changed), a surgical handrub with alcohol is the method of choice for both its effectiveness and time-saving.

To summarize the knowledge on the suitability of alcohols for surgical hand disinfection, it seems evident from the foregoing that alcohols are now the most effective chemicals in reducing the release of skin flora from the hands of the surgical team.

C. Skin Disinfection

The aim of skin disinfection is the quick removal and killing of as many of the skin flora as possible at the site of a planned invasive procedure, to prevent their translocation into the underlying tissues by a knife blade or a needle. This measure should be distinguished from procedures intended to treat skin areas for colonization with potential or true pathogens, a topic that will not be covered in this chapter.

In terms of strategies, preoperative skin disinfection differs from hand disinfection in that the aim is to hit all microorganisms in and on the skin, regardless of how easily they are released.

Thus, their actual number is of interest, rather than the number of detachable microorganisms. This causes a technical problem for their representative enumeration, unless biopsy techniques are used. Selwyn and Ellis [149] have shown that the sensitivity of most other sampling procedures was 0.5% and less than that of their biopsy method. Unfortunately, taking skin biopsies cannot be considered a routine sampling technique in tests for skin disinfectants. Therefore, many authors employed variations of the cylinder scrub method [147,218,219,220]. Others, such as Lowbury and the Birmingham group [205,210,216,221,222] used the skin of the hands and tested for effectiveness against the resident flora there. Evans and Stevens [157], investigating the flora of normal skin and the efficacy of alcohol disinfection of the antecubital fossa, the palm, and the forehead, produced their results by the combined use of the cotton swab and scrub methods [157,159,223]. Also, when using a combination of two consecutively employed sampling techniques—a modification of the detergent scrub method of Williamson and Kligman [220] and the cyanoacrylate method of Holland et al. [224]—Hartmann found a way to assess the bacterial population density in two skin compartments of different depth [225,226] and states that his technique could replace skin biopsies. Leyden and co-workers [227] proposed to test topical antimicrobial agents on the skin of the forearm after stimulating bacterial multiplication by an occlusive dressing.

From Table 22 it can be determined that on the finger surface, including the nails and the subungual spaces, a 2-min rub with 70% (v/v) ethanol produces a slight, but definite, reduction of resident flora by 0.5 log, when compared with a

TABLE 22 Efficacy of Various Treatments on Skin in Reducing the Content of Skin Bacteria at Various Body Sites and Assessed with Different methods

			Mean log reduction								
			Aerobic skin flora					Propionibacterium			
Site	Preparation	Conc. (%)[a]/ time (min)	Hand[b]	Biop	Swab	Scrub	CAM	Swab	Scrub	CAM	Ref.
Hand	Liquid soap	—/2	0.03								210
	Ethanol	70/2	0.5								153
	+ iodine	1	0.7								153
	+ CHG	0.5	0.7								153
	Ethanol	95/2									210
	+ CHG	0.5	1.6								
	Lugol	/2	0.5								153
	CHG aqueous	0.5/2	0.4								153
			0.2								221
Operation site	Ethanol	70/0.75									
	+ iodine	1.5		1.3		2.4					149
	+ CHG	0.5		0.9		2.2					149
Palm	Ethanol	70/4			2.3			3.0			223
Forehead	Ethanol	70/4			1.3	0.5		0.9	0.3		223
	Propanol-2	60/5				0.8	0.4		0.5	−0.1	226
	Propanol-1	60/5				2.1	1.4		0.6	0.4	226
	Povidone–iodine solution	10/5				1.8	0.4		1.3	0.3	226

[a]All alcohol concentrations are %v/v, all other agents are %w/v.
[b]Bowl-sampling of fingers [222].
CAM, cyanoacrylate-sampling; Biop, biopsy; CHG, chlorhexidine gluconate.

handwash with liquid soap [153,210]. This result can be slightly improved, by the addition of iodine (1%) or chlorhexidine gluconate (0.5%), to 0.7 log [153]. However, only an increase in alcohol concentration to 95% led to an augmentation of effectiveness worth mentioning, namely, to 1.6 log [210]. With nonalcoholic antiseptic solutions, such as Lugol's or an aqueous chlorhexidine gluconate (0.5%) solution, only marginal antimicrobial effects were observed [153].

These results cover only the effects of antiseptics on the skin of a very small and sebum-free part of the human body. Therefore, they may not be representative for regions typical for injections or for surgical operations. Furthermore, in these studies, the effect on the bacterial counts released from the skin, rather than actually residing there, was assessed. In contrast, Selwyn and Ellis [149] have shown, at various surgical sites, that combinations of 70% ethanol with iodine (1.5%) or chlorhexidine gluconate (0.5%) both act equally well on the surface: as found by scrub-sampling, the respective reductions were 2.4 and 2.2 log. However, biopsy samples lead to the conclusion that ethanol plus iodine causes a more pronounced effect (log reductions: 1.3 vs. 0.9).

In distinguishing between the more superficial flora, detachable by swabbing, and the anaerobic propionibacteria residing in the infrainfundibula of sebaceous glands and detectable by scrubbing, Evans and Mattern [223] showed that 70% ethanol within 4 min may considerably reduce the aerobic flora and propionibacteria from the palm. As there are no sebaceous glands, propionibacteria do not multiply on the palms, but exist there as superficial transient flora stemming from other skin regions. On the forehead, however, the deep flora, staphylococci or propionibacteria, were scarcely reduced in number, as demonstrable by the scrub technique (log reductions 0.5 and 0.3, respectively). From the more superficial layers sampled by swabbing, they were reduced by 0.9 and 1.3 log, respectively.

Hartmann's results [226], finally, confirm these findings and what has been said about the antimicrobial effectiveness of various alcohols; namely, that propan-1-ol is more active than propan-2-ol. His investigations also clearly show that disinfection of skin in regions of strong sebum production is difficult and that to attain sterility of the skin is impossible. Furthermore, interestingly, the reduction of more superficial propionibacteria is more readily achieved with povidone–iodine solution than with alcohol. As found with the cyanoacrylate technique, propionibacteria in the depth can seldom, or never, be attacked successfully by a single disinfection attempt. On the contrary, as a consequence of alcohol treatment a high rate of bacterial multiplication has been observed [226]. This was interpreted as a rebound effect attributable to the lipolytic action of the alcohol, which provoked an increased sebum production. However, the presence of salicylic acid and phenol, as contained in the isopropanol base of Fabry's tincture, obviously prevented this phenomenon and led to an increasing reduction of the bacterial skin population with long-lasting effect [226].

To make the best use of alcohols in skin disinfection, they should be applied by rubbing them onto the skin, applying a minimum pressure, since friction increases their antibacterial effect [228]. Whereas, for instance, chlorhexidine gluconate in 70 or 95% ethanol applied by gauze led to bacterial reductions of 1.0 and 1.2 log, the respective effects with the same preparations using the gloved hand as applicator were 1.9 and 3.0 log. In contrast, alcoholic sprays have relatively little effect [229].

When long-term effects of antiseptic skin care are needed, alcohols may not be the best choice. This applies to both treatment for carrier status or for colonization with unwanted or pathogenic organisms and for prevention of these conditions. For the prevention of vascular catheter-related septicemias in an intensive care unit, Maki and co-workers [230] compared the effect of three antiseptics for disinfection of patients' central venous and arterial catheter sites before insertion and for site care every other day. The use of aqueous chlorhexidine gluconate (2% w/v) was associated with the lowest incidence of local infection (2.3%), whereas with 70% (v/v) propan-2-ol and povidone–iodine solution, infections occurred in 7.1 and 9.3%, respectively; catheter-related bacteremias were seen in 0.5 versus 2.3 and 2.6%, respectively. These findings indicate that for this indication, the strong residual action of chlorhexidine gluconate on the skin flora in the vicinity of the catheter insertion site may be more important than the undoubtedly stronger immediate effect of alcohol, which does not possess residual bacteriostatic activity.

Alcohols for preoperative skin disinfection and for use before injections must be free of bacterial spores. This is usually achieved by filtration. In some countries it is routine to add 0.5% hydrogen peroxide to ethanol. This is reported to inactivate bacterial spores in the alcohol within 24 h [75,231].

To kill spores on the skin surface, for instance, before amputation of a malperfused leg, skin disinfection has been proposed with ethanol or propan-1-ol, with added 0.2% peracetic acid. This was reported to augment considerably both virucidal and sporicidal activity of alcohols in vivo [232]. On cadaver skin, 0.2% peracetic acid in 33% propan-1-ol has reduced spores of *Clostridium perfringens* by 3.0 log within 2 min [233].

VII. TOXICOLOGICAL ASPECTS

If disinfectants are used according to their intended purpose, only contact and inhalatory exposure need be considered. Oral toxicity of alcoholic disinfectants or antiseptics may be a problem in countries where alcohol (ethanol) is prohibited by religion or law, or when it is used with suicidal intent.

For most alcohols used in disinfection, standards exist for occupational exposure limits. Some of these are listed in Table 23.

TABLE 23 Occupational Exposure Limits of Some Standards for Various Alcohols

Alcohol	Type of limit	Concentration (ppm)	(mg/m³)		Standard: OSHA	ACGIH	NIOSH	DFG
Methanol	TWA/MAK	200	262		s	s	s	s
	STEL	250	328		*	*	*	
	MAK-ST	400	524	(30 min, 4×)				*
Ethanol	TWA/MAK	1000	1880		*	*	*	*
	STEL	2000	3760	(60 min, 2×)				*
Propan-1-ol	TWA	200	492		*	s	s	
	STEL	250	614		*	*	*	
Propan-2-ol	MWA/MAK	400	983		*	*	*	*
	STEL	500	1230		*	*	*	
	MAK-ST	800	1966	(30 min, 4×)				*
Butan-1-ol	MAX	50	150		s	s	s	
	TWA/MAK	100	300					*
	STEL	200	600	(5 min, 8×)				*
Butan-2-ol	TWA/MAK	100	303		*	*	*	*
	STEL	150	455				*	
	MAK-ST	200	606	(5 min, 8×)				*
Butanol, tert.	TWA/MAK	100	303		*	*	*	*
	STEL	150	455		*	*	*	
	MAK-ST	200	606	(5 min, 8×)				*
Isobutanol	TWA	50	152		*	*	*	
Pentan-1-ol	No limits				Established			
Pentan-2-ol	No limits				Established			
Pentan-3-ol	TWA/MAK	100	361			*		
	STEL	125	452			*		
Isoamyl alcohol	TWA/MAK	100	361		*	*	*	*
	STEL	125	452		*	*	*	
	MAK-ST	200	722	(30 min, 4×)				*

OSHA, Occupational Safety and Health Administration
ACGIH, American Conference of Government Industrial Hygienists
NIOSH, National Institute for Occupational Safety and Health
DFG, Deutsche Forschungsgemeinschaft
TWA, Time-weighted average
STEL, short-term exposure limit
MAX, ceiling concentration
MAK, maximum working place concentration
MAK-ST, MAK short-term

A. Percutaneous Absorption

Percutaneous absorption of alcohols does occur, but it is very small. It increases with growing chain length of the alcohol. Whereas, with ethanol and propanol, penetration of mouse or guinea pig skin was not detected by the methods employed, increased penetration was measured with higher alcohols [234]. With human abdominal skin, the following permeability constant (cm/h) were established in vitro [235]: water, 10^{-3}; methanol, 10^{-3}; ethanol, 1.2×10^{-3}; and propan-1-ol, 1.4×10^{-3}. Thus, the short-chain alcohols used for disinfection of hands and skin behave like water. The permeability of human skin for water was measured to be 0.2 mg/cm^2 h^{-1} [235].

Methanol may be absorbed through the skin. The consequences would be similar to those observed after ingestion; namely, metabolic acidosis and effects on sight (temporary or permanent blindness) and on other functions of the central nervous system.

Ethanol can also be absorbed through the skin. In mammals, however, it is quickly degraded so that below 10 mg/L the plasma concentration decreases exponentially [236]. At higher concentrations, the metabolizing enzyme systems are saturated. Consequently, the alcohol is eliminated at a rather constant rate of 0.5 mg/L min^{-1} [236]. Alcohol-soaked (200-ml) dressings placed on both legs in humans did not cause blood levels that were detectable after 3 h [237]. Therefore, the amount and rate of absorption must have remained well below the respective values for enzyme saturation and for elimination of the alcohol.

In very young ($<$ 1 year) children, however, abdominal alcohol dressings were reported to have caused intoxication [238]. However, 25 of the 28 children involved had skin lesions; therefore, it cannot be excluded that the alcohol at least partially penetrated the skin through these lesions.

As can be seen from Table 24, the acute toxicity of ethanol, propan-2-ol, and propan-1-ol—alcohols that are actually used on the skin as disinfectants, as antiseptics, or in cosmetics—is very low. Taken together with the small absorption rate through intact skin, it may be assumed, therefore, that toxicological arguments are of little importance in decisions concerning the choice of these alcohols for the foregoing purposes. And, indeed, this has been the general experience ever since these alcohols have been used by health care personnel, usually many times a day for many years. Contact with large areas of diseased or lesioned skin, however, should be permitted only with caution.

B. Inhalation Toxicity

According to Lester and Greenberg [239], ethanol is tolerated by the human respiratory tract without any reaction at 100 mg/m^3, whereas 300 mg/m^3 causes permanent irritations, and 400 mg/m^3 is intolerable. To give a picture of the whole range of harmful inhalatory exposure to alcohols, lethal concentrations (LCLO)

TABLE 24 Acute Oral Toxicity (Rat) of Alcohols

Alcohol	LD_{50} (g/kg)
Methanol	5.6
Ethanol	10.4–18.0
Propan-1-ol	1.9
Propan-2-ol	4.8–5.8
Butan-1-ol	0.8
Pentan-1-ol	2.2
Hexano-1-ol	0.7
Heptanol-1-ol	0.5
Benzyl alcohol	1.2

Source: Refs. 240, 251.

for mice are given as examples [240]: methanol, 50 g/m^3; ethanol, 39 g/m^3; and propan-2-ol, 31.5 g/m^3. As can be deduced from these values (a) toxicity increases with the growing chain length of the alcohol, and (b) lethal effects are observed only with enormously high concentrations of these alcohols.

Short-chain alcohols evaporate easily when used on the skin surface. Therefore, some inhalatory exposure is unavoidable. The following calculation may give an estimate of the effect of this exposure. Ten milliliters of 80% (w/w) ethanol consists of 6.3 g alcohol. When this amount has been rubbed onto the hands, has evaporated, and is completely distributed in a room of 10 m^3, the minimum volume usually required for a working person, the alcohol concentration will then be 0.63 g or 630 mg/m^3. Comparing this value with the occupational exposure limit (see TWA and MAK-values in Table 23), which is 1880 mg/m^3, it is evident that inhalation toxicity need not be considered as a problem in using ethanol for hand or skin disinfection. With 70% propan-2-ol, the corresponding values are 550 mg/m^3 and a TWA/MAK of 983 mg/m^3. Propan-1-ol 60% (w/w) produces a concentration of 480 mg/m^3, thus nearly reaching the TWA-limit of 492 mg/m^3 (OSHA). Because of natural or artificial ventilation, however, the estimated alcohol concentrations will be of only short duration. Therefore, the comparative basis should be the STEL/MAK-ST, rather than TWA/MAK values. For propan-1-ol the STEL is 614 mg/m^3.

Similar to absorption through the skin, the uptake of alcohol by the lungs is small. A concentration of 0.5 g/m^3 inhaled during 10 min causes an alcohol uptake of 10–100 mg. The blood level remains far below 0.1% [239]. Only concentrations of 15 g/m^3 inhaled for several hours at a breathing volume of 15 L/min will produce alcohol blood levels of 0.8–0.9% [239]. Thus, toxic effects by the inhalation of alcohol from disinfection of hands and skin with alcoholic solutions can reasonably be expected not to occur, as the amounts of alcohol employed are far below the thresholds of toxicity [54,240].

C. Acceptability

The assessment of acceptability of hand disinfectants and antiseptics is still a problem of methodology. Among other reasons, the small extent of objectively quantifiable changes in skin condition, the paucity of techniques for objective assessment, and the possible bias of prejudice in self-assessment may be listed. Therefore, the results of relatively few studies are available [35,217,241–243]. Usually, both methods of assessment, subjective and objective, are employed. The latter may be carried out by a dermatologist or by biophysical measurements of parameters, such as transepidermal water loss [242–244], skin temperature and skin blood flow [245], or the change in desquamation of the stratum corneum [246] as a consequence of skin disinfection. Another approach is the examination of silicone rubber templates of the skin by computer-assisted microimage analysis of the shadows generated from the skin profile by oblique illumination [247]. Because of the usually very slight deterioration of skin condition, the sensitivity of assessment by a dermatologist has not been found satisfactory [35,217,242,243]. In contrast, the results of semiquantitative self-assessment proved to be a sensitive and reliable indicator of product acceptability [35,217]. Evaluation of the acceptability of frequent application of alcohols on the hands, "intactness" (judging the roughness of skin), "turgor," and "sensation" proved to be very sensitive markers [217], whereas "appearance" was unaffected and, therefore, not usable as a distinguishing feature. To eliminate bias, however, an intelligent experimental design and blinding of the test subjects are of utmost importance. If it is not possible to conceal the identity of the test and placebo products from the subjects, objective methods of assessment are a better choice.

The most common effects of brief exposure of skin to alcohols are signs and symptoms of mild irritation, such as slight redness and sensations of heat and burning owing to removal of the water–lipid layer and dehydration, followed by toxic damage to living cells of the epidermis by the alcohol [248]. Symptoms of allergic contact dermatitis, such as contact urticaria at the exposed site, are extremely seldom [240,248]. Extended, repeated exposure to high concentrations of alcoholic solutions may cause drying of the skin, with roughness and reddening. In a comparative study involving daily surgical scrubbing for 5 consecutive days, Larson et al. [35] found preparations of unmedicated soap and of a triclosan-containing detergent rated as the mildest. Ethanol (70%) plus chlorhexidine gluconate (0.5%) was less harsh than a povidone–iodine liquid soap, but not as mild as a chlorhexidine detergent. Leyden [249] compared the acceptability of propan-1-ol and propan-2-ol with that of povidone–iodine soap. On the whole, he found that alcohols caused skin irritation more often than soap. The climate seems to play an important role, however. With low air temperatures and humidity, alcoholic handrubs seem to be better tolerated than the otherwise allegedly more accepted chlorhexidine gluconate preparation [241].

In a prospective, randomized, double-blind study, Rotter et al. [217] demonstrated that the only slight deterioration of skin condition diagnosed after 14 days of frequent (15 times per day) application of a high-concentration (50% + 30% v/v) mixture of propan-1-ol and propan-2-ol was significantly less if the antiseptic contained suitable emollients. Therefore, alcoholic solutions for frequent-use purposes, such as hand disinfection, should not be employed without emollients, thereby preventing skin problems and ensuring compliance of usage.

REFERENCES

1. A. Leo, C. Hansch, and D. Ilkins, Partition coefficients and their uses, *Chem. Rev. 71*:525 (1971).
2. A. Beyer, Alkoholdesinfektion, *Z. Hyg. Infektionskr. 70*:225 (1911).
3. P. B. Price, Reevaluation of ethyl alcohol as a germicide, *Arch. Surg. 60*:492 (1950).
4. F. Neufeld, and O. Schütz, Untersuchungen über die Sonderstellung des Alkohols als Desinfiziens und über die Rolle der Absorption bei Desinfektionsversuchen, insbesondere bei der Händedesinfektion, *Z. Hyg. Infektionskr. 124*:287 (1941).
5. W. C. Beck, Disinfection from antiquity to the present, *Guthrie J. 59*:191 (1990).
6. W. C. Beck, Benefits of alcohol rediscovered, *AORN J. 40*:57 (1984).
7. M. Dittrich, Geschichte der Antiseptik, *Handbuch der Antiseptik*, Vol. 1, Part 2 (W. Weuffen, A. Kramer, D. Gröschel, G. Berencsi, and E. Bulka, eds.), G. Fischer, Stuttgart, 1982, p. 13.
8. L. Buchholtz, Antiseptica und Bakterien, *Arch. Exp. Pathol. 4*:1 (1875).
9. R. Koch, Über Desinfektion, *Mitt. Kaiserlich. Gesundheitsam. 1*:234 (1881).
10. C. S. Stokvis, Alkohol- und Essigsäuretoleranz der Bakterien, *Zentralbl. Bakteriol. Abt. I Orig. 48*:436 (1909).
11. J. Christiansen, Zur Theorie und Praxis der Alkoholdesinfektion, *Z. Physiol. Chem. 102*:275 (1918).
12. F. Neufeld, and O. Schiemann, Über die Wirkung des Alkohols bei der Händedesinfektion, *Z. Hyg. Infektionskr. 121*:312 (1938).
13. E. H. Spaulding, Studies on the chemical sterilization of surgical instruments, *Surg. Gynecol. Obstet. 69*:738 (1939).
14. M. Witz, Experimentelle Untersuchungen über die entwicklungshemmende und keimtötende Wirkung des Äthylalkohols auf aerobe und anaerobe Sporenbildner unter besonderer Berücksichtigung menschenpathogener Anaerobier, *Arch. Hyg. 124*:243 (1940).
15. M. Seagesser, Die Gasbrandinfektion nach Injektionen, *Schweiz. Med. Wochenschr. 71*:552 (194).
16. H. E. Morton, The relationship of concentration and germicidal efficiency of ethyl alcohol, *Ann. N. Y. Acad. Sci. 53*:191 (1950).

17. R. W. Kolb, E. Schneiter, E. P. Floyd, and D. H. Byers, Disinfective action of methyl bromide, methanol and hydrogen bromide on anthrax spores. *Arch. Ind. Hyg. Occup. Med. 5*:354 (1952).
18. W. Adam, and G. Marcy, Untersuchungen über die Sporenwirksamkeit von Hautdesinfektionsmitteln, *Dtsch. Med. Wochenschr. 98*:2003 (1973).
19. P. Fürbringer, *Untersuchungen und Vorschriften über die Desinfektion der Hände des Arztes nebst Bemerkungen über den Bakteriologischen Charakter des Nagelschmutzes*, J. F. Bergmann, Wiesbaden, 1888.
20. C. Harrington, and H. Walker, The germicidal action of alcohol, *Boston Med. Surg. 148*:548 (1903).
21. V. Russ, Zur Frage der Baktericidie durch Alkohol, *Zentralbl. Bakteriol. Hyg. Abt. I. Orig. 37*:115; 280 (1904).
22. G. Wirgin, Vergleichende Untersuchungen über die keimtötenden und entwicklungshemmenden Wirkungen von Methyl-, Äthyl-, Propyl, Butyl- und Amylreihen, *Z. Hyg. Infektionskr. 46*:49 (1904).
23. G. Kabrhel, Über den Einfluß des Alkohols auf das Keimplasma, *Arch. Hyg. 71*:124 (1909).
24. A. Schinzel, and J. Lingnau, Studien zur Alkoholwirkung auf Mikroorganismen, *Arch. Hyg. 139*:265 (1955).
25. J. Schmidt, G. Naumann, and W. Horsch, *Sterilisation und Entwesung in der Medizinischen und Pharmazeutischen Praxis einschließlich Herstellung von Injektionsarzneien*, Thieme, Leipzig, 1968.
26. E. C. Hanse, Über die tötende Wirkung des Äthylalkohols auf Bakterien und Hefen, *Zentralbl. Bakteriol. Hyg. Abt. I Orig. 45*:466 (1908).
27. G. Sobernheim, Alkohol als Desinfektionsmittel, *Schweiz. Med. Wochenschr. 73*: 1280; 1304; 1333 (1943).
28. P. B. Price, The bacteriology of normal skin; a new quantitative test applied to a study of the bacterial flora and the disinfectant action of mechanical cleansing, *J. Infect. Dis. 63*:301 (1938).
29. P. B. Price, New studies in surgical bacteriology and surgical technique with special reference to disinfection of skin, *JAMA 111*:1993 (1938).
30. P. B. Price, Ethyl alcohol as a germicide, *Arch. Surg. 38*:528 (1939).
31. P. B. Price, Benzalkonium chloride (Zephiran Chloride) as a skin disinfectant, *Arch. Surg. 61*:23 (1959).
32. P. B. Price, Fallacy of a current surgical fad—the 3-minute scrub with hexachlorophene soap, *Ann. Surg. 134*:476 (1951).
33. P. B. Price, Present day methods of disinfecting the skin, *Arch. Surg. 61*:583 (1950).
34. T. C. King, and J. M. Zimmermann, Skin degerming practices: Chaos and confusion, *Am. J. Surg. 109*:695 (1965).
35. E. L. Larson, A. M. Butz, D. L. Gullette, and B. A. Laughon, Alcohol for surgical scrubbing? *Infect. Control Hosp. Epidemiol. 11*:139 (1990).
36. O. Kamm, The relation between structure and physiological action of the alcohols, *J. Am. Pharm. Assoc. 10*:87 (1921).
37. G. Sykes, *Disinfection and Sterilization*, E. & F. Spon, London, 1958.

38. H. E. Morton, Alcohols, *Disinfection, Sterilization and Preservation*, 3rd ed. (S. S. Block, ed.), Lea & Febiger, Philadelphia, 1983, p. 225.
39. K. H. Wallhäußer, *Praxis der Sterilisation-Desinfektion-Konservierung-Keimidentifizierung-Betriebshygiene*, 3rd ed., Thieme, Stuttgart, 1984.
40. G. Sykes, The influence of germicides on the dehydrogenases of *Bact. coli*. I. The succinic acid dehydrogenase of *Bact. coli*, *J. Hyg. Cambr. 39*:463 (1939).
41. S. Dagley, E. A. Dawes, and G. A. Morrison, Inhibition of growth of *Aerobacter aerogenes*. The mode of action of phenols, alcohols, acetone, and ethyl acetate, *J. Bacteriol. 60*:369 (1950).
42. F. Heinmets, W. W. Taylor, and J. J. Lehmann, The use of metabolites in the restoration of the viability of heat and chemically inactivated *Escherichia coli*, *J. Bacteriol. 67*:6 (1954).
43. J. Friedemann, and E. Stahl, Händeschnelldesinfektion mit Hilfe von Desinfektionsmittelkombinationen, *Wiss. Z. Humboldt-Univ. Math. Naturwiss. Reihe. 18*:1153 (1969).
44. A. Todrick, K. P. Fellowes, and J. P. Rutland, The effect of alcohol on cholinesterase, *Biochem. J. 48*:360 (1951).
45. H. Kirschhoff, Wirkungsmechanismen chemischer Desinfektionsmittel. I. Allgemeiner Reaktionsablauf, *Gesundheitwes. Desinfekt. 66*:125 (1974).
46. K. G. Lickfeld, Vergleichende Untersuchungen über bakteriostatische und bakterizide Wirkungen einiger chemischer Verbindungen auf *Pseudomonas aeruginosa*, unter besonderer Berücksichtigung des elektronenoptischen Äquivalenzbildes. IV. Die Wirkung von Laurylsufonat und von Äthylalkohol auf einen Pseudomonas aeruginosa-Stamm, *Arch. Hyg. (Berl.) 152*:468 (1968).
47. B. Pethica, Bacterial lysis. Lysis by physical and chemical methods, *J. Gen. Microbiol. 18*:473 (1958).
48. A. J. Isquith, and W. R. Chesbro, Pools, confluxes and transport of amino acids in *Streptococcus faecium*, *Biochim. Biophys. Acta 74*:642 (1963).
49. M. Craxi, M. G. Ferrara, and F. Ferina, Osservazione sulla resistenza di forme vegetative e cistiche di *Azotobacter* alla azione dell' alcohol, *Bull. Soc. Ital. Sper. 44*:1856 (1968).
50. R. J. V. Pulvertaft, and G. D. Lumb, Bacterial lysis and antiseptics, *J. Hyg. (Lond.) 46*:62 (1948).
51. S. Razin, and M. Argaman, Lysis of mycoplasma, bacterial protoplasts, spheroplasts and L-forms by various agents, *J. Gen. Microbiol. 30*:155 (19).
52. J. Friedemann, and I. Treuhoff, Ein- und mehrwertige Alkohole (einschließlich Glykoläther), *Handbuch der Desinfektion und Sterilisation*, Vol. 1 (W. Weuffen, ed.), Volk und Gesundheit, Berlin, 1972.
53. R. Franke, and A. Kramer, Quantitative structure–activity analysis for antimicrobial agents. *Handbuch der Antiseptik*, Vol. 1, Part 2 (W. Weuffen, A. Kramer, D. Gröschel, G. Berencsi, and E. Bulka, eds.), G. Fischer, Stuttgart, 1982, p. 117.
54. P. Heeg, D. Rehn, and U. Bayer, Alkohole, *Handbuch der Antisepsis*, Vol. 2, Antiseptika, part 3, Antibakterielle, Antifungielle und Antivirale Antiseptik—Ausgewählte Wirkstoffe (A. Kramer, W. Weuffen, A. P. Krasilmikow, D. Gröschel, E. Bulka, and D. Rehn, eds.), Volk und Gesundheit, Berlin, 1987, p. 215.

55. U. P. Kokko, Über die bakterientötende und ihr Wachstum hemmende Fähigkeit des einwertigen gesättigten Alkohols, *Arch. Hyg. (Berl.) 122*:44 (1939).
56. H. Eggensperger, Desinfektionswirkstoffe und ihre Wirkungsmechanismen, *Dtsch. Apoth. Ztg. 113*:785 (1973).
57. F. W. Tilly, and J. M. Schaffer, Relation between the chemical constitution and germicidal activity of the monohydric alcohols and phenol, *J. Bacteriol. 12*:303 (1926).
58. F. W. Tanner, and F. L. Wilson, Germicidal action of aliphatic alcohols, *Proc. Soc. Exp. Biol. 52*:138 (1943).
59. M. Rotter, Hygienic hand disinfection, *Infect. Control 5*:18 (1984).
60. F. W. Tilley, An experimental study of the influence of temperature on the bactericidal activity of alcohols and phenols, *J. Bacteriol. 43*:521 (1942).
61. W. Schürmann, and H. J. Eggers, Antiviral activity of an alcoholic hand disinfectant. Comparison of the in vitro suspension test with in vivo experiments on hands, and on individual fingertips, *Antiviral Res. 3*:25 (1983).
62. J. Friedmann, Zur Bestimmung des Eiweißfehlers von Desinfektionsmitteln, *Z. Gesamte Hyg. 7*:72 (1961).
63. E. C. McCulloch, *Disinfection and Sterilization*, 2nd ed., Kimpton, London, 1945.
64. L. Grün, and U. Heyn, Untersuchungen zur hygienischen Händedesinfektion mit einer Einwirkzeit von 30 Sekunden, *Hosp. Hyg. Desinfekt. 10*:307 (1976).
65. M. Rotter, Händedesinfektion, *Handbuch der Desinfektion und Sterilisation*, Vol. 5 (W. Weuffen, and E. Spiegelberger, eds.), Volk und Gesundheit, Berlin, 1984, p. 62.
66. K. Szerémi, Über die mikrobizide Wirkung verschiedener Alkohole und Alkoholkombinationen, *Wiss. Z. Humbolt-Univ. Berl. Math. Naturwiss. Reihe. 18*:455 (1969).
67. B. Wille, Praktische Laborversuche zur Resistenzentwicklung von Mikroorganismen gegen Desinfektionsmittel, *Hosp. Hyg. 1*:17 (1976).
68. H. Wigert, H. Schwotzer, and I. Feuerpfeil, Zur Resistenzentwicklung von Staphylokokken gegen Desinfektionsmittel, *Z. Gesamte Hyg. 25*:401 (1979).
69. I. Feuerpfeil, H.-J. Dobberkau, H. Schwotzer, and H. Wigert, Zum Resistenzverhalten gramnegativer Bakterienstämme gegen Desinfektionsmittel, *Z. Gesamte Hyg. 28*:781 (1982).
70. A. Kramer, W. Kezdia, G. Lebek, L. Grün, W. Weuffen, A. Poczta, In-vitro und in-vivo-Befunde zur Resistenzsteigerung bei Bakterien gegen Antiseptika und Desinfektionsmittel, *Handbuch der Antiseptik*, Vol. 1, Part 4 (A. P. Krasilnikow, A. Kramer, D. Gröschel, and W. Weuffen, eds.), G. Fischer, Stuttgart, 1985, p. 79.
71. R. Regamey, Les gangrénes gazeuses aprés injéctions medicamenteuses, *Schweiz. Med. Wochenschr. 67*:874 (1939).
72. G. Bernhard, Über Isopropylalkohol als Mittel zur Händedesinfekton, *Dtsch. Med. Wochenschr. 48*:68 (1922).
73. H. M. Powell, The antiseptic properties of isopropyl alcohol in relation to cold sterilization, *J. Indiana State Med. Assoc. 38*:303 (1945).
74. C. E. Coulthard, and G. Sykes, Germicidal effect of alcohol with special reference on bacterial spores, *Pharm. J. 137*:79 (1936).
75. W. Weuffen, R. Hetmanek, and H. Berling, Beitrag zur Eliminierung von Sporen aus

Ethanol, *Mikrobielle Umwelt und Antimikrobielle Massnahmen*, Vol. 8 (A. Kramer, H. Wigert, and B. Kemter, eds.), Barth, Leipzig, 1984, p. 246.
76. A. Kramer, G. Schuster, H. G. Hauthal, W. Weuffen, W.-D. Jülich, H. Höppl, B. Bandemir, and R. Hetmanek, Emulgator E 30–ein oberflächenaktiver Kombinationspartner für antimikrobielle Zubereitungen, *Handbuch der Antiseptik*, Vol. 2, Part 3 (A. Kramer, W. Weuffen, A. P. Krasilnikow, D. Gröschel, E. Bulka, and D. Rehn, eds.), Volk und Gesundheit, Berlin, 1987, p. 423.
77. E. A. Reinicke, Bakteriologische Untersuchungen über die Desinfektion der Hände, *Zentralbl. Gynäkol. 18*:1189 (1894).
78. B. Krönig, Versuche über Spiritusdesinfektion der Hände, *Zentralbl. Gynäkol. 18*: 1346 (1894).
79. G. L. Green, Versuche über die Spiritusdesinfektion der Hände, *Dtsch. Med. Wochenschr. 22*:361 (1896).
80. W. E. Post, and H. N. Nicoll, Comparative efficiency of some common germicides, *JAMA 55*:1635 (1910).
81. L. H. T. Bernstein, Technique for studying the epithelial cells of skin in relation to disinfection, *J. Invest. Dermatol. 67*:265 (1976).
82. M. Frobisher, Jr., and L. Sommermeyer, A study of alcohols on tubercle bacilli and other bacteria in sputum, *Am. Rev. Tuberc. 68*:419 (1953).
83. G. Lockemann, F. Bär, and W. Totzek, Über die keimtötende Wirkung von Alkoholen. *Zentralbl. Bakteriol. Abt. I Orig. 147*:1 (1941).
84. E. Heiler, and M. Heicken, Die Abtötung von Tuberkelbazillen in dünner Auswurfschicht (Händedesinfektion). *Beitr. Klin. Tuberk. 93*:1 (1939).
85. E. Heiler, Die Einwirkung keimtötender Stoffe auf Tuberkelbazillen des Typus humanus und bovinus, I, *Z. Hyg. Infektionskr. 110*:22 (1929).
86. R. Olivio, Ricerche sul potere battericida dell' alcohol denaturata. *Riv. Ital. Igiene 8*:430 (1948).
87. C. R. Smith, Alcohol as a disinfectant against the tubercle bacillus, *Public Health Rep. 62*:1285 (1947).
88. R. Hare, E. Raik, and S. Gash, Efficiency of antiseptics when acting on dried organisms, *Br. Med. J. 1*:496 (1963).
89. G. Stanek, A. Hirschl, and G. Laber, Sensitivity of various spiroplasma strains against ethanol, formalin, glutaraldehyde and phenol, *Zentralbl. Bakteriol. Hyg. Abt. I Orig. B 174*:348 (1981).
90. J. Christiansen, Zur chirurgischen Händedesinfektion, *Med. Klin. 57*:1076 (1956).
91. G. Wewalka, M. Rotter, W. Koller, G. Stanek, Wirkungsvergleich von 14 Verfahren zur Hygienischen Händedesinfektion, *Zentralbl. Bakteriol. Hyg. Abt. I Orig. B 165*: 242 (1977).
92. M. Rotter, W. Koller, M. Kundi, Eignung dreier Alkohole für eine Standarddesinfektionsmethode, *Zentralbl. Bakteriol. Hyg. Abt. I Orig. B 164*:424 (1977).
93. Austrian Society for Hygiene, Microbiology and Preventive Medicine, ed., Richtlinie der ÖGHMP vom. 4. November 1980 zur Prüfung der Desinfektionswirkung von Verfahren für die Chirurgische Händedesinfektion, *Hyg. Med. 6*:10 (1981).
94. German Society for Hygiene and Microbiology, ed., Richtlinien für die Prüfung und

Bewertung chemischer Desinfektionsverfahren—erster Teilabschnitt, *Zentralbl. Bakteriol. Hyg. Abt. I Orig. B 172*:528 (1981).

95. M. L. Tainter, F. H. Throndson, R. R. Beard, and R. J. Wheatlake, Chemical sterilization of instruments, *J. Am. Dent. Assoc. 31*:479 (1944).
96. D. H. Grant, Antiseptic and bactericidal properties of isopropyl alcohol, *Am. J. Med. Sci. 166*:261 (1923).
97. W. Westermann, Desinfektionsversuche mit Isopropylalkohol und Äthylalkohol unter Zusatz eines Seifenpräparates, *Z. Hyg. Infektionskr. 115*:154 (1933).
98. C. H. Shaffer, Jr., L. F. Ortenzio, and L. S. Stuart, Activity of phenol, ethyl alcohol, and isopropyl alcohol in the AOAC official, first action tuberculocidal test, *J. Assoc. Off. Anal. Chem. 50*:765 (1967).
99. H. A. Koch, and Y. Koch, Zur Wirkung von Desinfektionsmitteln auf Schimmelpilze, Dermatophyten und Hefen, *Wiss. Z. Humboldt-Univ. Berl. Math. Naturwiss. Reihe. 18*:1157 (1969).
100. R. H. Kruse, T. D. Green, R. C. Chambers, and M. W. Jones, Disinfection of aerosolized pathogenic fungi on laboratory surfaces. I. Tissue phase, *Appl. Microbiol. 11*:436 (1963); II. Culture phase, *12*:155 (1964).
101. C. W. Emmons, Fungicidal action of some disinfectants on two dermatophytes, *Arch. Dermatol. Syphilol. 28*:15 (1933).
102. K. Loewenthal, The antifungal effect of 70% ethyl alcohol, *Arch. Dermatol. 83*:803 (1961).
103. E. Gildemeister, E. Hailer, and G. Heuer, Das Verhalten des Vakzinevirus gegenüber keimtötenden Stoffen, *Arch. Hyg. 103*:132 (1930).
104. H. Lüthi, The resistance of conidia of *Penicillium* to alcohol, *Mitt. Lebensmitt. Hyg. 45*:26 (1954).
105. E. K. Kuwert, and O. Thraenhardt, Theoretische, methodische und praktische Probleme der Virusdesinfektion in der Humanmedizin, *Immun. Infekt. 4*:125 (1977).
106. M. H. Gordon, Studies of the viruses of vaccinia and variola. Privy Council, Med. Res. Council, Special Reports Series, 98.
107. K. F. Bingel, and C. Hermann, Die experimentelle Desinfektion des Vakzinevirus als Grundlage für die Klinische Pockendesinfektion, *Med. Welt 17*:76 (1966).
108. A. Kewitsch, and W. Weuffen, Wirkung chemischer Desinfektionsmittel gegenüber Influenza- Vaccinia- und Poliomyelitisvirus, *Z. Gesamte Hyg. 16*:687 (1970).
109. K. Großgebauer, Zur Desinfektion der mit Pocken kontaminierten Hand, *Gesundheitswesen Desinfekt. 59*:1 (1967).
110. V. Groupé, C. G. Engle, P. E. Gaffney, and R. A. Manaker, Virucidal activity of representative antiinfective agents against influenza A and vaccinia viruses, *Appl. Microbiol. 3*:333 (1955).
111. M. Klein, and A. Deforest, Antiviral action of germicides, *Soap Chem. Spec. 34*:70; 95 (1963).
112. K. Heicken, Desinfektion bei Pocken, *Ab. Bundesges. Amt. 7*:122 (1963).
113. M. Frobisher, Jr., Properties of yellow fever, *Am. J. Hyg. 11*:300 (1930).
114. M. A. Bucca, The effect of various chemical agents on Eastern equine encephalomyelitis virus, *J. Bacteriol. 71*:491 (1956).

115. O. Brahms, H. Lippelt, and F. Müller, Experimentelle Beiträge zur Frage der Virusinaktivierung, *Zentralbl. Bakteriol. Hyg. Abt. I Orig. 163*:425 (1955).
116. C. H. Cunningham, The effect of certain chemical agents on the virus of Newcastle disease of chickens, *Am. J. Vet. Res. 9*:195 (1948).
117. F. W. Tilley, and W. A. Anderson, Germicidal action of certain chemicals on the virus of Newcastle disease (avian pneumoencephalitis), *Vet. Med. 42*:229 (1947).
118. O. Jaeger, R. Barth, and W. Tusch, Zur Wirksamkeit von Desinfektionsmitteln gegenüber dem Tollwutvirus, *Zentralbl. Bakteriol. Hyg. Abt. I Orig. B 167*:183 (1978).
119. J. Platt, and R. A. Bucknall, The disinfection of respiratory syncytial virus by isopropanol and a chlorhexidine–detergent handwash, *J. Hosp. Infect. 6*:89 (1985).
120. E. W. Schulz, and F. Robinson, The in vitro resistance of poliomyelitis virus to chemical agents, *J. Infect. Dis. 70*:193 (1942).
121. W. Schürmann, and H. Eggers, Antiviral activity of an alcoholic hand disinfectant, comparison of the in vitro test with in vivo experiments on hands, and on individual fingers, *Antiviral Res. 3*:25 (1983).
122. J. B. Kurtz, T. W. Lee, and A. J. Parsons, The action of alcohols on rotavirus, astrovirus and enterovirus, *J. Hosp. Infect. 1*:321 (1980).
123. S. Stockman, and F. C. Minett, Researches on the virus of foot and mouth-disease, *J. Comp. Pathol. 39*:1 (1926).
124. J. M. Mbithi, S. Springthorpe, and S. S. Satar, Chemical disinfection of hepatitis A virus on environmental surfaces, International Symposium on Chemical Germicides, Atlanta, GA, July 1990.
125. W. W. Bond, M. S. Favero, N. J. Petersen, and J. W. Ebert, Inactivation of hepatitis B virus by intermediate-to-high-level disinfectant chemicals, *J. Clin. Microbiol. 18*:535 (1983).
126. H. Kobayashi, M. Tsuzuki, K. Koshimizu, H. Toyama, N. Yoshihara, T. Shikata, K. Abe, K. Mizuno, N. Otomo, and T. Oda, Susceptibility of hepatitis B virus to disinfectants or heat, *J. Clin. Microbiol. 20*:214 (1984).
127. D. Wachtel, H. Mücke, and M. Sprössig, Untersuchungen zur Desinfektion von *Toxoplasma gondii, Dtsch. Gesundheitwes. 24*:1571 (1969).
128. J. P. Duby, N. L. Miller, and J. W. Frenkel, The *Toxoplasma gondii* oocysts from cat feces, *J. Exp. Med. 132*:636 (1970).
129. J. P. Duby, N. L. Miller, and J. W. Frenkel, Characterization of the new fecal form of *Toxoplasma gondii*, *J. Parasitol. 56*:447 (1970).
130. R. C. S. Woodroffe, and D. A. Shaw, Natural control and ecology of microbial populations on skin and hair. *The Normal Microbial Flora of Man* (F. A. Skinner, and J. G. Carr, eds.), Academic Press, New York, 1974, p. 13.
131. G. A. J. Ayliffe, The effect of antibacterial agents on the flora of the skin, *J. Hosp. Infect. 1*:111 (1980).
132. R. R. Marples, and A. M. Kligman, Methods for evaluating topical antibacterial agents on human skin, *Antimicrob. Agents Chemother. 5*:323 (1974).
133. M. J. Marples, The normal microbial flora of the skin, *The Normal Microbial Flora of Man* (F. A. Skinner, and J. G. Carr, eds.), Academic Press, New York, 1974, p. 7.

134. C. A. Evans, W. M. Smith, E. A. Johnston, and E. R. Giblett, Bacterial flora of the normal human skin. *J. Invest. Dermatol. 15*:305 (1950).
135. T. Rosebury, *Microorganisms Indigenous to Man*, McGraw-Hill, New York, 1962.
136. M. J. Marples, *The Ecology of the Human Skin*, Charles C. Thomas, Springfield, IL, 1965.
137. H. I. Maibach, and G. Hildick-Smith, *Skin Bacteria and Their Role in Infection*, McGraw-Hill, New York, 1965.
138. E. J. L. Lowbury, Gram negative bacilli on the skin, *Br. J. Dermatol. 81* (Suppl. 1):55 (1969).
139. M. J. Marples, The normal flora of the human skin, *Br. J. Dermatol. 81* (Suppl. 2): (1969).
140. W. C. Noble, and D. A. Somerville, *Microbiology of Human Skin*, W. B. Saunders, Philadelphia, 1974.
141. M. E. McBride, W. C. Duncan, and J. M. Knox, The environment and the microbial ecology of human skin, *Appl. Environ. Microbiol. 33*:603 (1977).
142. A. M. Kligman, The bacteriology of normal skin, *Skin Bacteria and Their Role in Infection* (H. I. Maibach, and G. Hildick-Smith, eds.), McGraw-Hill, New York, 1965, p. 13.
143. J. A. Ulrich, Dynamics of bacterial skin populations, *Skin Bacteria and Their Role in Infection* (H. I. Maibach, and G. Hildick-Smith, eds.), McGraw-Hill, New York, 1965, p. 219.
144. D. J. Bibel, and D. J. Lovell, Skin flora maps: A tool in the study of cutaneous ecology, *J. Invest. Dermatol. 67*:265–269 (1976).
145. R. R. Marples, and P. Williamson, Effect of systemic dimethyltetracycline on human cutaneous microflora, *Appl. Microbiol. 18*:228 (1969).
146. J. Davies, J. R. Babb, G. A. J. Ayliffe, and S. H. Ellis, The effect on the skin flora of bathing with antiseptic solutions, *J. Antimicrob. Chemother. 3*:473 (1977).
147. J. Davies, J. R. Babb, G. A. J. Ayliffe, and M. D. Wilkins, Disinfection of the skin of the abdomen, *Br. J. Surg. 65*:855 (1978).
148. I. H. Blank, Survival of bacteria on the skin, *Skin Bacteria and Their Role in Infection* (H. I. Maibach and G. Hildick-Smith, eds.), McGraw-Hill, New York, 1965, p. 43.
149. S. Selwyn, and H. Ellis, Skin bacteria and skin disinfection reconsidered, *Br. Med. J. 1*:136 (1972).
150. R. J. Holt, The pathogenic role of coagulase-negative staphylococci, *Br. J. Dermatol. 86* (Suppl. 8):42 (1972).
151. P. R. Skinner, A. J. Taylor, and H. Coakham, Propionibacteria as a cause of shunt and postneurosurgical infection, *J. Clin. Pathol. 31*:1085 (1978).
152. E. L. Larson, K. J. McGinley, J. J. Leyden, M. E. Cooley, and G. H. Talbot, Skin colonization with antibiotic-resistant (JK group) and antibiotic-sensitive lipophilic diphtheroids in hospitalized and normal adults, *J. Infect. Dis. 153*:701 (1986).
153. E. J. L. Lowbury, H. A. Lilly, and J. P. Bull, Disinfection of the skin of operation sites, *Br. Med. J. 2*:1039 (1960).
154. J. N. Bruun, and C. O. Solberg, Hand carriage of gram negative bacilli and *Staphylococcus aureus*, *Br. Med. J. 2*:580 (1973).

155. D. G. Eckert, N. J. Ehrenkranz, B. C. Alfonso, and L. B. Moskowitz, Proteac groin skin carriage among nursing home residents—resistance to antiseptics, *Infect. Control Hosp. Epidemiol. 10*:155 (1989).
156. M. Casewell, and I. Phillips, Hands as route of transmission of *Klebsiella* species, *Br. Med. J. 2*:1315 (1977).
157. C. A. Evans, and R. L. Stevens, Differential quantitation of surface and subsurface bacteria of normal skin by the combined use of the cotton swab and the scrub methods, *J. Clin. Microbiol. 3*:576 (1976).
158. K. McGinley, G. F. Webster, and J. J. Leyden, Regional variations of cutaneous propionibacteria, *Appl. Environ. Microbiol. 35*:62 (1978).
159. C. A. Evans, Persistent individual differences in the bacterial flora of the skin of the forehead: Numbers of propionibacteria, *J. Invest. Dermatol. 64*:42 (1975).
160. M. J. Marples, The normal microbial flora of the skin, *The Normal Microbial Flora of Man* (F. A. Skinner, and J. G. Carr, eds.), Academic Press, New York, 1974, p. 7.
161. P. D. Meers, and G. A. Yeo, Shedding of bacteria and skin squames after handwashing, *J. Hyg. 81*:99 (1979).
162. H.-M. Beetz, Zur Tiefenverteilung der Hautbakterien im Stratum corneum, *Arch. Dermatol. Forsch. 244*:76 (1972).
163. B. M. Gibbs, and L. W. Stuttard, Evaluation of skin germicides, *J. Appl. Bacteriol. 30*:66 (1967).
164. L. F. Montes, and W. H. Wilbourne, Location of bacterial skin flora, *Br. J. Dermatol. 81* (Suppl. 1):23 (1969).
165. J. V. S. Pether, and R. J. Gilbert, The survival of salmonella on finger-tips and transfer of the organisms to foods, *J. Hyg. Cambr. 69*:673 (1971).
166. R. R. Marples, The effect of hydration on the bacterial flora of the skin, *Skin Bacteria and Their Role in Infection* (H. I. Maibach, and G. Hildick-Smith, eds.), McGraw-Hill, New York, 1965, p. 33.
167. R. W. Lacey, V. G. Alder, and W. A. Gillespie, The survival of *Staphylococcus aureus* on human skin. An investigation using mixed cultures, *Br. J. Exp. Pathol. 51*:305 (1970).
168. D. D. Marsh, and S. Selwyn, Studies on antagonism between human skin bacteria, *J. Med. Microbiol. 10*:161 (1977).
169. S. Selwyn, Natural antibiosis among skin bacteria as a primary defence against infection, *Br. J. Dermatol. 93*:487 (1975).
170. E. J. L. Lowbury, H. A. Lilly, and J. P. Bull, Disinfection of hands: Removal of transient organisms, *Br. Med. J. 2*:230 (1964).
171. H. Mittermayer, and M. Rotter, Vergleich der Wirkung von Wasser, einigen Detergentien und Äthylalkohol auf die transiente Flora der Hände, *Zentralbl. Bakteriol. Hyg. Abt. I Orig. B 160*:163 (1975).
172. G. A. J. Ayliffe, J. R. Babb, and A. H. Quoraishi, A test for "hygienic" hand disinfection, *J. Clin. Pathol. 31*:923 (1978).
173. M. Rotter, Are models useful for testing hand antiseptics? *J. Hosp. Infect. 11* (Suppl. A):236 (1988).
174. M. Rotter, H. Mittermayer, and M. Kundi, Untersuchungen zum Modell der künst-

lich kontaminierten Hand—Vorschlag für eine Prüfmethode, *Zentralbl. Bakteriol. Hyg. Abt. I Orig. B 159*:560 (1974).
175. M. Rotter, W. Koller, M. Kundi, G. Wewalka, and H. Mittermayer, Testmethode für die Wertbemessung von Verfahren für die Hygienische Händedesinfektion, *Zentralbl. Bakteriol. Hyg. Abt. I Orig. B 164*:498 (1977).
176. U.S. General Services Administration, O-T-C drugs generally recognized as safe, effective and not misbranded, *Fed. Reg. 43*:1210 (1978).
177. G. A. J. Ayliffe, J. R. Babb, and A. H. Quoraishi, A test for hygienic hand disinfection, *J. Clin. Pathol. 31*:923 (1978).
178. J. Ojajärvi, Effectiveness of hand washing and disinfection methods in removing transient bacteria after patient nursing, *J. Hyg. Cambr. 85*:193 (1980).
179. M. Rotter, W. Koller, G. Wewalka, H.-P. Werner, G. A. J. Ayliffe, and J. R. Babb, Evaluation of procedures for hygienic hand disinfection, controlled parallel experiments on the Vienna test model, *J. Hyg. Cambr. 96*:27 (1986).
180. G. A. J. Ayliffe, J. R. Babb, J. G. Davies, S. W. B. Newsom, C. Rowland, J. H. Platt, and B. Mason, Hygienic hand disinfection tests in three laboratories, *J. Hosp. Infect. 16*:141 (1990).
181. M. L. Rotter, and G. A. J. Ayliffe, Practical guide on rationale and testing procedures for disinfection of hands, *MIM 91.2*:1 (1991).
182. M. L. Rotter, and W. Koller, A European test for the evaluation of the efficacy of procedures for the antiseptic handwash? *Hyg. Med. 16*:4 (1991).
183. R. N. Michaud, M. B. McGrath, and W. A. Goss, Application of a gloved-hand model for multiparameter measurements of skin-degerming activity, *J. Clin. Microbiol. 3*:406 (1976).
184. Austrian Society for Hygiene, Microbiology and Preventive Medicine, Richtlinie der ÖGHMP vom 4. November 1980 zur Prüfung der Desinfektionswirkung von Verfahren für die Chirurgische Händedesinfektion. *Österr. Krankenhausztg. 22*:33 (1981); *Hyg. Med. 6*:10 (1981).
185. W. Koller, M. Rotter, H. Mittermayer, and G. Wewalka, Zur Kinetik der Keimabgabe von der künstlich kontaminierten Hand, *Zentralbl. Bakteriol. Hyg. Abt. I Orig. B 163*:509 (1976).
186. G. A. J. Ayliffe, J. R. Babb, J. G. Davies, and H. A. Lilly, Hand disinfection: A comparison of various agents in laboratory and ward studies, *J. Hosp. Infect. 11*:226 (1988).
187. R. R. Marples, and A. G. Towers, A laboratory model for the investigation of contact transfer of microorganisms, *J. Hyg. Cambr. 82*:237 (1979).
188. H. A. Lilly, and E. J. L. Lowbury, Transient skin flora—their removal by cleansing or disinfection in relation to their mode of deposition, *J. Clin. Pathol. 31*:919 (1978).
189. M. Rotter, W. Koller, and G. Wewalka, Über die Wirksamkeit von PVP-Jod-haltigen Präparationen bei der Händedesinfektion, *Hyg. Med. 5*:553 (1980).
190. M. Rotter, W. Koller, and G. Wewalka, Povidone–iodine and chlorhexidine gluconate-containing detergents for disinfection of hands, *J. Hosp. Infect. 1*:149 (1980).
191. M. Rotter, G. Wewalka, and W. Koller, Einfluß einiger Variablen auf die Ergebnisse von Prüfungen Hygienischer Händedesinfektionsverfahren, *Hyg. Med. 7*:157 (1982).

192. M. Rotter, W. Koller, and G. Wewalka, Eignung von Chlorhexidinglukonat- und PVP-Jod-haltigen Präparationen zur Händedesinfektion, *Hyg. Med. 6*:425 (1981).
193. N. J. Ehrenkranz, and B. C. Alfonso, Failure of handwashing to prevent hand transfer of patient bacteria to urethral catheters, *Infect. Control Hosp. Epidemiol. 12*:654 (1991).
194. H.-G. Sonntag, and S. Niesert, Untersuchungen zur Desinfektionswirkung auf Mykobakterien, *Mitt. Österr. Sanitaetsverwalt. 78*:186 (1977).
195. H.-G. Sonntag, Desinfektionsverfahren bei Tuberkulose, *Hyg. Med. 3*:322 (1978).
196. J. Steinmann, R. Nehrkorn, E. Lösche, E. Sasse, and B. Bogumil-Puchert, Viruswirksamkeit der hygienischen Händedesinfektion, *Hyg. Med. 15*:7 (1990).
197. K. O. Schümann, and K. Grossgebauer, Versuche zur Desinfektion von Vakziniaviren in Kaninchenborken bzw. an der Hand, *Zentralbl. Bakteriol. Hyg. Abt. I Orig. B 164*:45 (1977).
198. J. O. Hendley, L. A. Mika, and J. M. Gwaltney, Jr., Evaluation of virucidal compounds for inactivation of rhinovirus on hands, *Antimicrob. Agents Chemother. 14*: 690 (1978).
199. S. A. Ansari, S. A. Sattar, V. S. Springthorpe, G. A. Wells, and W. Tostowaryk, In vivo protocol for testing efficacy of hand-washing agents against viruses and bacteria: Experiments with rotavirus and *Escherichia coli*, *Appl. Environ. Microbiol. 55*:3113 (1989).
200. R. Tyler, G. A. J. Ayliffe, and C. Bradley, Virucidal activity of disinfectants: Studies with the poliovirus, *J. Hosp. Infect. 15*:339 (1990).
201. J. M. Gwaltney, Jr., and J. O. Hendley, Rhinovirus transmission one if by air, two if by hand, *Am. J. Epidemiol. 107*:357 (1978).
202. E. J. L. Lowbury, H. A. Lilly, and G. A. Ayliffe, Preoperative disinfection of surgeon's hands: Use of alcoholic solutions and effects of gloves on skin flora, *Br. Med. J. 4*:369 (1974).
203. M. L. Rotter, Povidone–iodine and chlorhexidine gluconate containing detergents for disinfection of hands [Letter to the Editor], *J. Hosp. Infect. 2*:273 (1981).
204. M. Furuhashi, and T. Miyamae, Effect of pre-operative hand scrubbing and influence of pinholes appearing in surgical rubber gloves during operation, *Bull. Tokyo Med. Dent. Univ. 26*:73 (1979).
205. H. A. Lilly, and E. J. L. Lowbury, Disinfection of the skin: An assessment of some new preparations, *Br. Med. J. 3*:674 (1971).
206. E. J. L. Lowbury, H. A. Lilly, and J. P. Bull, Disinfection of hands: Removal of resident bacteria, *Br. Med. J. 1*:1251 (1963).
207. M. L. Rotter, and W. Koller, Surgical hand disinfection: Effect of sequential use of two chlorhexidine preparations, *J. Hosp. Infect. 16*:161 (1990).
208. G. Wewalka, M. Rotter, and W. Koller, Wirksamkeit verschiedener Mittel zur Chirurgischen Händedesinfektion und zur präoperativen Hautdesinfektion, *10 Jahre Ludwig Boltzmann-Institut zur Erforschung der Infektionen und Geschwülste* (P. Porpaczy, ed.), H. Egermann, Wien, 1980, p. 9.
209. M. L. Rotter, and W. Koller, Effect of sampling method on the results of evaluating procedures for surgical hand disinfection, 24th congress of the Austrian Society of Hygiene, Microbiology and Preventive Medicine, Salzburg, 17–19 May 1994.

210. H. A. Lilly, E. J. L. Lowbury, and D. Wilkins, Detergents compared with each other and with antiseptics as skin "degerming" agents, *J. Hyg. Cambr. 82*:89 (1979).
211. J. Borneff, Die Testung chirurgischer Händedesinfektionsmittel nach der Handschuhmethode, *Arch. Hyg. 150*:281 (1966).
212. W. A. Altemeier, Surgical antisepsis, *Disinfection, Sterilization and Preservation*, 2nd ed. (S. S. Block, ed.), Lea & Febiger, Philadelphia, 1977, p. 641.
213. R. Aly, and H. I. Maybach, Comparative study on the antimicrobial effect of 0.5% chlorhexidine gluconate and 70% isopropyl alcohol on the normal flora of hands, *Appl. Environ. Microbiol. 37*:610 (1979).
214. M. L. Rotter, and W. Koller, Sequential use of chlorhexidine detergent and alcohol for surgical hand disinfection, Second International Conference of the Hospital Infection Society, London 2-6. September 1990, Poster 0073.
215. P. Heeg, W. Osswald, and N. Schwenzer, Wirksamkeitsvergleich von Desinfektionsverfahren zur chirurgischen Händedesinfektion unter experimentellen und klinischen Bedingungen, *Hyg. Med. 11*:107 (1986).
216. H. A. Lilly, E. J. L. Lowbury, and M. D. Wilkins, Limits to progressive reduction of resident skin bacteria by disinfection, *J. Clin. Pathol. 32*:382 (1979).
217. M. L. Rotter, W. Koller, and R. Neumann, The influence of cosmetic additives on the acceptability of alcohol-based hand disinfectants, *J. Hosp. Infect. 18* (Suppl. B):57 (1991).
218. P. Story, Testing of skin disinfectants, *Br. Med. J. 2*:1128 (1952).
219. E. A. Pachtmann, E. E. Vicher, and M. J. Brunner, The bacteriologic flora in seborrhoic dermatitis, *J. Invest. Dermatol. 22*:389 (1954).
220. P. Williamson, and A. M. Kligman, A new method for the quantitative investigation of cutaneous bacteria, *J. Invest. Dermatol. 45*:498 (1965).
221. E. J. L. Lowbury, Skin disinfection, *J. Clin. Pathol. 14*:85 (1961).
222. E. J. L. Lowbury, H. A. Lilly, and J. P. Bull, Methods for disinfection of hands and operation sites, *Br. Med. J. 2*:531 (1964).
223. C. A. Evans, and K. L. Mattern, The bacterial flora of the antecubital fossa, the efficacy of alcohol disinfection on this site, the palm and the forehead, *J. Invest. Dermatol. 75*:140 (1980).
224. K. T. Holland, C. D. Roberts, K. T. Cunliffe, and C. D. Williams, A technique for sampling micro-organisms from the pilosebaceous ducts, *J. Appl. Bacteriol. 37*:289 (1974).
225. A. A. Hartmann, A comparison of the effect of povidone–iodine and 60% *n*-propanol on the resident flora using a new test method, *J. Hosp. Infect. 6* (Suppl. A):73 (1985).
226. A. A. Hartmann, C. Pietzsch, P. Elsner, T. Lang, H. Hackel, P. Fischer, and T. Bertelt, Antibacterial efficacy of Fabry's tinctura on the resident flora of the skin of the forehead. Study of bacterial population dynamics in stratum corneum and infundibulum after single and repeated applications, *Zentralbl. Bakteriol. Hyg. B 182*:499 (1986).
227. J. J. Leyden, R. Stewart, and A. M. Kligman, Updated in vivo methods for evaluating topical antimicrobial agents on the human skin, *J. Invest. Dermatol. 72*:165 (1979).

228. E. J. L. Lowbury, and H. A. Lilly, Gloved hand as applicator of antiseptics to operations sites, *Lancet 1*:153 (1975).
229. E. J. L. Lowbury, H. A. Lilly, and J. P. Bull, Methods for disinfection of operation sites, *Br. Med. J. 2*:531 (1964).
230. D. G. Maki, M. Ringer, and C. J. Alvarado, Prospective randomised trial of povidone–iodine, alcohol and chlorhexidine for prevention of infection associated with central venous and arterial catheters, *Lancet 338*:339 (1991).
231. W. Weuffen, H. Berling, and R. Hetmanek, Verwendung von Äthanol für Desinfektionszwecke, *Krankenhaushygiene*, 2nd ed. (W. Weuffen, F. Oberdoerster, and A. Kramer, eds.), Barth, Leipzig, 1981, p. 518.
232. M. Sprössig, and H. Mücke, Steigerung der virusinaktivierenden Wirkung von Peressigsäure durch *n*-Propyl-alkohol, *Acta Biol. Med. German. 14*:199 (1965).
233. H. P. Schau, Zum Problem der Gasbrandspritzeninfektion [cited by 54, p. 229] (1973).
234. F. Meyer, and J. Ziegenmeyer, Resorptionsmöglichkeiten der Haut, *J. Soc. Cosmet. Chem. 26*:93 (1975).
235. R. Scheuplein, and L. Ross, Effects of surfactants and solvents on the permeability of epidermis, *J. Soc. Cosmet. Chem. 21*:853 (1970).
236. E. J. Ariens, E. Mutschler, and A. M. Simonis, *Allgemeine Toxikologie*, Thieme, Stuttgart, 1978.
237. R. V. Bowers, W. D. Burleson, and J. F. Blades, Alcohol absorption from the skin in man, *Q. J. Stud. Alcohol 3*:31 (1942).
238. E. R. Gimenez, N. E. Vallejo, E. Roy, M. Lis, E. M. Izurieta, S. Rossi, and M. Cappuccio, Percutaneous alcohol intoxication, *Clin. Toxicol. 1*:39 (1968).
239. D. Lester, and L. A. Greenberg, The inhalation of ethyl alcohol by man, *Q. J. Stud. Alcohol 12*:167 (1951).
240. H.-P. Harke, Zur Toxikologie der in Desinfektionsmitteln als Wirkstoffe eingesetzten aliphatischen Alkohole, *Hyg. Med. 3*:78 (1978).
241. J. Ojajärvi, P. Mäkelä, and I. Rantasalo, Failure of hand disinfection with frequent hand washing, a need for prolonged field studies, *J. Hyg. 79*:107 (1977).
242. P. J. Kolari, J. Ojajärvi, J. Lauharanta, and P. Mäkelä, Cleaning of hands with emulsion—a solution of skin problems of hospital staff? *J. Hosp. Infect. 13*:377 (1989).
243. E. Larson, J. J. Leyden, K. J. McGinley, and G. H. Talbot, Physiologic and microbiologic changes in skin related to frequent handwashing, *Infect. Control 7*:59 (1986).
244. G. E. Nilsson, On the measurement of evaporative water loss. Methods and clinical application, Thesis No 47, University of Linköping, 1977.
245. G. E. Nilsson, T. Tenland, and P. A. Öberg, A new instrument for continuous measurement of blood flow by light beating spectroscopy, *IEEE Trans. Biomed. Eng. 27*:12 (1980).
246. K. J. McGinley, R. R. Marples, and G. Plewing, A method for visualizing and quantitating the desquamating portion of the human stratum corneum, *J. Invest. Dermatol. 53*:107 (1969).

247. P. Corcuff, F. Chatenay, and J. L. Leveque, A fully automated system to study skin surface patterns, *Int. J. Cosmet. Sci. 6*:167 (1984).
248. H. Tronnier, Die Wirkung von Desinfektionsmitteln auf die Haut und Möglichkeiten von Hautschutz und -pflege, *Händedesinfektion*, International Seminar, München, September 1981, Informed, Gräfelfing, 1982, p. 53.
249. J. J. Leyden, Hautirritationen durch Desinfektions-und Waschmitteln, *Händedesinfektion*, International Seminar, München, September 1981, Informed, Gräfelfing, 1982, p. 51.
250. G. Hommel, *Handbuch der Gefährlichen Güter*, 4th ed., Springer, Berlin, 1980.
251. J. F. Threon, Alcohols, *Industrial Hygiene and Toxicology* (F. A. Patty, ed.), 2nd ed., Vol. 2; *Toxicology* (D. W. Fassett, and D. D. Irish, eds.), Interscience Publishers, John Wiley & Sons, New York, 1963.

11

Chlorhexidine

N. S. Ranganathan
Amway Corporation
Ada, Michigan

I. INTRODUCTION

The observations of antimalarial activity in certain substituted diguanides [1] led to the search for other therapeutically useful members of this series. Chlorhexidine (Hibitane) was first synthesized in 1954 in the laboratories of ICI England [2]. The bactericidal activity of chlorhexidine is much greater than that of monomeric biguanides [3]. Recognized as an effective antimicrobial agent for over 30 years, a wide range of antimicrobial products containing chlorhexidine have been developed and marketed in Europe, Canada, Australia, New Zealand, and United States. The products include surgical scrubs, health care personnel handwash, preoperative skin preparations, skin antiseptics, skin cleansers, wound protectants, urinary irritants, burn ointments, acne creams, cosmetic preservatives, ophthalmic solution, and oral products such as mouth rinses. During its 30 years of marketing, the safety and efficacy of chlorhexidine has been widely reported in the medical literature. Because of the extensive nature of the data available, this chapter will attempt to cover only the major areas of interest in relation to chlorhexidine and its role in infection control. In the United States, approval for topical antiseptic use was granted in 1976 by the U.S. Food and Drug Administration (FDA), based on its high level of antimicrobial activity, low mammalian toxicity, and a strong affinity for binding to skin and mucous membranes. In addition, chlorhexidine has been used as a preservative in certain ophthalmic solutions. As the digluconate salt, chlorhexidine has been extensively studied for its antimicrobial properties, its absorption, excretion, metabolism by various routes of administration, and its toxicological properties.

II. CHEMICAL PROPERTIES

The observation of marked bacteriostatic action was made with the mixture of polymeric diguanides by Curd [1] as well as Davies [2]. These observations were made from the fusion of 1,6 di-(N^3-cyano-N'-guanidino)hexane with hexamethylenediamine dihydrochloride. Attempts to determine actual chain lengths, or to separate the mixture into homogeneous fractions, were unsuccessful. However, it is known that molecules carrying only one diguanide residue are weak in antimicrobial potency. A stepwise synthesis of polydiguanides indicated that full biological activity could be obtained when two diguanides are incorporated in to each molecule [1,2]. Further studies indicated that lower aliphatic acids were more soluble, which led to the finding of chlorhexidine diacetate. However, in aqueous solution it gave an odor of acetic acid. Several polyhydroxy acids were tested for their solubility, and the digluconate (from 1,5-D-gluconolactone) was the best [4]. A strength of 20% w/v chlorhexidine gluconate was the most convenient concentration to handle in terms of formulating the agent. Acetate and gluconate salts of chlorhexidine may be used interchangeably in aqueous preparations; however, because of its economy and convenience, the gluconate is much preferred over the acetate. Acetate is a better choice if an alcoholic solution of minimum water content is required. Precipitation may occur if chlorhexidine gluconate solution is added directly to alcohol. When used in oral preparations, it renders an extremely bitter taste, which can be masked by using flavoring agents.

The active ingredient chlorhexidine gluconate has the following structural formula:

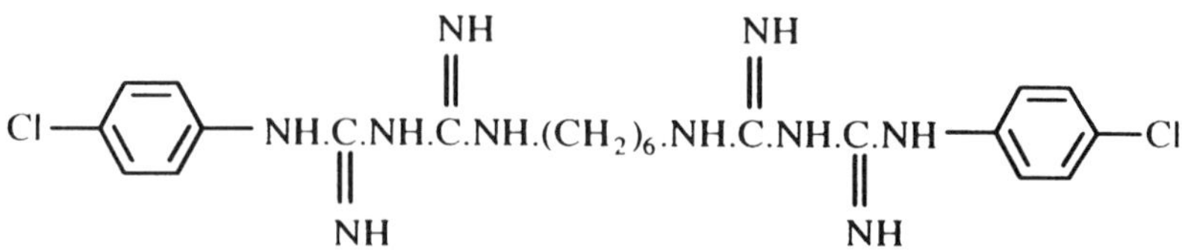

The chlorhexidine molecule possesses strongly cationic groups and, therefore, binds to surfactants, detergents, and other excipients, reducing the availability of the free compound to exert its antimicrobial effect in their presence. Various inorganic anions (chloride, sulfate, carbonate, and nitrate) and organic ions (natural soaps and synthetic detergents) are incompatible with chlorhexidine. Very high concentrations of nonionic surfactants reduce its activity. In certain cases, incompatibility may occur owing to chlorhexidine being incorporated in to micelles, resulting in a reduced antimicrobial activity. Chlorhexidine is moderately surface-active and forms micelles in solution [5]. A pH range between 5 and 8 is ideal for the stability of aqueous solutions of chlorhexidine, whereas, at extreme acidic conditions, there is a gradual decrease in activity. Also, a high alkaline pH (above 9.0) will result in precipitation of the active component. Chlorhexidine hydrolyzes at elevated temperatures (above 40°C), yielding *p*-chloroaniline (PCA). Stability studies show a higher rate of hydrolysis at elevated temperature and pH conditions [6].

III. CHLORHEXIDINE GLUCONATE AND INFECTION CONTROL

Despite advances in hospital design and medical care, serious nosocomial infections remain a major hazard in hospital environments. Their incidence is increasing together with continued high morbidity and mortality. The Communicable Disease Center of the United States Public Health Service estimates that an average of 5% of all patients in a short term, acute care hospital develop nosocomial infections. Of patients having surgery, it has been reported that 2–20% develop wound infections [7]. Meanwhile, the cost of hospital associated infection in terms of mortality, morbidity, and economics continues to mount. Today, nosocomial infections represent approximately 11.5 million additional patient days of hospitalization each year [8].

Recently, the anaerobic *Bacteroides* spp., accounts for well over half of all hospital acquired infections. Gram-negative infections are common, not only among surgical patients, but medical patients as well, and they now account for increasing morbidity. Many epidemiologists and infection control personnel believe that despite the availability of many new antibiotics aimed specifically at treating gram-negative infections, hard to eradicate flora will continue to dominate the hospital environment and be the leading cause of nosocomial infections. To help combat this growing problem, bioengineers have altered the hospital environment to include such innovations as clean rooms, improved ventilating systems, and room layouts aimed at separating operating rooms from other sections of the hospital. Isolating susceptible patients from the general hospital population also reduces the possibility of infection. Presently, the emphasis is on deterring the source of gram-negative infections, which usually can be traced to contact with moisture-retaining or containing objects. Among the main sources of contamination are the hands, sinks, in-dwelling catheters and intravenous lines, and the bodies and clothes of surgical and other hospital personnel.

The initial contamination of skin, mucosal, or wound surfaces with potentially pathogenic bacteria is a fundamental step in the pathogenesis of most hospital-acquired infections. Because the patient's skin and the surgeon's hands harbor a large population of microbial flora and are major factors in transmission of infection, medical research continues to strive for new antimicrobial agents for use as presurgical hand scrubs and washing solutions as skin decontaminants in the preoperative preparation of the patient. Chlorhexidine gluconate formulations provide a rapid bactericidal action, along with a persistent antimicrobial effect against a wide range of organisms. Chlorhexidine formulated in a detergent base is used extensively for disinfection of the hands of surgeons and nurses and also for whole-body skin disinfection of patients undergoing surgery. Alcohol-based chlorhexidine gluconate solutions with emolients are also used by surgeons and nurses for hand disinfection. Alcohol-based chlorhexidine solutions are particularly suitable for final preparation of the operative site. The immediate bactericidal action of chlorhexidine gluconate surpasses that of similar preparations containing povidone–iodine, triclosan, hexachlorophene or *p*-chloro-*m*-xylenol (PCMX). Its valuable persistent (residual) effect, which prevents regrowth of organisms on the skin, is comparable with that of hexachlorophene or triclosan, although chlorhexidine has a broader spectrum of activity, particularly against gram-negative organisms.

IV. MECHANISM OF ACTION OF CHLORHEXIDINE

Chlorhexidine is both inhibitory and lethal to vegetative gram-positive and gram-negative bacteria at relatively high dilutions according to Davies et al. [2]. Heat-resistant spores are not susceptible to its lethal action at ambient temperature, even at high concentrations. At 98–100°C, however, it destroys mesophilic spores; hence, its use in low concentrations (0.01%) in the sterilization of certain eye drops. Laurence [9] found pathogenic fungi to be as susceptible to its inhibitory and lethal action as vegetative bacteria. The susceptibility is in the following ascending order: enteric viruses, bacterial spores, fungal spores, fungi, yeasts, vegetative bacteria [10–12].

Chlorhexidine is rapidly absorbed by bacterial cells and accompanied by other cytological changes that include changes in the permeability of the cells as well as their optical properties. The amount of drug absorption causing maximum leakage of cell constituents and changes of extinctions was equivalent to *Escherichia coli* and *Staphylococcus aureus*. Higher doses of chlorhexidine gluconate, resulting in a higher level of drug absorption, caused correspondingly less leakage and change in extinction, although such higher doses were more rapidly bactericidal. Hugo and Longworth [13] conducted a series of experiments determining drug uptake, leakage of cell constituents, and turbidometric changes designed to elucidate the mode of action of chlorhexidine gluconate. They concluded that the mode of action of the drug is to react with the cell, causing a disorientation of the lipoprotein membrane, by virtue of the lipophilic groups of the drug molecule, so that the membrane no longer fulfills its function as an osmotic barrier. Once this initial reaction has occurred, the subsequent events depend on the concentration of the drug present, which may prevent leakage from the damaged cell by physically sealing the cell, owing to formation of a complete layer, or layers, of drug on the cell surface. This is demonstrated by the inactivation of autolytic enzymes or by denaturation of the cytoplasmic membrane or cytoplasm. Its possible mode of action, referred to previously, of exerting a lethal action by combining with the cell surface, disorganizing permeability barriers, and coagulating the cytoplasmic contents, was thought to be similar to that of quaternary ammonium salts, such as cetrimide [14]. Activity is reduced in the presence of organic matter, such as blood, serum, and milk, although considerable activity still remains even at dilutions down to 0.02%. The adequacy of residual activity in the presence of blood or protein was demonstrated by Calman and Murray [15], who chose chlorhexidine as the best antibacterial agent for use in obstetrics, basing their recommendation on both long- and short-contact time tests. The mode of action for chlorhexidine as it relates to membrane disruption has been studied by Hugo and Longworth [16–18]. At low concentrations (up to 200 μg/ml), chlorhexidine inhibits membrane enzymes and promotes leakage of cellular constituents thus suggesting its bacteriostatic property. At higher concentrations, cytoplasmic constituents are coagulated and a bactericidal effect is seen. In the case of chlorhexidine-induced leakage of intracellular material from *E. coli* and *S. aureus*, a diphasic leakage/concentration pattern is found. The first part of the curve shows increasing leakage with increasing concentration of the active. Over a period of time, at higher concentrations of the active, the protoplasmic contents became progressively lower. Chlorhexidine also damages the cytoplasmic membrane of yeasts [19] and prevents the outgrowth, but not the germination of bacterial spores [20]. Chlorhexidine has been claimed to be an inhibitor of both membrane-bound and soluble adenosine triphosphatase (ATPase) and also of net K^+ uptake in *S. faecalis*. However, it is believed that inhibition of membrane-bound ATPase is not a primary target of chlorhexidine action, since the activity is inhibited only at high biguanide concentrations [21]. It is possible that chlorhexidine phosphate is insoluble and may have acted as an ATPase inhibitor by precipitating its substrate.

The mechanism of action of chlorhexidine as an anti-fungal agent has been studied. The action of chlorhexidine on yeast was studied by Elferink and Booij [22]. They found that the biguanide induced the rapid release of K^+ ions, indicative of membrane damage. Similar findings were made by Walters et al. [19] who observed that pentose release from *Saccharomyces cerevisiae* was induced maximally at a concentration of 50 μg/ml (when exposed at 30°C for 3 h) as opposed to minimal inhibitory concentration of 7.0 μg/ml. There was no evidence of lysis. However, the highest chlorhexidine level (1000 μg/ml) induced some clumping.

Bobichon and Bouchet [23] investigated the action of chlorhexidine at a sublethal concentration on the ultrastructure of budding *Candida albicans* and observed a loss of cytoplasmic components indicative of plasma membrane damage, together with coagulation of nucleoprotein.

Studies on membrane-associated enzyme systems have failed to reveal specific effects. Inhibition of ATPase and of the active transport systems for amino acids has been reported, whereas dehydrogenase activity is stimulated by a low concentration of chlorhexidine, possibly because the damaged membrane is more permeable to their substrates. The behavior of chlorhexidine at an oil–water interface suggests that the molecules may become oriented within the lipid component, causing general disruption of membrane structure and function.

V. IN VITRO STUDIES

The minimum inhibitory concentration (MIC) for an antimicrobial agent is an in vitro measurement of the bacteriostatic and fungistatic effects of the compound. It is the lowest concentration, expressed in micrograms per milliliter (μg/ml) or parts per million (ppm), at which the antibacterial agent will inhibit the growth of a specific organism. Molds and yeasts were tested on Sabouraud's agar incubated at 30°C for 24–72 h. Anaerobes were incubated anaerobically for 2–3 days on agar containing 5% lysed blood. Tables 1 and 2 [24] show the MIC values for chlorhexidine gluconate. The MIC is the lowest chlorhexidine concentration that prevented both bacterial and fungal growth. There are numerous publications that refer to the various methods to measure the efficacy of active compounds under different end-use conditions. Most antiseptics are not simple solutions, but complex formulations of one or more antimicrobial agents, in combination with excipients, such as detergents, perfumes, and colorants. The method of formulation can profoundly affect the physical and biological properties of the active agent [25]. Therefore, it is not adequate to assume that an antiseptic formulation will be effective merely because of a specific amount of active agent has been included. Testing of antiseptics by microbiological methods, as opposed to chemical assay, therefore, is of vital importance. Another type of assay, which, similar to the MIC, relates to bacteriostatic activity, is the inhibition zone test, or diffusion assay. In this type of test the antiseptic, or a dilution of it, is introduced into a central well in an agar plate. It diffuses into the surrounding agar, which has been inoculated with the test organism. The growth of the organism is inhibited (if it is sensitive), and seen as zones of inhibition that are formed and can be quantified after an appropriate incubation period. Thus, a compound giving a small zone of inhibition may not be intrinsically less active than a different type of chemical compound giving a large zone of inhibition when tested at the same concentration under similar conditions. It is really a qualitative, rather than a quantitative test, although it has been adapted to be fully quantitative for the assay of antibiotics. Cationic compounds in general do not give large diffusion zones because of interaction with the growth medium and their poor diffusibility.

An extension of this sort of test is an in vitro measure of skin persistence. Instead of antiseptic dilutions being put into a well in an agar plate, the fingertips of volunteers, or pieces of skin that have previously been dipped in antiseptic solution and then washed with water, are laid on the agar surface. A persistent effect is noted if an inhibition zone occurs around the skin or around the area touched by the fingertips when the plate is incubated. This test can be quite useful, but the size of the inhibition zone is meaningful only if the test is properly calibrated. Comparisons of zone sizes between different active agents is beyond the scope of this test owing to differences in diffusibility.

The in vitro bactericidal and fungicidal activity of 0.05% chlorhexidine gluconate was determined using a procedure based on British Standard 3286 (1960), and is presented in Tables 3 and 4, respectively [24]. One milliliter of a 24-h broth culture of the test organism was added to 10 ml of aqueous 0.05% (w/v) chlorhexidine gluconate solution, which was maintained at ambient temperature (18–21°C). One-milliliter aliquots of the mixture were removed after 20 s, 1 min, and 10 min, and transferred to inactivator broth containing 1.5% soya lecithin and 10% polysorbate 80. A viable count was performed on appropriate further dilutions and, by comparison with an untreated control, a 10-log reduction factor was calculated.

Chlorhexidine has little sporicidal activity, except at elevated temperatures [26]. For viruses, the activity of chlorhexidine is restricted to the nucleic acid core or the outer coat. The outer coat may be entirely protein or a combination of lipoprotein or glycoprotein. Chlorhexidine has been effective against respiratory viruses, herpes virus, and cytomegalovirus. Other viruses, such as poliomyelitis and papilloma (warts) virus are not affected. Viruses are generally very simple structures, consisting of a core of nucleic acid, either RNA or DNA, surrounded by a thin coat of protein or lipoprotein. The strong disinfectants, such as glutaraldehyde, formaldehyde, iodine, and chlorine, act by destroying nucleic acids and denaturing proteins, thereby inactivating all viruses at a certain time period. However, detergent-based antibacterials act mainly by destroying membranes and, therefore, are active against those viruses that have lipoprotein coats. Chlorhexidine gluconate being cationic, has much of its antibacterial activity because of its ability to interfere with the metabolism of a cell. This is of no value against viruses, which do not possess a defined metabolic system.

Human immunodeficiency virus (HIV), the organism responsible for acquired immunodeficiency syndrome (AIDS), has been studied for its sensitivity to chlorhexidine [27]. Published data on the activity of chlorhexidine against a wide range of viral agents are summarized in Table 5 [24]. A series of in vitro studies

TABLE 1 Bacteriostatic Activities of Chlorhexidine Gluconate

Test organism	No. of strains	MIC (mg/L) Mean	MIC (mg/L) Range
Gram positive cocci	—		
Micrococcus navus	1	0.5	
M. lutea	1	0.5	
Staphylococcus aureus	16	1.6	1–4
S. epidermidis	41	1.8	0.25–8
Streptococcus faecalis	5	38	32–64
S. mutans	2	2.5	
S. pneumoniae	5	11	8–16
S. pyogenes	9	3	1–8
S. sanguis	3	9	4–16
S. viridans	5	25	2–32
Gram-positive bacilli			
Bacillus cereus	1	8	
B. subtilis	2	1	
Clostridium difficile	7	16	8–32
C. welchii	5	14	4–32
Corynebacterium spp.	8	1.6	0.5–8
Lactobacillus casei	1	128	
Listeria monocytogenes	1	4	
Propionibacterium acne	2	8	
Gram-negative bacilli			
Acinetobacter anitratus	3	32	16–64
A. lwoffii	2	0.5	
Alcaligenes faecalis	1	64	
Bacteroides distasonis	4	16	
B. fragilis	11	34	8–64
Campylobacter pylori	5	17	8–32
Citrobacter freundii	10	18	4–32
Enterobacter cloacae	12	45	16–64
Escherichia coli	14	4	2–32
Gardnerella vaginalis	1	8	
Haemophilus influenzae	10	5	2–8
Klebsiella aerogenes	5	25	16–64
K. oxytoca	2	32	
K. pneumoniae	5	64	32–128
Proteus mirabilis	5	115	64–>128
P. morganii	5	73	16–128
P. vulgaris	5	57	32–128
Providencia stuartii	5	102	64–128
Pseudomonas aeruginosa	15	20	16–32

TABLE 1 Continued

Test organism	No. of strains	MIC (mg/L)	
		Mean	Range
Gram-negative bacilli (continued)			
P. cepacia	1	16	
P. fluorescens	1	4	
Salmonella bredeney	1	16	
S. dublin	1	4	
S. gallinarum	1	8	
S. montivideo	1	8	
S. typhimurium	4	13	8–16
S. virchow	1	8	
Serratia marcescens	10	80	16–64

performed at Vanderbilt University indicate the sensitivity of AIDS virus to chlorhexidine. However, considering the mode of transmission of AIDS by sexual contact or by parenteral inoculation, topically applied antimicrobials, in general, may have a less significant role to play in its prevention.

VI. IN VIVO TESTS

The most meaningful evaluation of the potency of an active compound must be relevant to in-use conditions as close as possible to those likely to be encountered in practice. For instance, an antiseptic solution or cream to be used on the skin or on wounds can be tested on experimentally contaminated intact human or animal skin, or on a wound model in animals. An antibacterial agent incorporated in a soap can be evaluated by determining its effect on the normal flora of the skin in handwashing experiments. The testing of skin antiseptic products provides a constant subject for debate among clinical microbiologists. When assessing the effectiveness of an agent for skin disinfection, several properties of the product must be taken into consideration. Rapid broad-spectrum bactericidal action against both the resident and transient microflora, followed by residual activity to prevent regrowth of skin organisms, will be the basic requirements. Cumulative efficacy of the active agent as result of repeated use will be an added benefit. The product must be compatible with body fluids and have cosmetic acceptability for the user. The relative importance of these individual properties varies, to some degree, according to the particular application of the antiseptic. The methods of assessing the efficacy have been defined by the regulatory agency (*Federal Register* 21, parts 333 and 369).

Table 2 Fungistatic Activity of Chlorhexidine

Organism	No. of strains	Mean MIC (mg/L)
Mold fungi		
Aspergillus flavus	1	64
A. fumigatus	1	32
A. niger	1	16
Penicillium notatum	1	16
Rhizopus spp.	1	8
Scopulariopsis spp.	1	8
Yeasts		
Candida albicans	2	9
C. guillermondii	1	4
C. parapsilosis	2	4
C. pseudotropicalis	1	3
Cryptococcus neoformans	1	1
Saccharomyces cerevissiae	1	1
Torulopsis glabrata	1	6
Dermatophytes		
Epidermophyton floccosum	1	4
Microsporum canis	2	4
M. fulvum	1	6
M. gypseum	1	6
Trichophyton equinum	1	4
T. interdigitale	2	3
T. mentagrophytes	1	3
T. quinkeanum	1	3
T. rubrum	2	3
T. tonsurans	1	3

A. The Hygienic-Handwashing Test

This test is performed according to the method described by Ayliffe et al. [28]. The fingertips are contaminated with a bacterial culture, and the hands are washed for the recommended time. The numbers of viable bacteria are determined by sampling the fingertips in neutralizer broth containing glass beads. By comparing bacterial counts pre- and posttreatment, an assessment of the activity of the handwash against transient bacterial contamination can be made. Maki et al. [29] evaluated several antibacterial handwashing agents in a clinical setting, using 15-s handwash contact time. Chlorhexidine skin cleanser, a liquid soap with alcoholic rinse, and a povidone–iodine skin cleanser were compared. The chlorhexidine preparation was the most effective agent tested and provided adequate protection.

B. The Surgical-Handwashing Test

This test is designed to mimic a ward handwash. Volunteers washed their hands using chlorhexidine skin cleanser. Hands are sampled to see how many of the resident flora had survived the antiseptic handwash. A loose-fitting glove is placed over the hands and 50 ml of liquid is poured into the glove. The glove is sealed at the wrist by using a rubber band and the hand massaged for 1 min. The resident bacteria that survive the handwash procedure are washed into the sampling liquid. The numbers of surviving bacteria are then counted and the antiseptic efficacy is determined by comparing pre- and posttreatment counts. The glove juice test [30] could be modified to have a certain handwashing period, followed by extraction of the hands to determine the residual antimicrobial efficacy of the formulation, as applicable under FDA guidelines. More demanding test procedures are used to evaluate the effect of the normal resident flora [31–34]. The Peterson's glove juice method [30] and the method described by Ayliffe [28] may be preferred because they sample the hands and, therefore, are affected by the handwashing technique, a factor of considerable importance.

VII. CLINICAL ASPECTS OF CHLORHEXIDINE

Developments in the medical field are progressing faster than ever before in terms of equipment, drugs, and surgical techniques, and yet, cross-infection in hospitals remains a major problem. There is no single answer, as hospital-acquired infection has many causes and routes of transmission. Consequently, its control must rely on the combined effect of numerous established procedures throughout the hospital. It is now recognized, however, that hand hygiene is the single most important measure in the control of hospital-acquired infection. Three basic levels of hand hygiene are recognized in infection control practice: (a) social handwashing with nonmedicated soap and water; (b) hygienic hand disinfection with an antiseptic handwash or handrub in situations for which a significant risk of cross-infection exists; (c) surgical hand disinfection with an antiseptic handwash or handscrub used by the operating room staff. The social handwashing is recommended before starting the work, handling food, leaving for home, after visiting the toilet, handling bedding, and so forth. However social handwash has little effect on the microbial flora and, therefore, will not be discussed in this chapter.

A. Surgical Hand Disinfection

For surgical hand disinfection, immediate antimicrobial action is required to eliminate acquired transient pathogens and also to reduce, as far as possible, the resident microflora on the hand. In addition, the treatment should be persistent on the hands, thereby preventing the regrowth of the survivors in the warm and moist conditions that exist inside the surgeon's gloves. The activity should also be retained in the presence of organic matter.

TABLE 3 Bactericidal Activity of 0.05% Chlorhexidine Gluconate

Test organism	No. of strains	Mean $\log_{10}$ reduction at		
		1/3 min	1 min	10 min
Gram-positive				
Micrococcus navus	1	0.1	0.4	2.1
M. lutea	1	0.2	0.7	2.9
Staphylococcus aureus	16	0.4	0.7	2.5
S. epidermidis	41	2.2	3.4	>5.1
Streptococcus faecalis	5	0.4	0.4	1.1
S. mutans	2	0.8	>4.6	5.8
S. pneumoniae	5	0.8	1.5	>3.5
S. pyogenes	9	1.2	1.8	>3.7
S. sanguis	3	1.1	2.2	>3.9
S. viridans	5	0.4	0.8	2.3
Gram-positive bacilli				
Bacillus cereus	1	2.0	2.0	4.7
B. subtilis	2	0.5	0.5	0.3
Clostridium difficile	7	0.2	0.3	0.3
C. welchii	5	2.1	3.1	>4.8
Corynebacterium spp.	8	1.1	1.4	3.7
Lactobacillus casei	1	0.2	0.2	4.1
Listeria monocytogenes	1	0.6	2.2	4.8
Propionibacterium acne	2	0.7	1.8	3.6
Gram-negative bacilli				
Acinetobacter anitratus	3	1.4	2.6	>5.3
A. lwoffii	2	>4.0	>4.3	>4.8
Alcaligenes faecalis	1	1.5	2.7	4.1
Bacteroides distasonis	4	0.9	2.7	>4.9
B. fragilis	11	3.0	4.2	5.2
Campylobacter pylori	5	N.T.	2.8	>4.0
Citrobacter freundii	10	3.4	4.9	>6.0
Enterobacter cloacae	12	3.5	4.5	>6.3
Escherichia coli	14	3.2	5.0	>6.4
Gardnerella vaginalis	1	2.3	3.3	>5.8
Haemophilus influenzae	10	>4.1	>4.1	>4.1
Klebsiella aerogenes	5	2.7	3.9	>5.9
K. oxytoca	2	3.2	5.2	>6.4
K. pneumoniae	5	3.0	4.8	>6.2
Proteus mirabilis	5	0.8	0.9	2.9
P. morganii	5	1.0	1.5	4.2
P. vulgaris	5	0.8	1.0	4.1
Providencia stuartii	5	0.6	0.9	1.8
Pseudomonas aeruginosa	15	1.7	2.7	4.9

TABLE 3 Continued

Test organism	No. of strains	Mean $\log_{10}$ reduction at		
		1/3 min	1 min	10 min
Gram-negative bacilli (continued)				
P. cepacia	1	1.1	1.3	>4.6
P. fluorescens	1	3.8	5.0	>6.7
Salmonella bredeney	1	1.6	3.4	>6.4
S. dublin	1	1.5	2.9	3.2
S. gallinarum	1	2.5	4.0	>6.2
S. montivideo	1	2.4	3.8	>6.3
S. typhimurium	4	2.0	3.7	>6.0
S. virchow	1	1.9	3.9	>6.2
Serratia marcescens	10	1.5	3.7	>5.9

Chlorhexidine is normally incorporated as an active agent in a detergent base or an alcoholic base. The cleansing products containing the chlorhexidine are used for the rapid disinfection of hands, percutaneous and intracutaneous operative sites, as well as intravascular catheter sites. The types of products chosen will depend on their intended use—hygienic washing in general patient care areas, for skin antisepsis before high-risk procedures, or for surgical scrubbing. Alcohol-based chlorhexidine products are generally used for hygienic hand disinfection of nursing staff on the ward. The most important feature is immediate elimination of transfer of the transient bacteria from one patient to the another. The immediate bacterial action of chlorhexidine seems to be better than similar preparations containing other active components, such as povidone–iodine, triclosan, *p*-chloro-*m*-xylenol (PCMX), or others. The valuable persistent activity of chlorhexidine on skin in preventing regrowth of organisms is well documented by various workers [35–37].

Antimicrobial activity of chlorhexidine gluconate against natural and artificial contamination was studied by simulating in-use environmental contamination [38]. Commercial skin cleanser or antiseptic preparations containing chlorhexidine gluconate were experimentally contaminated with *S. aureus*, *Proteus mirabilis*, *Pseudomonas aeruginosa*, or *Pseudomonas cepacia*. To simulate what often happens in hospitals when residual product from old bottles are transferred to new, several transfers with further bacterial challenges were made over a period. After the test period the product and samples from the delivery spouts were cultured and found to have no bacterial growth.

The cumulative effect of bactericidal skin cleansers over a number of usage

TABLE 4 Fungicidal Activity of 0.05% Chlorhexidine Gluconate

Test organism	No. of strains	Mean $\log_{10}$ reduction after		
		1/3 min	1 min	10 min
Mold fungi				
Aspergillus flavus	1	0.4	0.8	1.7
A. fumigatus	1	0.7	1.2	2.4
A. niger	1	0.7	1.2	3.0
Penicillium notatum	1	0.6	2.0	3.5
Rhizopus spp.	1	0.4	0.4	0.5
Scopulariopsis spp.	1	0.6	1.1	2.3
Yeasts				
Candida albicans	2	2.8	>4.1	>4.2
C. guillermondii	1	3.5	>4.3	>4.3
C. parasilosis	2	2.1	3.4	>4.2
C. pseudotropicalis	1	3.6	>4.4	>4.4
Cryptococcus neoformans	1	4.0	>4.2	>4.2
Saccharomyces cerevissiae	1	3.7	>3.7	>3.7
Torulopsis glabrata	1	1.3	2.2	>4.4
Dermatophytes				
Epidermophyton floccosum	1	0.7	0.5	>1.8
Microsporum canis	2	0.4	1.0	>2.0
M. fulvum	1	0.2	0.6	>2.4
M. gypseum	1	0.1	0.3	2.0
Trichophyton equinum	1	0.5	1.1	>2.1
T. interdigitale	2	0.4	0.9	>2.4
T. mentagrophytes	1	1.3	>2.1	>2.1
T. quinkeanum	1	0.2	0.9	>2.8
T. rubrum	2	0.3	0.6	>2.4
T. tonsurans	1	0.4	0.3	1.6

periods studied by several workers [39,40], indicated that chlorhexidine wash has a better kill rate after several applications over a time period, similar to povidone–iodine. However, user compatibility of a skin disinfectant is considered important, especially when staff are subjected to frequent use of the product. The importance of providing an agent pleasing to users has been recognized [41]. A study identified skin damage as a major deterrent to handwashing among patient care personnel [42]. Another in-use ward study found no single preparation to be superior in terms of user acceptability for skin mildness over continued use [40]. Contrary to popular belief, there is no direct correlation between an agent's antiseptic effectiveness and its effect on the skin; thus, indicating that an antiseptic is not

TABLE 5 Virucidal Activity of Chlorhexidine Gluconate

Virus	Viral family	Activity[a]	Conc. (%)
Respiratory syncytial virus	Paramyxovirus	+	0.25
Herpes hominis/simplex	Herpesvirus	+	0.02
Poliovirus type 2	Enterovirus	−	0.02
Adenovirus type 2	Adenovirus	−	0.02
Equine infectious anemia virus	Retrovirus	+	2.0
Variola virus (smallpox)	Poxvirus	+	2.0
Herpes simplex type 1/type 2	Herpesvirus	+	0.02
Equine influenza virus	Orthomyxovirus	+	0.001
Hog cholera virus	Togavirus	+	0.001
Bovine viral diarrhea	Togavirus	+	0.001
Parainfluenza virus	Paramyxovirus	+	0.001
Transmissible gastroenteritis virus	Coronavirus	+	0.001
Rabies virus	Rhabdovirus	−	0.001
Canine distemper virus	Paramyxovirus	+	0.01
Infectious bronchitis virus	Coronavirus	+	0.01
Newcastle virus	Paramyxovirus	+	0.01
Pseudorabies virus	Herpesvirus	−	0.01
Cytomegalovirus	Herpesvirus	+	0.1
Coxsackievirus	Picornavirus	−	0.4
Echovirus	Picornavirus	−	0.4
Human rotavirus	Reovirus	−	1.5
Human immunodeficiency virus type 1	Retrovirus	−	0.2

[a]+ = Active in vitro at the concentration stated; − = not active in vitro at the concentration stated.

necessarily more drying or damaging to skin than a nonmedicated soap [42]. The real challenge in developing topical antimicrobials will be to select a system with adequate skin protection as well as satisfactory skin disinfection. It has been demonstrated that frequent use of chlorhexidine skin cleanser increased counts on the fingertips [43]. This was attributed to skin damage, and chlorhexidine alcohol was suggested as an in-between use product, to reduce the number of handwashings. A comparative study on hand disinfectants indicated that undesirable skin effects were reported most often for povidone–iodine, followed by hexachlorophene and chlorhexidine, respectively [44]. Hence, the study concluded that, although chlorhexidine skin cleanser is not totally free from causing skin irritation, it appears to be better tolerated than other handwashing agents containing antimicrobials. Maki et al. [29] found chlorhexidine skin cleanser to be better tolerated than a soap handwash with alcohol rinse or with a povidone–iodine handwash.

By following the FDA guidelines, a large volunteer study was undertaken to evaluate surgical hand scrubs [30]. A chlorhexidine handwash produced significantly greater reductions in numbers of resident bacteria than either hexachlorophene or povidone–iodine preparations and maintained these low numbers for up to 6 h under gloves. Povidone–iodine was not persistent and allowed the grow-back of microflora with time. This particular chlorhexidine handwash (Hibiclens) was also the first surgical handscrub to be approved as safe and effective by the Topical Antimicrobials Committee of the FDA.

The immediate and persistent antibacterial efficacy of chlorhexidine gluconate incorporated in a sponge–brush was assessed under simulated operating room conditions [45]. During the test period, the subjects cleansed their hands with water and the chlorhexidine gluconate-impregnated brush by initially using the sponge and brush sides then, after rinsing, the sponge side only for an additional 3 min. Since there was no significant increase in bacterial numbers up to 6 h after scrubbing, it was concluded that the chlorhexidine gluconate sponge–brush produce a highly effective means for presurgical scrubbing. The effectiveness of sponge–brushes impregnated with either chlorhexidine gluconate or povidone–iodine was studied for immediate and persistent antimicrobial activity and also the effect of blood, which commonly penetrates through punctured surgical gloves [46]. Chlorhexidine gluconate was significantly more effective than povidone–iodine in both the presence and absence of blood, thereby confirming an earlier study [47,48]. The effect of 4% chlorhexidine detergent solution and 2% triclosan formulation were compared for initial and prolonged effects of the agents. Chlorhexidine gluconate was significantly better than triclosan formulation [49].

The problems associated with ineffective surgical hand disinfection is illustrated in a clinical report of an epidemic of prosthetic valve endocarditis [50]. Serious postoperative infection and rejection of the prosthesis occurred in patients who had undergone heart surgery. At investigation, the outbreak was found to be due to a specific strain of *Staphylococcus* that was part of the resident microflora of one particular surgeon. It was found that after routine handwashing with povidone–iodine, the surface of this surgeon's hands became recolonized with significant numbers within a short time, thereby confirming the lack of persistence of povidone–iodine efficacy. Carriage of microorganisms on the hands has been implicated in more than 40 nosocomial outbreaks, primarily of infections of the newborn and in catheter-related urinary tract infections [51]. Most nosocomial infections are caused by gram-negative rods, and it has been asserted that hand carriage of gram-negative rods may not be transient [52]. With the development of intensive care units to care for those who are seriously ill, a population of patients has emerged that is at greater risk for contracting nosocomial infections.

In some European countries, alcoholic preparations are recommended as surgical hand scrubs. Indeed, 60–90% *n*-propanol, isopropanol, and ethanol have been repeatedly demonstrated in European and American studies to have superior antimicrobial activity when compared with detergent-based formulations [53–57]. Despite the demonstrated superiority of alcoholic solutions, they are not widely used for scrubbing in the United States, perhaps because of the problems associated with flammability and their drying effect on skin. However, an enclosed dispensing system containing alcohol could minimize the flammability issue. The additions of emollients to alcoholic formulations will minimize the drying effects of the skin. A recent study recommends the consideration of alcoholic preparations and concludes that 70% ethyl alcohol with 0.5% chlorhexidine gluconate could be an efficacious and acceptable alternative for surgical scrubbing [58].

B. Hygienic Hand Disinfection

The primary objective of hygienic hand disinfection is to eliminate the spread of transient organisms between patients. The washing process eliminates, or greatly reduces, the role of the physician or medical health care worker as a vector for disseminating microbiological diseases between patients. Although a surgeon may scrub his or her hands two or three times a day in the surgical setting, it is not uncommon for health care workers to wash their hands 25–30 times a day before examining a patient in a hospital unit. User acceptability of a formulation is of particular importance in the ward situation, where frequent handwashing is often necessary. Health care workers have a strong preference for products that are mild and nonirritating to their skin. Factors such as soreness and dryness of the hands can have more influence on the use of a product than its antimicrobial efficacy [42].

The action of chlorhexidine on test sites artificially contaminated with *Serratia marcescens* showed a 99.9% reduction of transient contaminants after a single 15-s handwash [31]. In a prospective study in a large intensive care unit, the efficacy for prevention of nosocomial infection was followed using chlorhexidine, povidone–iodine, and a nonmedicated liquid soap. Chlorhexidine was considered acceptable by most personnel [59]. The cumulative irritancy potential of the chlorhexidine solution was compared with those of hexachlorophene, povidone–iodine, and nonmedicated cleanser. Chlorhexidine showed no potential for skin irritation [35,60]. Handwashing often has been singled out as the most important procedure in preventing nosocomial infections. Disinfection of hands after every patient contact has been recommended by both the Centers for Disease Control [61] and the American Hospital Association [62]. Several factors have been related to handwashing efficacy, including the microbicidal agent, the duration of its use, and the techniques employed [29,32,40,63]. Recommendations on the acceptable duration of handwashing vary, but a minimum of 10 s is recommended, with vigorous lathering and rubbing of the hands before rinsing [64]. A study correlating handwashing practices and nosocomial infection rates were carried out through educational programs [65,66]. These studies stress the underlying fact that such educational programs could be successful only if there is constant monitoring and feedback, followed by continuous reinforcement of infection control principles. The primary purpose of infection control programs is to prevent hospital-acquired infections. Several patient care techniques have been modified to decrease the incidence of infection. However, many of the risk factors predisposing to infection are unavoidable. For instance, host factors, such as debilitating underlying disease, old age, organ failure, and necessary treatments that may compromise host defenses, often cannot be modified. Strategies for infection control focus on interventions, practices, and procedures that modify correctable risks. However most programs emphasize the importance of handwashing practices. The external routes of transmission of nosocomial infection are well established. Handwashing, particularly with an antiseptic solution, has long been demonstrated to be an effective control measure. Recently, high-technology automated, handwashing systems have been tried for possible commercialization.

VIII. OTHER APPLICATIONS OF CHLORHEXIDINE

The potential uses of chlorhexidine for infection prevention are not limited to preoperative skin disinfection and hand cleansing. Preoperative total body bathing to reduce postoperative infection has been studied by several workers [67–73]. These studies concluded that 4% chlorhexidine solution used for whole-body disinfection is a valuable adjunct to existing antiseptic and aseptic procedures that contribute in reducing the microflora of the skin. The disinfecting efficacy of preoperative total body washing using chlorhexidine could be correlated with a significant reduction in the postoperative infection rates. Early postoperative wound infections complicating orthopedic surgery are thought to be derived from the patient's own resident bacterial flora. Chlorhexidine when used as a postoperative wash reduced the bacterial flora considerably, thereby minimizing infection [68].

Patients who require bone marrow transplants or intensive chemotherapy are at high risk of infection and require special protective measures to prevent potentially fatal infections. Long-term daily applications of chlorhexidine for skin decontamination of isolated patients undergoing marrow transplantation, has been effective in reducing complications from infection [74].

Since chlorhexidine gluconate displays a wide spectrum of antimicrobial activity and is not appreciably absorbed through intact skin, its ability to disinfect groin and perineum was studied by Byatt and Henderson [75]. A 0.2% chlorhexidine formula was used as vaginal washing to minimize the neonatal colonization of group B streptococci [76]. A significant reduction in colonization by this organism was noted in the infants of treated mothers, thus preventing serious infection in the early neonatal period. For perigenital skin disinfection of predelivery, chlorhexidine appears to be the agent of choice, because of its immediate and sustained antimicrobial activity.

Nasal and cutaneous carriage of *S. aureus* is a causative factor of morbidity and mortality in chronic hemodialysis patients. Since skin disinfection at the vascular access site may reduce the incidence of staphylococcal infections, the efficacy of chlorhexidine in skin disinfection was studied [77]. It reduced total microflora and cutaneous *S. aureus* and maintained these reductions throughout the duration of the hemodialysis procedure (4 h). The effectiveness of chlorhexidine gluconate in the prevention of urinary tract infections has been reported [78]. A formulation containing low levels of chlorhexidine was effective in reducing the incidence of bacterial infection in catheterized patients [79,80]. Mitchell and his co-workers [81], using 0.05% chlorhexidine gluconate in glycerin as a urethral disinfectant and catheter lubricant, showed falling infection rates in indwelling catheter-related infection.

The use of chlorhexidine as a therapeutic or prophylactic drug in oral hygiene has been followed for a long time. Chlorhexidine showed substantial antimicrobial activity against *Bacteroides gingivalis*, *B. intermedius*, and black pigmented gram-negative obligate anaerobes associated with periodontal disease in adults [82]. The in vivo plaque-inhibitory activity of chlorhexidine is attributed to the combination of its antimicrobial activity and its ability to adsorb to the surface of the oral cavity [83]. According to Davis et al. [84], tissue binding and slow release probably accounted for the sustained antibacterial action, as well as the therapeutic effect in gingivitis. A histological study of the oral mucosa of hamsters [85] led to the observation that chlorhexidine does not penetrate the undamaged oral epithelium. Use of a mouth rinse containing chlorhexidine digluconate (0.2%) showed a bacteriostatic action against salivary organisms [86]. Pre- and postoperative use of chlorhexidine mouthwash has reduced the oral microflora and the incidence of postextraction bacteremia [87,88]. A mouthwash and gel formulation have been successfully used in immunocompromised patients to control candidiasis and mucositis [89,90]. Soaking dentures in chlorhexidine has been effective in reducing colonization with *Candida* species [91,92]. One disadvantage of using chlorhexidine for oral hygiene is the development of brown stains on teeth, with some, but not all users. The highly reactive molecule of chlorhexidine reacts with dietary chromogens, which are then precipitated onto the tooth surface to form a stain [93].

With the limited scope the chlorhexidine offers as a hard surface disinfectant, there are some European products with a combination of ethyl alcohol, chlorhexine gluconate, and quaternary ammonium compound. The product is being offered as a disinfectant wipe and is used for the disinfection of clinical hard surfaces in all hospital departments and laboratories. Chlorhexidine has been shown to disinfect alginate impression material when used as a soaking solution

[94]. Chlorhexidine along with thimerosal is commonly used as a disinfectant soaking solution for contact lenses [95]. When the cleaning and rinsing steps were included in the disinfection regimen of the contact lenses, the soaking solution containing chlorhexidine was effective for 6-hour soaking period with lenses challenged with *S. marcescens* and *Candida parapsilosis* [96]. However, Richardson et al. [97] reported that the inactivation of the preservatives in contact lens solutions during storage was dependent upon the nature of the preservative system as well as the type of plastic material used to fabricate the container system.

A number of investigators have reported that various contact lens solutions, particularly certain formulations with chlorhexidine, are susceptible to contamination by *S. marcescens* [98,99]. Adaptation of isolates of *S. marcescens* with persistence to disinfecting solutions containing chlorhexidine as preservative, was suggested as a possible epidemiology of contact lens associated eye infections by Gandhi et al. [100]. In-use contact lens cases are frequently contaminated with microorganisms potentially hazardous to the eye [101]. This contamination was reported to be the result of imporoper hygienic practices on the part of the contact lens wearer [96]. However, the inability of some contact lens solutions to kill *S. marcescens* is of particular concern because of the association of contaminated solutions with ocular infections [102].

IX. IN USE SAFETY DATA

Chlorhexidine has a long consumer use history of 30 years, having been used in a wide variety of clinical situations. During the extensive use of chlorhexidine, very few adverse reactions to the active component have been reported. Systemic (oral, intramuscular, intravenous) and topical routes of administration have been investigated for short- and long-term adverse effects. Adverse reactions, such as dermatitis and photosensitivity, are relatively rare. However, few hypersensitivity and generalized allergic reactions have been reported. Oral LD_{50} values are high because chlorhexidine is poorly absorbed from the gut and is excreted mainly in the feces. Laboratory experiments in humans have failed to produce evidence of irritation or sensitization [60].

The earliest toxicological evaluation of chlorhexidine showed a marked difference between intravenous and oral LD_{50} values [103]. Systemic administration of formulations containing chlorhexidine has been performed in rats and mice. The potential of chlorhexidine gluconate as an irritant was the subject of a series of tests during which water and hexachlorophene were used as controls [104]. The 21-day cumulative irritancy assay, which employed occlusive dressings, showed that chlorhexidine gluconate is a compound with extremely low irritancy potential for the human skin.

Carcinogenicity studies have been performed in both rats and mice using chlorhexidine and one of its main degradation products *p*-chloroaniline (PCA). No evidence of carcinogenicity was found in rats after 2 years of oral feeding.

Although there is no evidence that chlorhexidine gluconate is toxic if absorbed through the skin, studies were conducted to determine the degree of absorption. Chlorhexidine gluconate was tagged with radioactive ^{14}C and applied to the skin of volunteers. Blood and urine samples were analyzed after 3–6 h for radioactivity, which was undetectable. However, a fecal sample showed traces of radioactivity. A study to determine percutaneous absorption over a 3-week period in volunteers undergoing a handscrub regimen, found no detectable levels of chlorhexidine gluconate in the blood. This result was confirmed by yet another study conducted among hospital staff members who have used chlorhexidine gluconate for 6 months. This study indicated that the degree of percutaneous absorption, if any at all, is extremely low [105]. Neonatal rhesus monkeys that were bathed daily for 90 days with an 8% (w/v) chlorhexidine gluconate formulation produced no adverse findings in the brain or other tissues [106].

X. ENVIRONMENTAL FATE OF CHLORHEXIDINE

Chlorhexidine is extensively used in medical institutions and hospitals. The environmental impact of chlorhexidine containing waste water could be important as part of the ecological system. Biological treatment of chlohexidine gluconate containing waste water has been studied [107–109]. It has been suggested that when the wastewater containing chlohexidine gluconate is treated with batch activated sludge process, longer time aeration would be required to obtain high treatment efficiency [110].

XI. MICROBIAL RESISTANCE TO CHLORHEXIDINE

There has been renewed interest in the action of chlorhexidine and of other cationic-type bactericides. This has stemmed partly from the work of Nikaido [111], Hancock [112], and their co-workers on the role of gram-negative bacterial membrane acting as a permeability barrier to the entry of antibacterial agents. The findings of Stickler et al. [113,114], that many clinical isolates of gram-negative organisms are resistant to chlorhexidine, provides an opportunity for a rational approach to new antiseptic and disinfectant formulations. The combination of an agent that opens up the outer membrane with compounds that can attack, for example, the cytoplasmic membrane could produce a range of new products with improved activity against refractile organisms. It has been shown that the combination of EDTA and chloroxylenol potentiates the activity of this phenolic compound against *P. aeruginosa* [115].

A general review of disinfectant policies for hopsitals were suggested by several workers [116,117] in order to reduce the use of disinfectants in circumstances where heating or thorough cleaning will be sufficient. This makes good economic sense and is also to be approved of on the general grounds that the more frequently an antimicrobial agent is used, the more likely it will become that a resistant microbial flora will emerge. An ideal situation could be to get the hospital's control infection committee to think carefully about their antiseptic or disinfectant policies in situations where intrinsically resistant bacteria are producing infections. This is further supported by the observations made on the mode of development of urinary tract infections in paraplegic patients enduring long-term intermittent bladder catheterization [118,119]. Catheterization that was performed three or four times daily involved the washing of the periurethral area with chlorhexidine prior to insertion of the catheter. The effect of this repeated application of antiseptic on the bacterial flora of the urethral meatus was examined in the patients from the date of injury and admission to the spinal unit up to the time they developed urinary tract infection. The urethral flora was examined daily before and after the application of chlorhexidine. The general pattern that emerged was that for a few days after trauma, the meatal skin carried a gram-positive flora which was greatly reduced by the application of the antiseptic. A gram-negative flora usually developed by about day 4 and proved to be more refractory to chlorhexidine. In particular, *P. mirabilis*, *P. aeruginosa*, *Providencia stuartii*, and *Klebsiella* spp. frequently survived the meatal cleansing and proceeded to infect the bladder.

Stickler and Thomas [120] examined large collection of isolates of gram-negative bacilli causing urinary tract infections in patients to ascertain whether the resistance to chlorhexidine was a general phenomenon or was restricted to individual cases. It was observed that chlorhexidine resistance was not a widespread phenomenon. It was limited to *P. mirabilis*, *P. aeruginosa*, *P. stuartii*, and *S. marcescens*. These observations suggest that an antiseptic policy involving the long-term and extensive use of chlorhexidine in clinical situations such as the catheterized urinary tract could be counterproductive. As a means of controlling urinary tract infections in patients with indwelling catheters, a model was developed to examine the effect of chlorhexidine on urinary pathogens [121]. The studies revealed that *P. stuartii*, *P. aeruginosa*, *P. mirabilis*, *E. coli*, *K. pneumoniae*, and *S. faecalis* growing in the urine in the bladder model rapidly recovered from the initial bactericidal effect of the antiseptic. The experiments also showed that the films of bacterial growth developed on the walls of the model were resistant to antiseptics and initiated the recovery of the cultures after the instillation. More recently, biofilms of *E. coli* and other urinary tract pathogens established on silicon discs have also been shown to survive well in chlorhexidine for up to 2 h, where as in urine suspension the cells were rapidly killed by the same concentration of the antiseptic [122,123]

When bacterial resistance to biocides are reported, cationic-type bactericides are often grouped together, despite the fact that marked differences exist in their chemical structure. It is possible that the less susceptible strains of gram-negative bacteria possess some means of limiting the concentration of biguanide that passes across the outer membrane [124]. Chemical analysis of the cell envelope of different strains provide useful data but do not necessarily explain different degrees of susceptibility. More subtle differences might have to be followed, for example, resistance in *Proteus* spp. to polymyxin has been attributed to a less acidic type lipopolysaccharide [125]. There is no evidence that chlorhexidine resistance in some gram-negative bacteria is plasmid-borne or that it is transferable [126–128].

XII. OTHER BIGUANIDES

The bactericidal activity of chlorhexidine gluconate is much greater than its respective monomers [3]. This stimulated the development of other polymeric biguanides. Picloxidine has been used in combination with quaternary ammonium compounds for surface disinfection. A polymeric form of biguanide particularly polyhexamethylene biguanide (PHMB) hydrochloride is used in a variety of antimicrobial applications. Cosmocil CQ is the cosmetic grade used as preservative in cosmetic formulations. Another bisbiguanide, alexidine has been widely recommended for use as an oral antiseptic and antiplaque compound [129]. Alexidine differs from chlorhexidine in that it possesses ethylhexyl rather than chlorophenol end groups.

REFERENCES

1. F. H. S. Curd and F. L. Rose, Synthetic antimalarials, Part X. Some aryldiguanide derivatives, *J. Chem. Soc.* Part II:729–737 (1946).
2. G. E. Davies, J. Francis, A. R. Martin, F. L. Rose, and G. Swain, 1:6-Di-(4′-chlorophenyl)diguanide-hexane (Hibitane)—laboratory investigation of a new antibacterial agent of high potency, *Br. J. Pharmacol.* 9:192–196 (1954).
3. G. E. Davies and B. S. Field, Action of biguanides, phenols, and detergents on *E. coli* and its spherplasts, *J. Appl. Bacteriol.* *32*:233–243 (1969).
4. N. Senior, Some observations on the formulation and properties of chlorhexidine, *J. Soc. Chem.* *24*:259–278 (1973).
5. D. D. Heard and R. W. Ashworth, The colloidal properties of chlorhexidine and its interaction with some macromolecules, *J. Pharm. Pharmacol.* *20*:505–518 (1969).
6. R. R. Goodall, J. Goldman, and J. Woods, Stability of chlorhexidine solutions, *Pharm J.* *200*:33–34 (1968).
7. R. W. Haley, D. H. Culver, and J. W. While, The nationwide nosocomial infection rate. A new need for vital statistics, *Am. J. Epidemiol.* *121*:159–162 (1985).
8. F. Daschner, Economic aspects of hospital infections, *J. Hosp. Infect.* *3*:1–4 (1982).
9. C. A. Laurence, In-vitro antimicrobial activity of chlorhexidine, *J. Am. Pharm. Assoc. (Sci. Ed.)* *49*:731–735 (1960).
10. P. M. Woodcock, Biguanides as industrial biocides. *Industrial Biocides* (K. R. Payne, ed.), John Wiley & Sons, Chichester, 1988, pp. 19–36.
11. L. A. Shaker, A. D. Russell, and J. R. Furr, Aspects of the action of chlorhexidine on bacterial spores, *Int. J. Pharm.* *34*:51–56 (1986).

12. T. L. Pitt, M. A. Gaston, and P. N. Hoffman, In-vitro susceptibilities of hospital isolates of various bacterial genera to chlorhexidine, *J. Hosp. Infect. 4*(2):173–176 (1983).
13. W. Hugo and A. R. Longworth, Effect of chlorhexidine on protoplasts and spheroplasts of *E. Coli*, *J. Pharm. Pharmacol. 16*:751–755 (1964).
14. A. Davies, M. Bentley, and B. S. Field, Comparison of the action of Vantocil, cetrimide, and chlorhexidine on *E. coli* and its spheropasts and the protoplasts of gram positive bacteria, *J. Appl. Bacteriol. 31*:448–461 (1968).
15. R. M. Calman and J. Murray, Antiseptics in midwifery, *Br. Med. J. 2*:200–204 (1956).
16. W. B. Hugo and A. R. Longworth, Some aspects of mode of action of chlorhexidine, *J. Pharm. Pharmacol. 16*:655–662 (1964).
17. W. B. Hugo and A. R. Longworth, Cytological aspects of the mode of action of chlorhexidine, *J. Pharm. Pharmacol. 17*:28–32 (1965)
18. W. B. Hugo and A. R. Longworth, The effect of chlorhexidine on the electrophoretic mobility, cytoplasmic constituents, dehydrogenase activity and cell walls of *E. coli* and *S. aureus*, *J. Pharm. Pharmacol. 18*:569–578 (1966).
19. T. H. Walters, J. R. Furr, and A. D. Russell, Antifungal action of chlorhexidine, *Microbios. 38*:195–204 (1983).
20. L. A. Shafer, A. D. Russell, and J. R. Furr, Aspects of the action of chlorhexidine on bacterial spores, *Int. J. Pharm. 18*:225–231 (1986).
21. I. Chopra, S. C. Johnson, and P. M. Bennett, Inhibition of *Providencia stuartii* cell envelope enzymes by chlorhexidine, *J. Antimic. Chemo. 19*:743–751 (1987).
22. J. G. R. Elferink and H. L. Booij, Interaction of chlorhexidine with yeast cells, *Biochemical Pharmacology 23*:1413–1419 (1974).
23. H. Bobichon and P. Bouchet, Action of chlorhexidine on budding *Candida albicans* scanning and transmission electron microscopic study, *Mycopathol. 100*:27–35 (1987).
24. G. W. Denton, Chlorhexidine. *Disinfection, Sterilization, and Preservation* (S. S. Block, ed.), Lea & Febiger, Philadelphia, 1991, pp. 278–281.
25. B. J. Meakin, Some physico-chemical factors affecting the antimicrobial activity of antiseptic formulations, *Br. J. Clin. Pract. 125* (Suppl.):12–22 (1983).
26. L. A. Shaker, A. D. Russell, and J. R. Furr, Aspects of the action chlorhexidine on bacterial spores, *Int. J. Pharm. 34*:51–56 (1986).
27. M. A. Harbison and S. M. Hammer, Inactivation of human immunodeficiency virus by Betadine products and chlorhexidine, *J. AIDS 2*:16–20 (1989).
28. G. A. Ayliffe, J. R. Babb, Jr., and A. M. Quoraishi, A test for hygienic hand disinfection, *J. Clin. Pathol. 31*:923–928 (1978).
29. D. G. Maki, M. A. Zilz, and C. J. Alvarado, Evaluation of antibacterial efficacy of four agents for hand washing, *Current Chemotherapy and Infectious Disease*, Proc. 11th Int. Cong. Chem. *11*:10 –1090 (1979).
30. A. F. Peterson, Comparative evaluation of surgical scrub preparations, *Surg. Gynecol. Obstet. 146*:63–65 (1978).
31. M. A. K. La Rocca, P. T. La Rocca, and R. La Rocca, Comparative study of three handwash preparations for efficacy against experimental bacterial contamination of human skin, *Adv. Ther. 2*:269–274 (1985).

32. E. J. L. Lowbury and H. A. Lilly, Use of 4% chlorhexidine detergent solution and other methods of skin disinfection, *Br. Med. J. 1*:510–515 (1973).
33. E. J. L. Lowbury, H. A. Lilly, and G. A. J. Ayliffe, Pre-operative disinfection of surgeon's hands, use of alcoholic solutions and effects of gloves on skin flora, *Br. Med. J. 4*:369–374 (1974).
34. K. Askgaard, Some actions of hand disinfection agents, *Ugeskrift. Laeger. 137*: 2515–2518 (1975).
35. D. G. Maki and J. A. Hecht, Comparative study of handwashing with chlorhexidine, povidone–iodine, and non-germicidal soap for prevention of nosocomial infection, *Clin. Res. 30*:Abstr. 303A (1988).
36. G. Stanley, C. Sheetz, M. Pfaffer, and R. Wenzel, The comparative effect of alternative handwashing agents on nosocomial infection rates, Proceeding 29th ICAAC, Houston, Abstr. 662, 1989.
37. S. A. Hammond, J. R. Morgan, and A. D. Russel, Comparative susceptibility of hospital isolates of gram-negative bacteria to antiseptics and disinfectants, *J. Hosp. Infect. 9*:255–264 (1987).
38. R. Aly, Antimicrobial activity of chlorhexidine gluconate against natural and artificial contamination during simulation of in-use conditions, *J. Pharm. Sci. 70*:964–968 (1981).
39. J. D. Jarvis, C. D. Wyne, L. Enwright, and J. D. Williams, Handwashing and antiseptic containing soaps in hospital, *J. Clin. Pathol. 32*:732–736 (1979).
40. J. Ojajarvi, An evaluation of antiseptics used for hand disinfection in wards, *J. Hyg. (Cambr.) 76*:75–81 (1976).
41. J. Ojajarvi, The importance of soap selection for routine hand hygiene in hospital, *J. Hyg. (Cambr.) 86*:275–283 (1981).
42. E. Larson, J. J. Leyden, and K. J. McGinley, Physiologic and microbiologic changes in skin, related to frequent handwashing, *Infect. Control 7*:18–24 (1986).
43. J. Ojajarvi, P. Makela, and I. Ransalo, Failure of hand disinfection with frequent handwashing. A need for prolonged field studies. *J. Hyg. (Cambr.) 79*:107–113 (1977).
44. H. G. Smylie, J. R. C. Logie, and G. Smith, From Phisohex to Hibiscrub, *Br. Med. J. 4*:586–593 (1973).
45. M. A. K. La Rocca and P. T. La Rocca, An evaluation of the antimicrobial effect of a hand sponge–brush impregnated with 4% CHG (Hibiclens), *Dev. Ind. Microbiol. 23*:543–546 (1982).
46. R. Aly and H. L. Maibach, Comparative evaluation of CHG (Hibiclens) and povidone–iodine (E-Z scrub) sponge–brushes for presurgical hand scrubbing, *Curr. Ther. Res. 34*:740–745 (1985).
47. W. R. Cole and H. R. Bernard, Inadequacies of present methods of surgical skin preparation, *Arch. Surg. 89*:215–222 (1964).
48. E. J. L. Lowbury and H. A. Lilly, The effect of blood on disinfection of surgeon's hands, *Br. J. Surg. 61*:19–21 (1974).
49. J. W. A. Bendig, Surgical hand disinfection—comparison of 4% chlorhexidine detergent solution and 2% triclosan detergent solution, *J. Hosp. Infect. 15*:143–148 (1990).
50. P. G. Vanden Broek, A. S. Lampe, G. A. M. Berbee, J. Thompson, and R. P. Mouton,

Epidemic of prosthetic valve endocarditis caused by *S. epidermidis*, *Br. Med. J. 291*:949–950 (1985).

51. D. Maki, Control of colonization and transmission of pathogenic bacteria in hospital, *Ann. Intern. Med. 89*:33–46 (1978).
52. B. G. Adams and T. J. Marrie, Hand carriage of aerobic gram negative rods may not be transient, *J. Hyg. 89*:33–46 (1982).
53. G. A. J. Ayliffe, Surgical scrub and skin disinfection, *Infect. Control 5*:23–27 (1984).
54. E. L. Larson, P. I. Eke, and B. E. Laughen, Efficacy of alcohol based hand rinses under frequent use conditions, *Antimicrob. Agents Chemother. 30*:542–544 (1986).
55. E. J. L. Lowbury, H. A. Lilly, and G. A. J. Ayliffe, Preoperative disinfection of surgeon's hand, use of alcoholic solutions and effects of gloves on skin flora, *Br. Med. J. 4*:369–372 (1974).
56. M. L. Rotter, Hygienic hand disinfection, *Infect. Control 5*:18–22 (1984).
57. J. A. Murie and S. G. Macpherson, Chlorhexidine in methanol for the preoperative cleansing of surgeon's hands: A clinical trial, *Scot. Med. J. 25*:309–311 (1980).
58. E. L. Larson and A. M. Butz, Alcohol for surgical scrubbing, *Infect. Control Hosp. Epidemiol. 11*:139–143 (1990).
59. D. Maki and J. Hecht, Antimicrobial agents and chemotherapy, Proceedings of the 22nd Interscience Conference on Antimicrobial Agents and Chemotherapy, 1981, Abstr. 699.
60. A. S. D. Rosenberg, S. D. Alatary, and A. F. Peterson, Safety and efficacy of the antiseptic chlorhexidine gluconate, *Surg. Gynecol. Obstet. 143*:789–792 (1976).
61. Centers for Disease Control, *Isolation Techniques for Use in Hospitals*, U.S. Government Printing Office, Washington, 1970, p. 9.
62. American Hospital Association, *Infection Control in the Hospital*, 3rd ed., AHA, Chicago, 1974, p. 111.
63. Z. A. Quraishi, M. McGuckin, and F. X. Blais, Duration of hand washing in intensive care units—a descriptive study, *Am. J. Infect. Control 11*:83–87 (1984).
64. B. P. Simmons, S. Hill, and G. F. Mallison, Guidelines for hospital environmental control, U.S. Department of Health and Human Services, October 1981, p. 3.
65. J. M. Conly, S. Hill, J. Ross, J. Lertzman, and T. J. Louie, Handwashing practices in an intensive care unit: The effects of an educational program and its relationship to infection rates, *Am. J. Infect. Control 17*:330–339 (1989).
66. R. K. Albert and F. Condie, Handwashing patterns in medical intensive care units, *N. Engl. J. Med. 304*:1465–1466 (1981).
67. A. Brandberg, J. Holm, and J. Hammerstein, Effect of preoperative whole-body disinfection by shower-bath with chlorhexidine, *Problems in the Control of Hospital Infection*, Royal Society of Medicine International Congress and Symposium Series 23, Academic Press, London, 1979, pp. 71–75.
68. S. Seeberg, A. Lindberg, and B. R. Bergman, Pre-operative shower bath with 4% chlorhexidine detergent solution, *Skin Microbiology* (H. I. Maibach, and R. Aly, eds.), Springer-Verlag, New York, 1981, pp. 86–91.
69. A. B. Kaiser, D. E. Kernodle, N. L. B. Barg, and M. R. Petracek, Influence of preoperative showers on *Staphylococcus* skin colonization, *Ann. Thorac. Surg. 45*:35–38 (1988).

70. L. J. Hayek, J. M. Emerson, and A. M. N. Gardner, Clinical trial of "Hibiscrub" whole-body disinfection in elective surgery, *J. Hosp. Infect. 10*:165–172 (1987).
71. L. J. Hayek and J. M. Emerson, Pre-operative whole body disinfection—a controlled study, *J. Hosp. Infect. 11* (Suppl. B):15–19 (1988).
72. P. E. Randall, L. A. Ganguli, and R. W. Marcuson, Wound infection following vasectomy, *Br. J. Urol. 55*:564–567 (1983).
73. P. E. Randall, L. A. Ganguli, M. G. L. Keaney, and R. W. Marcuson, Prevention of wound infection following vasectomy, *Br. J. Urol. 57*:227–229 (1985).
74. A. S. D. Spiers, S. F. Dias, and J. A. Lopez, Means of preventing bacterial infection, *Skin Microbiology* (H. I. Maibach, and R. Aly, eds.), Springer-Verlag, New York, 1981, pp. 305–316.
75. M. E. Byatt and A. Henderson, Pre-operative sterilization of the perineum, A comparison of six antiseptics, *J. Clin. Pathol. 26*:921–924 (1973).
76. K. K. Christensen, P. Christensen, A. K. Dykes, and G. Kahlmeter, Chlorhexidine for prevention of neonatal colonization with group B streptococci, *Eur. J. Obstet. Gynecol. Reprod. Biol. 19*:231–236 (1985).
77. S. E. Goldblum, J. A. Ulrich, and R. S. Goldman, Comparison of 4% CHG in a detergent base for skin preparation of hemodialysis patients and personnel, *Am. J. Kidney Dis. 11*:548–552 (1983).
78. A. Miller, Prevention of urinary tract infection after prostatectomy, *Lancet 2*:886–888 (1960).
79. A. J. Ball, Bladder irrigation with chlorhexidine for the prevention of urinary infection after transurethral operations. A prospective controlled study, *J. Urol. 138*:491–494 (1987).
80. D. Kirk, Hibitane bladder irrigation in the prevention of catheter associated urinary infection, *Br. J. Urol. 51*:528–531 (1979).
81. J. P. Mitchell, N. Slade, and K. B. Linton, Instrumental bacteraemia and its prevention, *Br. J. Urol. 34*:454–458 (1962).
82. P. J. Baker, R. A. Coburn, R. J. Genco, and R. T. Evans, Structural determinants of activity of chlorhexidine and alkyl biguanides against the human flora, *J. Dent. Res. 66*:1099–1106 (1987).
83. P. Gjermo, P. Bonescoll, and G. Rolla, Relationship between plaque inhibitory effect and retention of chlorhexidine in the human oral cavity, *Arch. Oral Biol. 19*:1031–1034 (1974).
84. R. M. Davies, S. Jensen, C. R. Schlott, and H. Loe, The effect of topical application of chlorhexidine on the bacterial colonization of the teeth and gingiva, *J. Periodont. Res. 5*:96–100 (1970).
85. J. Lindhe, G. Heyden, G. Svanberg, H. Loe, and C. R. Schlott, Effect of local applications of chlorhexidine on the oral mucosa of the hamster, *J. Periodont. Res. 5*:177–181 (1970).
86. T. D. Hennessey, Some antibacterial properties of chlorhexidine, *J. Periodont. Res. 8*:61–67 (1973).
87. M. V. Martin and D. Nind, Use of chlorhexidine gluconate for preoperative disinfection of apicectomy sites, *Br. Dent. J. 162*:459–461 (1987).
88. E. A. Field, D. Nind, E. Varga, and M. V. Martin, The effect of chlorhexidine

irrigation on the incidence of dry socket, a pilot study, *Br. J. Oral Maxillofac. Surg. 26*:395–401 (1988).
89. G. A. Ferretti, Therapeutic use of chlorhexidine in bonemarrow transplant patients, *Oral Surg. Oral Med. Oral Pathol. 63*:683–687 (1987).
90. S. Jacobsen, I. L. Brynhni, and P. Gjermo, Oral candidiasis—frequency, treatment, and relapse tendency in a group of psychiatric inpatients, *Acta Odontol. Scand. 37*:353–357 (1979).
91. I. Olsen, Denture stomatitis: The clinical effects of chlorhexidine and amphotericin B, *Acta Odontol. Scand. 33*:47–52 (1975).
92. E. Budtz-Jorgensen, Hibitane in the treatment of oral candidiasis, *J. Clin. Periodontol. 4*:117–128 (1977).
93. M. Addy, J. Moran, A. A. Griffiths, and N. J. Willis-Wood, Extrinsic tooth discoloration by metal and chlorhexidine, *Br. Dent. J. 159*:281–285 (1985).
94. L. Z. G. Touyz and M. Rosen, Disinfectin of alginate impression material using disinfectants as mixing and soak solutions, *J. Dent. 19*:255–257 (1991).
95. C. A. Penley, R. C. Schlitzer, D. G. Ahearn, and L. A. Wilson, Laboratory evaluation of chemical disinfection of soft contact lenses, *Intraocular Lens Med. J. 7*:101–110 (1981).
96. C. A. Penley, D. G. Ahearn, R. C. Schlitzer, and L. A. Wilson, Disinfection of contact lenses, *Intraocular Lens Med. J. 7*:197–204 (1981).
97. N. E. Richardson, D. J. G. Davies, B. J. Meakin, and D. A. Norton, Loss of antibacterial preservatives from contact lens solutions during storage, *J. Pharm. Pharmacol. 29*:717–722 (1977).
98. P. B. Donzis, B. J. Mondino, B. A. Weissman, and D. A. Bruckner, Microbial contamination of contact lens care systems, *Am. J. Ophthalmol. 104*:325–333 (1987).
99. M. S. Mayo, R. L. Schlitzer, M. A. Ward, L. A. Wilson, and D. G. Ahearn, Association of *Pseudomonas* and *Serratia* corneal ulcers with use of contaminated solutions, *J. Clin. Microbiol. 25*:1398–1400 (1987).
100. P. A. Gandhi, A. D. Sawant, L. A. Wilson, and D. G. Ahearn, Adaptation and growth of *S. marcescens* in contact lens disinfectant solutions containing chlorhexidine gluconate, *Appl. and Environ. Microb. 59*:183–188 (1993).
101. L. A. Wilson, A. D. Sawant, R. B. Simmons, and D. G. Ahearn, Microbial contamination of contact lens storage cases and solutions, *Am. J. Ophthalmol. 110*:193–198 (1990.
102. P. A. Parment, R. Ronnerstam, and M. Walder, Persistance of *S. marcescens* and *E. coli* in solutions for contact lenses, *Acta. Ophthalmol. 64*:456–462 (1986).
103. D. M. Foulkes, Some toxicological evaluations on chlorhexidine, *J. Periodont. Res. 8*:55–57 (1973).
104. D. Hubben, Safety testing of chlorhexidine gluconate. Animal data, Paper presented at symposium entitled "Skin Antisepsis: Chlorhexidine Gluconate," 1977.
105. D. E. Case, Chlorhexidine: Attempts to detect percutaneous absorption in man, *The Control of Hospital Infection* (S. W. B. Newsom, and A. D. S. Caldwell, eds.), Royal Society of Medicine International Congress and Symposium Series, 23, London, RSM/Academic Press, London, 1980, pp. 39–43.
106. L. E. Gongwer, The effects of daily bathing of neonatal rhesus monkeys with an

antimicrobial skin cleanser containing chlorhexidine gluconate, *Toxicol. Appl. Pharmacol. 52*:255–258 (1980).
107. H. Kodama, T. Hashimoto, M. Tsuruoka, Y. Kido, M. Uyeda, and M. Shibata, *EISEI KAGAKU 34*: 408–413 (1988).
108. N. Sugio and S. Kojima, Biological treatment of chlorhexidine digluconate containing waste water, *Jap. J. Toxicol. Environ. Health. 38*:1–11 (1992).
109. N. Sugio and S. Kojima, Biological treatment of chlorhexidine digluconate containing waste water through tubular contact aeration treatment, *Jap. J. Toxicol. Environ. Health. 39*:242–246 (1993).
110. N. Sugio and S. Kojima, Treatment of chlorhexidine digluconate containing waste water, *Jap. J. Toxicol. Environ. Health. 38*:329–333 (1992).
111. H. Nikaido and M. Vaara, Molecular basis of bacterial outer membrane permeability, *Microbiol. Rev. 49*:1–32 (1985).
112. R. E. W. Hancock, Alterations in outer membrane permeability, *Annu. Rev. Microbiol. 38*:237–264 (1984).
113. D. J. Stickler and B. Thomas, Intrinsic resistance to nonantibiotic antimicrobial agents. *Principles and Practice of Disinfection, Preservation, and Sterilization* (A. D. Russell, W. B. Hugo, and G. A. J. Aylifee, Eds.), Blackwell Sci. Publ. Oxford, 1982, pp. 186–198.
114. D. J. Stickler, B. Thomas, C. L. Claton, and J. C. Chawla, Studies on the genetic basis of chlorhexidine resistance, *Br. J. Pract. Symposium Number 25*:23–28 (1983).
115. A. D. Russell and J. R. Furr, The antibacterial activity of a new chloroxylenol preparation containing EDTA, *J. Appl. Bact. 43*:253–260 (1977).
116. J. C. Kelsey and I. M. Maurer, The use of chemical disinfection in hospitals, Public health monograph series-2, London, HMSO.
117. E. J. L. Lowbury, G. A. J. Ayliffe, A. M. Geddes, and J. D. Williams, In *Control of Hospital Infection*, Published by Chapman & Hall.
118. D. J. Stickler, C. B. Wilmot, and J. D. Flynn, The mode of development of urinary infection in intermittently catheterized male paraplegics, *Paraplegia 8*:243–252 (1971).
119. J. D. O'Flynn and D. J. Stickler, Disinfectants and gram-negative bacteria, *Lancet 1*:489–490 (1972).
120. D. J. Stickler and B. Thomas, Antiseptic and antibiotic resistance in gram-negative bacteria causing urinary tract infection, *J. Clin. Pathol. 33*:288–296 (1980).
121. D. J. Stickler, C. L. Clayton, and J. Chawla, The resistance of urinary pathogens to chlorhexidine bladder washouts, *J. Hospit. Inf. 10*:28–39 (1989).
122. D. J. Stickler, J. Dolman, S. Rolfe, and J. Chawla, Activity of antiseptics against *E. coli* growing as biofilms on silicone surfaces, *Eur. J. Clin. Microbiol. 8*:974–978 (1990).
123. D. J. Stickler, J. Doleman, S. Rolfe, and J. Chawla, Activity of some antiseptics against urinary tract pathogens growing on biofilms on silicone surfaces, *Eur. J. Clin. Microbiol. Infec. Diseases 10*:410–415 (1991).
124. M. J. Norris, D. T. Rogers, and A. D. Russell, Cell envelope composition and sensitivity of *Pseudomonas mirabilis*, *Pseudomonas aeuroginosa* and *Serratia marcescens* to polyxin and other antibacterial agents, *Lett. Appl. Microbiol 1*:3–6 (1985).

125. K. Sidorczyk, U. Zahringer, and E. T. Rietschel, Chemical structure of the lipid A component of the lipopolysaccharide from *Pseudomonas mirabilis*, *Eur. J. Biochem.* *137*:15–22 (1983).
126. I. Ahonkai and A. D. Russell, Response of RPI^+ and RPI^- strains of *Escherichia coli* to antibacterial agents and transfer of resistance to *Pseudomonas aeruginosa*, *Curr. Microbiol.* *3*:89–94 (1979).
127. I. Ahonkai, W. J. Pugh, and A. D. Russell, Sensitivity to antimicrobial agents of some mercury sensitive and mercury resistant strains of gram-negative bacteria, *Curr. Microbiol* *11*:183–185 (1984).
128. A. D. Russell, The role of plasmids in bacterial resistance to antiseptics, disinfectants, and preservatives, *J. Hosp. Inf.* *6*:9–19 (1985).
129. P. Gjermo, G. Rolla, and L. Arskang, The effect on dental plaque formation and some in vivo properties of 12 bis-biguanides, *J. Periodont. Res.* *8*:81–88 (1973).

12

Chloroxylenol: An Old–New Antimicrobial

Mary K. Bruch
Mary Bruch Micro-Reg, Inc.
Leesburg, Virginia

I. INTRODUCTION AND HISTORY

Chloroxylenol (PCMX) has a rather enigmatic position in the array of antimicrobial chemicals in current use. A number of uses for PCMX were tested in the 1950s as the inclusion of antimicrobials in our commercial products greatly expanded after World War II. The increasing uses diminished, however, as the popularity of a new chemical, hexachlorophene, became directly competitive with PCMX. This situation reversed when most uses of hexachlorophene were prohibited by the U.S. Food and Drug Administration (FDA) in 1972. During this past 20 years of post-hexachlorophene formulation, many companies have chosen to use chloroxylenol in place of hexachlorophene in products from surgical hand scrubs and handwashing products to preservatives in cosmetics and cutting oils.

Chloroxylenol is one of our oldest chemical entities being used for its antimicrobial properties. The first widely used disinfectants were black and white solutions made from coal tar extracts [1]. German patents No. 300321, 1913 and 302013, May 7, 1913 to Schulke and Mayr and Fleming, described the enhancement of bactericidal effectiveness with mixtures of cresols and symmetrical xylenols.

One group of compounds found in the extracts was halogen-substituted xylenols. During the late teens and 1920s, German chemists formulated a series of substituted xylenols (substituted in all possible permutations) with chlorine and bromine. The very precise testing of these chemical substitutes was done with exquisite precision, using equimolar amounts of chemicals against the test organism [2].

Several substituted xylenols showed good activity, but were discarded for different reasons. A selection of one of the best compounds to be used as a disinfectant was *p*-chloro-*m*-xylenol [3]. One company interested in PCMX left Germany for England to eventually establish a company and a product still used and recognized throughout the English-speaking world as Dettol, produced by Reckitt and Colman. Before proceeding with this backward look at how the use of this chemical developed, a description is in order.

Chemically, the compound can be described as a substituted phenol or 3,5-dimethyl-4-chlorophenol (Fig. 1), with a molecular weight of 156.6. Chloroxylenol is soluble in alcohol, ether, benzene, terpenes, fixed oils, and solutions of alkali hydroxides. Chloroxylenol is a white to off-white crystalline powder, depending on synthesis, and it has a slight phenolic odor. It has only slight solubility in water, which affects its formulation significantly. The UV light absorption spectra occur at a maxima of 282 nm in ethanol, 279 nm in 0.1 *N* hydrochloric acid, and 296 nm in 0.1 *N* sodium hydroxide. This compound reacts with oxidizing agents and does not readily form insoluble salts.

Although the patent was not issued until 1940 to George Gladden and W. W. Cocker, it was synthesized by at least two chemists in 1923 and 1924 [4,5].

Most readers are familiar with the early use of phenol by Lister (1867), applying the persuasive lessons from the publications of Pasteur to surgery, he selected phenol because, as a surgeon, he was familiar with its use to cover foul odors. He literally used saturation disinfection in his surgical theater, encompassing the air, drapes, bandages, and surgical equipment, and so, inspired the world to revolutionize surgical procedure [6].

Lister, as with many innovators in the last half of the 19th century, built on the work of observers in the early years of the century. Prevost, Kuchenmeister, and Jaques Bechamp described the uses of chemicals for their antiseptic action. Many times these historic figures have relied on intuition and their own observations. It is not clear how much communication occurred between them. We certainly know that the years between 1850 and 1900 produced remarkable advances in science and saw the birth of microbiology; witnessing very quickly, innovation to eliminate the microorganisms only recently identified.

We have observed that phenol was used early in efforts to reduce the frustrating effects of infection. Soon attempts were made to find means for

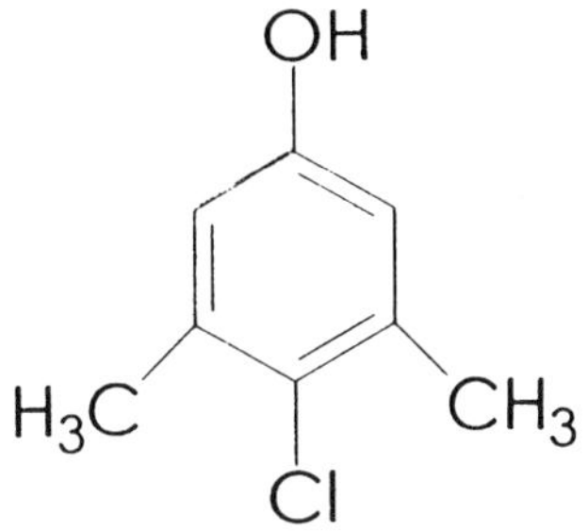

FIGURE 1 Chloroxylenol is also known as *p*-chloro-*m*-xylenol (PCMX), 4-chloro-3,5-xylene, 2-chloro-*m*-xylenol, 2-chloro-*m*-xylene, 2-chloro-5-hydroxy-1,3-dimethyl-benzene.

commercializing these findings and to produce a formulation less toxic than phenol [7–9]. One source of phenolic chemicals was coal tar extracts, which had been formulated by emulsification by LeBeuf in 1850. With the recognized insolubility of creosotes and xylenols, Jeyes (Jeyes Fluid, 1890s) made an emulsified product based on solubilized tar compounds with pine soap, which became the forerunner of Lysol and Dettol. The British used the German technology to produce what have been named Black Fluids and White Fluids; the first, solubilized with soap, and, the second, with other emulsifiers such as casein or extracts from seaweed.

A highly refined examination of halogenated derivatives of phenol was published in 1932 by Etinger-Tulczynska and Ulrich [10], who reported on investigators who had found that the halogen substitution of methyl derivatives of phenol and of xylenols were more active than phenol. Chloroxylenol was identified as one of the most active from this work. Cocker used this compound by emulsifying it with pine oil (terpineol) to produce Dettol. Many imitators have produced similar products, but Dettol remains a product used worldwide both institutionally and by consumers. Today, a variety of topical and industrial products are marketed with chloroxylenol as an active or preservative. In addition to worldwide acceptance of PCMX as a safe product, based on almost 60 years of extensive use, extensive oral and systemic exposure to humans, human topical application, and animal studies have been accumulated in the United States.

There have been several government-initiated reviews of chloroxylenol. The FDA began their review in 1972 and, in this same year, took action to prohibit marketing of hexachlorophene. Chloroxylenol was reviewed only cursorily in the initial Antimicrobial I Review in the Over-The-Counter Drug Review because only scanty information was submitted and, at that time, it was not formulated into either hospital or consumer handwashing products. The initial purview of the Antimicrobial I Panel was topical products without specific indications in infections. When the Antimicrobial II Panel reviewed PCMX, more data were submitted, including British data, and at that time, a prominent use was in an antifungal athlete's foot product. Most ingredients in the Antimicrobial I Review were placed in category III (not enough information available to make a decision as category I, or "generally recognized as safe and effective") [11,12].

Subsequently, the Cosmetic, Toiletries, and Fragrance Association (CFTA) and FDA again reviewed chloroxylenol, using much of the data in the OTC review as part of the Cosmetic Ingredient Review (CIR). They concluded that PCMX was safe to include in cosmetic products [13]. Chloroxylenol has been used in Europe as a preservative in cosmetics [14], in cleaning solutions as a disinfectant [15], and as an antiseptic for topical use [16–18].

In the years from 1980 to the present, a list of submissions of data has been made to FDA to form the basis for making PCMX category I [19]. Some questions were raised in the OTC review about toxic effects, but these have been resolved, and chloroxylenol has been categorized as I for safety by CIR and for use in athlete's foot and other fungal infections. There is little or no evidence of toxic effects from dosing with chloroxylenol, but the FDA has continued to request a long-term study more in conformity to new drug standards, rather than those of the OTC review for the repeatedly used handwashing products. This Category III status has continued into the Tentative Final Monograph, June 17, 1994, Federal Register.

Guess and Bruch [20] reviewed the toxicity data available on chloroxylenol, and their conclusions that the accumulated data do provide a basis for general recognition of safety were based on these data.

II. TOXICITY STUDIES

Animal studies have long been the basis of prediction of toxicity in humans, often based on dosing with very high levels of a chemical. The systemic toxicity of topically absorbed drug chemicals was virtually ignored until the hexachlorophene disaster reported in France from the contamination of baby powder with 6% hexachlorophene. Following these revelations, other antimicrobials were examined to see if the minute levels absorbed could be toxic, as they were with hexachlorophene. Although other chemicals are absorbed after topical application, there does not appear to be another "hexachlorophene" among the antimicrobials reviewed.

A. Animal Studies

There is one name that is paramount in the early animal and human studies with PCMX, Bernard Zondek [21]. A physician, he began investigating the potential of chloroxylenol for systemic users in infection, since he was familiar with descriptions of the antiseptic properties of PCMX as early as 1912, and had read of the use of Dettol in the control of puerperal fever, by Colebrook and Maxted [22] in 1933 in England, with handwashing and obstetrical use.

In initial animal studies Zondek injected chloroxylenol dissolved in olive oil subcutaneously (SC) into rabbits and collected urine over 5 days, recovering 15% of the administered dose and observing no toxic effects. He also applied 1 g of PCMX in tinctures and ointments to the shaved backs of rabbits and recovered 12–16% of the dose in 7 days. The goal of his studies was to use chloroxylenol in the treatment of human urogenital infections. Since he observed no toxic effects after the administration of a high dose in animals, he concentrated on human studies.

Zondek [23] repeated the subcutaneously injected chloroxylenol (1 g) in rabbits to examine metabolic fate. As will be described later, he did determine metabolites in humans, but not in this test. There were further metabolic studies in rats [24] injected with 100–200 mg chloroxylenol, and the urine, feces, and sacrificed body were analyzed at various times for PCMX. Again, about 15% of the dose was recovered in the feces. Yet, after 30 min 85–90% was recovered in the body in the free form, but after 48 h, only 10% was recovered from the body. He accounted for 55% of the dose and concluded that metabolites were formed that were not detected by his analytic methods. There was no reference to any toxic effects from these large doses.

Viewing these 50-year-old studies from 1995, one may criticize their analytic procedures or other details, but in reading them one is filled with admiration of the level of sophistication for their time, for instance, in using a high-dose administration and examining feces, urine, and the body for excreted chemical. Later we will see in the human studies, further refinements in his tests. These pioneers have directed our work in the channels we think we have selected ourselves.

Zondek's work was coincidental with the development of antibiotics and sulfanilamide, so that chloroxylenol did not develop as a systemic drug, since these newer ones appeared to be more effective in urogenital infections.

Further animal studies were delayed until 1952, and the publication of the first median lethal dose (LD_{50}) study. This study by Joseph [25] in mice was done by intraperitoneal (IP) injection of aqueous suspension, after which he found a minimum lethal dose of 115.2 g/kg. Calculations gave a very steep LD_{50} curve. Other studies described here are unpublished, but were submitted to the OTC or CIR reviews.

Further studies performed for Ottawa Chemical Co. (long-time producers of chloroxylenol in the United States) included an acute inhalation study in rats that showed no gross signs of systemic toxicity, and the authors concluded that, under their exposure conditions of 205 mg/L for 1 h, chloroxylenol was not toxic by inhalation.

Another LD_{50} was performed in rats resulting in an oral LD_{50} of 3.83 g/kg.

The authors recorded that no confidence limits could be calculated because of the very steep curve and the all-or-nothing effect of chloroxylenol dosing. The effect of dosing described in the study included depression, depressed righting, and other reflex depressions, indicating central nervous system toxicity as the cause of death.

Many studies submitted to the OTC review were performed with the Dettol formulation. This formulation contains 4.8% chloroxylenol, 10% alcohol, and 20% terpineol in a castor oil-based soap. The contribution of the other ingredients complicates the conclusions concerning chloroxylenol, but it should also be recognized that these additional ingredients would contribute to greater toxicity. However, even when the additional potential from added ingredients is factored, a very low order of toxicity emerges from a profile projected by these studies.

A variety of animal studies with Dettol were reported in the OTC material and were mostly submitted by the English manufacturer, Reckitt and Colman.

A 4-week study in rats was conducted with varying concentrations from 25 to 100% of Dettol administered by oral gavage. Undiluted Dettol produced emesis in the animals, so that only a 2 ml/kg per day was tolerated. Another dose-ranging study used 4 ml/kg per day for 4 weeks and then 5 ml/kg per day of a 50% dilution for 4 more weeks or 8 ml/kg per day. There were significant problems with tolerance of the doses, and the highest dose showed gross macroscopic abnormalities, whereas, although the animals could not tolerate dosing well, there were no other abnormalities on posting. It was concluded that a 13-week study at a dose of 5 ml/kg per day could be done. A 13-week follow-up study was conducted with the 5-ml dose as the highest dose. At doses ranging as high as 60 and 120 mg/kg per day of chloroxylenol, there were no significant gross or histopathological effects related to dosing. No target organ could be identified [26].

The metabolic rate of chloroxylenol in Dettol was examined by Havler [27], in several experiments after oral and percutaneous absorption of ^{14}C-labeled chloroxylenol in Dettol. Absorption, tissue distribution, and excretion data were obtained after dosing rats. Chloroxylenol at a dose of 48 mg/kg was given orally and 4 and 48 mg/kg, percutaneously.

Beagle dogs were also dosed orally with labeled Dettol at a 12-mg/kg dose of chloroxylenol, and its excretion was followed for 24 h. The results of these metabolic–excretion studies showed absorption and almost complete excretion in 24 h. Peak levels were found in the bloodstream 30 min after oral dosing and 2 h after percutaneous administration. Similar pathways and metabolites were found in the rat and the dog and were identified as a 6:1 ratio of glucuronide and sulfate conjugate. These findings will be reflected in the human studies described later.

A major study evaluating the toxicity potential of PCMX performed with Dettol was a 4-week dose-ranging study in beagle dogs [28], conducted to determine appropriate dose levels for a larger oral study. The dose of PCMX in Dettol amounted to 96 mg/kg, 192 mg/kg, and 384 mg/kg. An intermediate dilution of 120 mg PCMX per kilogram was given to a group of dogs that had already received 152 mg/kg for 4 weeks. This additional dosing continued another 4 weeks. The major event seen after dosing was occasional vomiting, which was seen at all levels, except when the additional time group received a 50% dilution of Dettol. This vomiting effect was probably vehicle-induced. Macroscopic examination of organs revealed edema of the pancreas and congestion of the kidneys in one animal in the high-dose group. Lower thymus weights of the high-dose group were noted, as it was for the spleen and pancreas. No other abnormalities were seen in any of the other dogs except overall weight loss in the high-dose group. On the basis of this range-finding experiment, it was concluded that a dose of 5 ml/kg per day of a 50% solution of Dettol would be a suitable high-dose level (120 mg/kg per day).

A second study in beagle dogs [29] was conducted over a 13-week period, using diluted Dettol solutions that gave PCMX concentrations of 1.25 mg/kg, 60 mg/kg, and 120 mg/kg daily by oral gavage. Four groups of six dogs were used; three groups received test compound, and the fourth group received a water control. Each individual group was composed of three male and three female dogs. The results of this study revealed a general lack of significant toxicity from PCMX in a solubilizing vehicle. There were no deaths in any animals. Clinically, the only signs noted was occasional vomiting in an animal receiving a 25 or 50% dilution of Dettol.

There were no dose-related body weight changes, no food nor water intake changes, and no macroscopic abnormalities at postmortem. Although most individual organ weight values were within normal ranges, the mean liver weights for dosed groups were greater than control, and the differences were dose-related. Even though liver weights were higher in the dosed groups, there were no histological changes noted in the liver nor any other tissues examined.

There was very little toxicity over a 13-week period at oral doses to 120 mg/kg. No target organ was identified except, possibly, the liver, which showed some slight weight increase, but no histological change. This very good subchronic exposure in dogs over 13 weeks would translate to the safe ingestion of 8.4 g daily in an adult human.

A subchronic 90-day oral gavage experiment using rats as the test animal was reported by Pennwalt Company [30]. In this study, propylene glycol served as the vehicle and control solution. The experiment compared PCMX in propylene glycol with Fresh Soap (3.75% PCMX in 50% propylene glycol), with hexachlorophene. The volume of each dose was maintained at 1 ml/kg. The doses of PCMX administered by gavage were 8 mg/kg, 24 mg/kg, and 75 mg/kg. A total of 280 rats or 40 rats per group were used. At the end of 90 days, 140 rats were sacrificed. The remaining animals were to be continued as a part of a 2-year exposure program. At the end of 90 days, an evaluation of many parameters, including weight, pathology, clinical signs, and hematology were conducted. The results clearly indicated that at the 8-mg/kg dose level there were no treatment effects. The effects from both the 25-mg/kg and 75-mg/kg doses were very mild and essentially the same: slight hemoconcentration, leukocytosis, and monocytosis. There was no dose–effect relationship. At the highest dose of 75 mg/kg, there was a fairly nonspecific response of some animals that showed nasal and ocular exudates. Even with this length of exposure, no specific target organ could be identified.

Although it is known that absorption from topical application is not as rapid or as complete as from an oral dose; nevertheless, it is essential to have topical exposure data just in case an aberrant chemical gave the opposite results.

Three topical exposure experiments, one in rabbits over a 30-day period, one in rabbits over a 90-day period, as well as one in dogs over an 8-month period, demonstrated no surprises in the evaluation of the toxicity potential of PCMX. In the first rabbit study [31], seven groups of rabbits were exposed to PCMX at dose levels of 1200 mg/kg, 360 mg/kg, and 112.5 mg/kg. Other groups were exposed to Fresh Soap solutions, hexachlorophene, or vehicle. Extensive serum chemistry, hematology studies, and weight gain profiles were performed. The test solutions were placed into a gauze pad fixed to either thorax or lumbar regions for ten doses, then moved. Clinical observations were made twice daily. One rabbit at the 1200-mg/kg dose and one at the 360-mg/kg group died during the experiment. None of the animals in the PCMX group experienced weight loss problems or dose-related changes in serum chemistry or hematology.

A 3-week study in abraded skin in rabbits and a 90-day intact skin study in rabbits was conducted for the Ottawa Chemical Company in 1965 [31]. A control group received 1.0 mg/kg per day of propylene glycol vehicle, a second group received 18 mg/kg per day of PCMX in propylene glycol, and a third group received 180 mg/kg per day of PCMX in propylene glycol. The skin of three rabbits from each of the three groups was abraded initially and at the beginning of the second and third week. The skin of the remaining rabbits was left intact. Dosing was 5 days per week, with the test solutions massaged with a glass rod into the shaved skin until dry. At the end of 3 weeks, three animals from each group were allowed to go dose-free for 2 weeks, sacrificed, and examined by gross and microscopic analysis for tissue or organ change. The animals remaining on treatment for 13-weeks were also sacrificed at the beginning of the 14th week and also autopsied. Initially, and just before sacrifice, a complete blood count and urinalysis were performed on each animal. The results showed absolutely no treatment-related effects, as evidenced by gross pathology, microscopic examination of representative tissues, blood or urine changes.

Chloroxylenol has been studied after chronic topical exposure in beagle dogs [30]. In this study, dogs were exposed daily to 40% solutions, 12% solutions, or 3.75% solutions of PCMX, to 20% Fresh Soap with PCMX or soap placebo, or to 1.6% hexachlorophene. The test material was placed under a gauze pad held in place by a canvas jacket, and the sites were therefore occluded. After a 10-day exposure, the application site was changed from the thoracic to the lumbar region and then back again after another 10 days. Changes were made in the original protocol because of serious erythema at the skin site and weight loss. For the first 123 days of the experiment, the exposure dosages of PCMX under occlusion were 600 mg/kg, 180 mg/kg, and 56.2 mg/kg. At day 124 through day 242, the dosage rate was reduced to 200 mg/kg, 60 mg/kg and 19 mg/kg, and 37.5 mg/kg. The change in dosing schedules on day 124 caused all animals to regain weight loss and largely reversed the cutaneous clinical signs. The original signs of cutaneous damage and weight loss did not recur when the dosage was increased. One dog in the 1.6% hexachlorophene group and one dog in the 40% PCMX group died spontaneously. All other animals survived the 8-month period. As might be expected, the most severe clinical signs of erythema and cutaneous damage were among the dogs receiving the 40% PCMX solution, with decreasing signs of irritation with the other PCMX doses. All dogs exhibited erythema, some with fissuring and exudate, but it should be recalled that all dosing sites were also occluded. When the dosing volume was reduced from 1.5 to 0.5 ml/kg, there was a significant decrease in these clinical signs. The average body weight of dogs receiving the 40% PCMX was significantly lower than the other groups receiving test material, and all of these were below vehicle control. When dosing rate was decreased, the weight loss among the 40% PCMX animals was reversed. However, since there were no clinical signs of toxicity, and the hematology and clinical chemistry of test dogs was not altered, the weight loss may have been associated with the extreme irritation at the site of application. Additionally, since the skin at the site was irritated to the point of fissuring in some cases, there is a possibility of greatly enhanced absorption of the test chemical from the site. Some lethargy was seen in groups receiving Fresh Soap vehicle (no PCMX) as well as in some PCMX groups. There was no gross sign of systemic toxicity attributed to PCMX after 8 months of daily dosing with rather massive doses under occlusion. It should be remembered that these doses were daily under occlusion and cumulative in the sense that they were never removed by rinsing or washing.

The effect of Dettol on wound healing was evaluated in rats following a midline incision compared with controls. Tests of wound strength showed that 5% Dettol had no significant effect on the rate of wound healing. Undiluted Dettol did show delay in healing, but the delay was overcome in 10 days [32].

Pennwalt also planned a four-generation reproductive study [30], but it was discontinued just after the end of the F_1 and F_2 generations. Animals were treated with oral gavage with PCMX for 100–150 days before mating. The F_1 rats were bred at 120–150 days of age after dosing for 99–129 days. After observation of the parameters selected to evaluate this test, the litter weight was reduced at the dose level of 75 mg/kg per day for the F_1 and F_2 generations for chloroxylenol. The chloroxylenol formulated soap and the 75 mg/kg per day chloroxylenol had lower numbers of pups per female bred.

B. Human Studies

Not only have animal studies been conducted to evaluate the toxicity potential of PCMX, but humans were used, in 1942 when Zondek [21] began to use massive doses by various routes of administration to treat urogenital infections. In this early paper, Zondek reported on 150 patients to whom he administered doses of 6 g PCMX orally and intramuscular (IM) injections of 1 g PCMX in olive oil. He did not specify how many patients were in each dose group. One female received 20 g PCMX mixed with PCMX benzoate intragluteally. One week later 8 g was administered topically in the form of a 25% solution in alcohol and glycerin. To treat urogenital infections, Zondek recommended that on day 1 a patient receive (a) twice that day, a 10-ml injection of a 10% solution (2 g IM); (b) 33 g of a 30% PCMX ointment on the right arm in the morning and on the left arm in the evening (20 g topically) for a total of 22 g PCMX. This was to be repeated on day 2, with one injection and one ointment application; on day 3, one ointment application only; day 4, an injection and ointment treatment daily. If the urogenital infection persisted, treatment with injections and ointment could continue. The total dose of PCMX administered to the patient in 1 week amounts to 84 g of PCMX with no systemic reactions.

In another experiment Zondek [23] concentrated on locating PCMX after injection or inunction. He found that 50 g applied topically resulted in approximately 3 mg/dl in blood 4–5 h after application. After 1 g was injected IM, 1 mg/dl could be detected in the blood. In another study, he injected 1 or 2 g PCMX intramuscularly and found that in three patients no demonstrable amount could be found in the blood up to 48 h later, but in other cases, 1 mg% was found 1–2 h after injection. The highest value found was 4 mg%, 12 h after injection in a case of hepatitis. When given orally or rectally in doses of 10 and 2 g, respectively, the highest blood values found were 2 mg% orally and 1 mg% rectally. Doses of 4 g rectally gave blood values of 3 mg%. Topical administration of doses of 2 g showed no blood values, 5 g gave only traces in the blood and 8 g resulted in 1–4 mg%, whereas 20 g resulted in a maximum of 4 mg% (40 μg/ml) [24].

Even with the massive doses of PCMX used by Zondek in humans, he reported no adverse systemic reactions, even when these doses were administered over several days. In later years, the lack of toxicity from PCMX was confirmed in the animal studies reported in the foregoing. Additionally, Zondek reported that blood concentrations of PCMX rapidly declined, showing rapid metabolism and elimination by the urine as glucuronates and sulfates. This too was later confirmed in animal studies. The insertion of metabolic animal studies in this discussion of human testing serves as the preliminary work for the human-bathing studies performed by Reckitt and Colman. For example, in a study performed by Reckitt and Colman [33], the metabolism of [^{14}C]PCMX was examined in two strains of rats after IM, SC, and oral administration of Dettol and intravenous (IV) administration of PCMX. Despite being genetically deficient in UDP-glucuronyl-transferase, the Gunn–Wistar rats appeared to metabolize PCMX as rapidly as did the Sprague-Dawley rats. In addition to the metabolic studies, tissue distribution studies on [^{14}C]PCMX were performed on a variety of tissues at 15-, 30-, 60-, and 120-min intervals. It is interesting to note that brain levels of labeled PCMX were insignificant in either rat strain. As might be expected, there was a rapid appearance of PCMX in plasma, whole blood, and kidney following IM injection. Following IV injection, blood levels were essentially zero by 1 h.

In a second experiment conducted by Reckitt and Colman [27] to study the metabolism of PCMX, there were extensive absorption, tissue distribution, and excretion data generated after oral and percutaneous administration of [^{14}C]PCMX to female rats. Dosing was with a 25% solution of Dettol orally at a rate of 4 mg/kg, which is a dose of 48 mg of PCMX per kilogram. For percutaneous absorption studies, a 25% solution of Dettol, 4 mg/kg, was applied to a lint dressing, with aluminum foil backing, to the shaved and abraded backs of the rats. The occluded pad was removed after 6 h and the skin swabbed with cotton wool soaked in detergent. Both the dressing and cotton swabs were extracted in methanol and analyzed for residual labeled PCMX. Male beagle dogs were dosed orally with 25% Dettol in water at a rate of 1 mg/kg (12 mg/kg). Blood samples were taken up to 24 h after dosing. Urine and feces were collected for 5 days, after which the animals were sacrificed and selected tissue analysis performed.

The results of this study showed that practically all of the oral, labeled PCMX was recovered in the first 24 h, indicating complete absorption. In the percutaneous study, approximately half of the administered dose was excreted, with most of the remainder being on the lint dressing. Most of the label was recovered in the urine with both routes of administration. The plasma levels following oral administration reached a peak 30 min after dosing, declining steadily until very little remained at 24 h. After percutaneous dosing, peak plasma levels occurred 2 h after dosing, declining steadily up to the 6-hour period, when the dressing was removed. Following removal, there was a rapid plasma decline, with very little PCMX in the plasma by 24 h postdosing. In the plasma, the level of free PCMX was essentially undetectable, all the label being in the form of polar metabolite whether the route of administration was oral or percutaneous.

The study showed that PCMX is well absorbed after oral administration, rapidly metabolized, and eliminated by the urinary system. After percutaneous administration, absorption is somewhat slower and amounted to approximately 50% of the applied dose. It should be borne in mind, however, that the site of application was abraded skin under occlusion. Even so, the PCMX is rapidly metabolized and eliminated by the urinary system. With a plasma half-life after oral dosing of 60 min in the rat and 50 min in the dog, there is little chance of free PCMX accumulating in the blood or tissue from multiple exposure to smaller topical doses in humans.

Two human bathing studies have been conducted and both showed that from intact human skin, very little PCMX would be absorbed under bathing conditions. In the first study [34], 25 ml of Dettol was added to 125 L of water and the volunteers bathed for 10 min in this solution. Total PCMX in the bath water amounted to approximately 120 mg. Blood and urine analysis were performed, and peak urine concentration of metabolite occurred between 1 and 4 h. The calculated total amount absorbed after analysis of urine, bath mats, towels, or other materials, was approximately 0.5%.

In a second bathing study [35], a single experiment was conducted in which a single human volunteer had 1-ml of Dettol applied to a 20-m^2 area of the back, and occluded with foil for 30 min. The Dettol applied contained 46 mg PCMX. Analysis of skin swabs and patch revealed 38.2 mg, showing 7.2 mg on or in the skin, or 15.7% of the amount applied. Analysis of urine for up to 48 h showed a recovery of 4.4% PCMX (as conjugate), which indicates that approximately 10% of the applied dose had actually been absorbed. When the four healthy male volunteers bathed daily for 10 min in a solution of 25 ml Dettol in 100 L of water, after analysis, it was calculated that from 0.3 to 0.9% of the available PCMX had been absorbed and excreted.

The following facts should help put the question of safety of PCMX into appropriate perspective:

1. The Antimicrobial Panel II has declared PCMX in 3.75% concentrations safe for direct application to fungal infections for periods up to 1 month. Since efficacy studies have been done and when finally recognized as effective, PCMX would also be allowed to be used prophylactically twice-a-day for an indefinite time, based on safety decisions of the panel.
2. In November 1984, an expert panel published their final report for the Cosmetic Ingredient Review in which they found chloroxylenol, in concentrations up to 5%, to be safe as a cosmetic ingredient. It should be remembered that most cosmetic ingredients remain on the skin for rather lengthy periods during a day.
3. Ninety-day oral toxicity studies in rats at dose levels up to 75 mg/kg per day did not reveal significant toxicity that could be traced to treatment, and no target organ could be identified, even in this study. Thirty-day studies in rats, using Dettol sufficient to yield PCMX doses of 120 mg/kg, indicated no treatment-related effects. This general lack of toxicity is supported by an LD_{50} value of 3.83 g/kg.

4. A subchronic topical study of 90 days in rabbits at a high-dose level of 180 mg/kg per day revealed no signs of toxicity, except to the skin at the site of application. There was a complete blood, urine, and organ evaluation by microscopic histology, and even this, revealed no evidence of toxicity to any organ or tissue.
5. Eight-month topical studies in beagle dogs, with doses as high as 600 mg/kg, showed only topical irritation and weight loss at the high-dose level, and this was reversed when the dose was reduced to 200 mg/kg. There were no other clinical or hematological signs of toxicity from PCMX.
6. Massive doses of PCMX administered to humans orally, parenterally, topically, and rectally by Zondek yielded blood concentrations far in excess of what could be realistically expected from the application of a 3% product. Yet Zondek reported no side effects from the repeated daily use of these massive doses in humans.
7. Metabolic studies have shown that there is little chance of accumulation in the body from either topical or oral administration. The PCMX is rapidly metabolized, with a half-life in dogs and rats of approximately 1 h. It is completely excreted in the urine, primarily as the glucuronate and sulfate, even when hepatitis is present.

The human data from the just-described bathing studies has been the subject of a computer-modeling study by Stavchansky [36], who used permutations of the derived data to model what potential might exist if dosing were exaggerated in time and amount. References in the literature show that this procedure has been used to predict half-life after exaggerated use. From work by Guy [26], Stavchansky showed that PCMX had little potential for accumulation, based on rapid metabolism. A half-life of 2–5 h was predicted from the computer modeling under exaggerated dose potential, compared with a calculated value of 1 h from the animal metabolism data.

Accidental exposure, and exaggerated dosing from PCMX, has been analyzed from British use data [37]. Of the approximately 130,000 emergency calls reported from 1960 to 1976, 1819 were reports from chlorinated xylenol products/pine product. Almost 700 were specifically related to Dettol. There were four deaths. Many cases were exposed to large amounts, as high as 350 ml of Dettol (16.8 g PCMX). Toxic effects have been noted from the pine extracts added to xylenol formulations. Generally, patients had little effect from what was frequently, an exaggerated dose of chloroxylenol.

C. Skin and Eye Studies

When dealing with topically applied products, a major consideration with an antimicrobial formulation is sensitization and irritation. Calnan [38] has made a survey of positive drug sensitization tests. Chloroxylenol was the third highest on the list of chemicals tested. At 4.8% PCMX in Dettol (with no dilution), it contains a level that could easily be an inducer. Chlorocresol and resorcinol cross-react with PCMX in skin testing and have been much more commonly used in England than in the United States. More often, much more dilute concentrations are used in the United States, and, most frequently, are washed off of the skin in use. It is not surprising, then, that there have been no verified cases of sensitization in the United States. Certainly, the potential is there. Schorr [39] has also reported the sensitizing potential of chloroxylenol as well as other antimicrobials with phenolic groups. Marzulli and Maibach conducted a repeated insult patch test in an all-male population using chloroxylenol at 0.5% in petrolatum. There were no positive reactions in this population [40].

Irritation studies have been performed in rabbits [31] in subacute and chronic exposure. Moderate to extreme erythema was found in high-dose exposure (18% chloroxylenol). Dettol (4.8% chloroxylenol) has also been tested in rabbits with abraded and nonabraded skin and had a moderate irritant reaction. Various other irritation studies [26] in rats and guinea pigs using various Dettol-like products bear out the findings that chloroxylenol is a mild irritant, with test results very dependent on the concentration tested.

In human skin patch tests, conclusions from animals were borne out in tests when Ottosept Extra (chloroxylenol) showed results resembling chemical burns encouraged with the repeated applications of concentrations (0–1%) in the particular vehicle tested (propylene glycol and sodium lauryl sulfate). As we might also expect, chloroxylenol in high concentrations (30%) is a severe eye irritant [20]. Irrigation reduces the effect in animal tests. When the concentration is reduced to a use level (3%), only conjunctival redness was observed at 24 h, and the product was not scored as an eye irritant.

III. EFFECTIVENESS

A. Historical Perspective and Context

The second aspect of any antimicrobial review is its antimicrobial profile and range of effectiveness. Two major use areas have developed since the 1940s, until today, when most products are handwashing and surgical scrubs or it is used as a preservative in numerous types of products.

B. Minimal Inhibitory Concentration

The minimal inhibitory concentration (MIC) data, in most circumstances, have dealt with antibiotics, and the adaptation of these data to zone-of-inhibition interpretation and correlation is widely used clinically. It is arguable whether the differences in concentrations, availability, and type of conventional application make comparison of MIC data from topically applied antimicrobials reasonable.

In this test, a known inoculum of microbial cells is diluted twofold in a growth medium, incubated and the tubes or wells, and positive or negative growth responses are read. The antimicrobial remains in contact with the cells as a lethal or inhibitory agent. No subculturing is usually done. When the negative tubes are subcultured for viability, the test becomes the determination of the minimal bactericidal concentration (MBC).

The MBC determination has more relevant bearing in situations for which topical products are used and in application to skin microbiology because of the shorter contact usually encountered and the intent-in-use for kill rather than inhibition. Because the contact time is extended in MIC determinations, slower kill times are often reflected as apparently greater activity than is warranted from the data. Precise time–kill data are more meaningful for these short exposure times. The MBC values can be calculated from data derived from kill in 10 min, but not in 5 min (often found in antimicrobial testing related to phenol coefficient testing). Therefore, the exposure time for kill is significantly shorter for MBC than the standard incubation period routinely used in MIC tests. Time–kill analysis was performed in the extensive in vitro Dettol data submitted to FDA.

A significant amount of data have been recorded with MIC values that deviate from each other. The true values without the influencing factors lie somewhere in the range of these reported values. For topical application where the site is known and visualized, these values have meaning only as a theoretical basis for the determination that the chemical is antimicrobial, and to give some idea of the concentrations required in use.

It is always true that there will be variation when many strains are tested, depending on the source of the strains selected for test and in their cultural history.

Chloroxylenol has a solubility of 300 ppm (μg/ml or mg/L) in water. If chloroxylenol (PCMX) or a formulated emulsion is used for MIC or MBC determination and casually added to water solutions without solubilization, the PCMX will precipitate out and the emulsion is difficult to uniformly redisperse in water. The emulsion will break, and, again, the excess PCMX will precipitate.

Some MIC values have been published by the manufacturer of chloroxylenol, Ottawa Chemical Company (Ferro Corporation, now NIPA/Hardwicke). Both the values for the dilution used and the conversion value derived from the dilutions have been published (Technical Brochure). There is no detailed description of how these values were derived, but standard dilution procedures in water are assumed.

MBC data were submitted to the FDA by Reckitt and Colman, the manufactures of Dettol [41]. Dettol is a disinfectant–antiseptic product that has been marketed in England and British Empire countries since 1933. These tests were performed using a widely accepted testing procedure determining the dilution of Dettol required to kill a standardized inoculum after a 10-min exposure, but not after 5. The effect of serum (organic load) and hard water (which might possibly be used in the dilution of this product) was determined in this series of studies.

The effect of the formulated product in vivo (on skin) was determined. Even though the sampling procedure was done with a velvet block transfer, the relative effectiveness remains demonstrable. This type of impression plating transfers superficial bacteria and samples only a small proportion of the organisms on the skin.

It is clear from the results that, given an organism such as *Staphylococcus aureus*, *Escherichia coli*, *Pseudomonas aeruginosa*, or *Proteus* spp., the results show a variation in dilution values that are quite easily converted to parts per million (ppm) of the active ingredient (here, for PCMX in Dettol, 48,000 ppm). It should be kept in mind that these data, whether recorded as the specific dilution or in parts per million, are *kill* or MBC data and not MIC or inhibitory data. Also, the difference between 50 and 100 or 100 and 200 is one tube or well. The actual value is in between the dilution of active agent in the tubes and is not actually determined.

Dettol is formulated as described generically in the *British Pharmacopoeia*, with approximately 10% alcohol and 20% terpineol. Products may contain 10–20% terpineol. The alcohol at 10% is not sufficient to add much, if anything, as an antimicrobial. Terpineol is present as a solubilizer for chloroxylenol and does possess some antimicrobial activity. Russell and colleagues state the following about pine oil and terpineol [42].

> The chief constituent of turpentine is the cyclic hydrocarbon pinene. The odor of pinene, whether in turpentine or from pine made by distilling wood chips or leaves (needles) from various coniferous trees, has long held an association in the public's mind with freshness, cleanliness, and a safe, disinfected environment, but the terpene hydrocarbons of which pinene is but one example, have little or no biocidal activity.
>
> The terpene alcohol terpineol, which may be produced synthetically from pinene or turpentine via terpin hydrate, or in 80% purity by steam distilling pine-wood fragments, is another ingredient of pine disinfectants and has been exploited as an ingredient of the Colebrook and Maxted [22] chloroxylenol formulation. Unlike pinene, it possesses antimicrobial activity in its own right and it shares with pinene the property of modifying the action of phenols in solubilized disinfectant formulations, although not in the same ways for all microbial species. Varying amounts of pine oil in a PCMX/soap formulation modify the phenol coefficient of the preparation depending on the test organism used. Soap in these disinfectant formulations brings about changes in activity by aiding formation of ionic micelles [43].

> Unfortunately there has been a tendency to ignore or be unaware of the above biocidal trends when labelling these varied products, and preparations containing small amounts of pine oil or pinene have been described as disinfectants. Attempts to remedy this situation have been made through the publication of a British Standard entitled "Aromatic Disinfectant Fluids" (BS 5197: 1976). The standard makes it clear that it specifies the requirements for a general-use disinfectant and does not imply use in hospitals or in other situations where there is a risk of infectious disease. The active ingredients, as stated in the standard, are substituted phenols together with pine oil or related terpenes and aromatic oils.

There is some thought that the terpineol may stimulate PCMX activity or enhance it; however, there is not much evidence to support this hypothesis. Certainly there is no evidence or reports in the literature that chloroxylenol is not the ingredient with the major antimicrobial activity in Dettol.

When one is dealing with exposure to hundreds of times the MIC–MBC concentrations with a topically applied chemical antimicrobial, and this must be contrasted with perhaps a two to ten times the multiple with antibiotics given systemically, very small MIC values as an attribute or, indeed, the use of MIC determination as a basis for decisions about activity is questionable.

A tremendous variety of clinical strains with many species were repeatedly tested. Some representative values are included. The basic testing is to find the dilution that kills the standardized inoculum in 10 min, but not in 5 min.

Dettol was evaluated against a total of 168 strains of microorganisms using a modified Association of Official Analytical Chemists (AOAC) phenol coefficient technique. Of these organisms 70 were gram-positive, 89 gram-negative, 8 strains of fungi, and 1 protozoan were tested. Forty of the strains were obtained as standard cultures from various official culture collections, whereas the remainder were from hospital sources of infection [41]. Dettol was bactericidal against the wide range of microorganisms tested, achieving a complete kill of the inoculum within 10 min at dilutions in excess of the recommended use-dilution. This activity, although slightly reduced, is maintained at dilutions in excess of the use-dilution when Dettol is evaluated in the presence of hard water and organic matter.

Pseudomonas aeruginosa was the organism most resistant to the activity of Dettol, certain strains required concentrations of Dettol in excess of the use-dilution to achieve complete kill. However, dried cultures of *P. aeruginosa* were rapidly killed by a 1:1000 dilution of Dettol, and at 37°C, Dettol demonstrated a killing dilution of 1:60 against this organism in an aqueous system. Higher temperatures (37°C) increased activity, and dried organisms were particularly sensitive to treatment for all cultures evaluated.

Evaluations of 0.25% PCMX (Dettol) against laboratory strains of *S. aureus*,

P. aeruginosa, *Proteus mirabilis*, *Enterococcus faecium*, and *Candida albicans* indicate that, in the presence of hard water, greater than 6 $\log_{10}$ of each bacterium, except *P. aeruginosa*, were killed. Greater than 5 $\log_{10}$ *C. albicans* were killed in the 5-min exposure time [44]. When tested in the distilled water, all bacteria showed greater than 6 $\log_{10}$ reduction in 5 min. The presence of organic soil, in the form of albumin (1.0%), did not seem to affect the activity of 0.25% PCMX in the 5-min exposure time.

For use as a topical antimicrobial, exposure times of 5 min is appropriate; however, for use as a surface disinfectant, an exposure of 1 min is considered appropriate for testing purposes. With a 1-min exposure of the foregoing laboratory strains, PCMX (0.25%) was not effective against *P. aeruginosa* or *E. faecium*. The relative resistance of *P. aeruginosa* to lower concentrations (0.1–0.25%) of PCMX has been reported [45,46]. The activity of PCMX against *P. aeruginosa* is potentiated by the addition of ethylene-diaminetetraacetic acid (EDTA) [46–48].

Minimum inhibitory concentrations and minimum bactericidal concentrations of PCMX solubilized in polyethylene glycol (PEG 400) were determined as part of an evaluation for handscrub solutions and presented to the FDA [49]. These values are found in Tables 1 and 2. The MBC values are low, with the exception of those for *Pseudomonas*, *Penicillium*, and *Mucor*. The MIC determinations made with PCMX (3%) formulations [50] are presented in Table 3. These values were determined by agar dilution, a slightly different method than used in determining values presented in Tables 1 and 2. The data indicate that the formulations are more effective against gram-positive and acid-fast bacteria and fungi than they are against the gram-negative bacteria studied.

Challenge testing of Ultradex formulation and of the formulation without the antibacterial, chloroxylenol, clearly shows the rapid activity of chloroxylenol, even against the fungus *Candida* [51]. Ultradex does not contain either alcohol or terpinol as solubilizing agents in the formula. To determine the contribution of vehicle to Ultradex, the formula with and without chloroxylenol was challenged in a time series, sampling with several different microorganisms (Table 4). Aliquots were withdrawn at 1, 2.5, and 5 min after inoculation and immediately neutralized with an appropriate neutralizer. Organisms tested were *S. aureus*, *P. aeruginosa*, *E. coli*, *C. albicans*, *Aspergillus niger*, and *Streptococcus faecalis*. These selected microorganisms were grown in appropriate media and adjusted to given an inoculum level of approximately 1×10^6 CFU/ml. The study concluded that the chloroxylenol present in the Ultradex formulation is an active bactericidal chemical, showing quick kill of at least 10^6 cells in one minute.

An MIC determination was made with a PCMX producer, Ottawa Chemical Company, using PCMX, solubilized in propylene glycol, against *C. albicans* and *Trichophyton mentagrophytes*. The results of the twofold tube dilution test indicated a MIC of 125 ppm for *C. albicans* and 1000 ppm for *T. mentagrophytes* [52]. Another study was submitted showing the activity of chloroxylenol (PCMX)

TABLE 1 Microbial Effectiveness of PCMX
Inhibiting dilutions of PCMX[a]

Organism	Minimal inhibitory concentration (MIC) ppm
Candida albicans	50
Pseudomonas aeruginosa	1000
Proteus vulgaris	100
Escherichia coli	50
Bacillus subtilis	50
Penicillium chrysogenum	1000
Alternaria solani	200
Aspergillus niger	200
Ehisopus nigricans	100
Aerobacter aerogenous	50
Staphylococcus aureus	10
Bacillus cereus	50
Streptococcus faecalis	20
Mucor species	1000
Oidium lactis (*Candida*)	50
Saccharomyces pastorianus	50
Phoma roseus	100

The broad spectrum microbiocidal activity of PCMX has proved it effective against a variety of vegetative microorganisms. Fungi, gram-positive bacteria, and gram-negative bacteria were tested
[a]Subject to carrier used.
Source: This table was summarized from data published by Ottawa Chemical Co.

in a modified AOAC killing dilution test, using recently isolated pathogenic strains of fungi [53]. Ten strains of yeast and dermatophytes were tested with appropriate neutralizers used in the recovery medium. The results shown in Table 5 indicate the fungicidal activity of Dettol (4.78% PCMX) against *Candida* spp. and dermatophytes. The effective dilutions ranged from 1:130 (370 ppm) for *C. albicans* to 1:300 (160 ppm) for *T. rubrum*. Rickett and Colman [54] demonstrated the trichomonacidal activity of Dettol against *Trichomonas vaginalis*.

Reports of antiviral and anti-rikettsial activity have been made; however, extensive data do not exist. Anasari et al. [55] have shown that Dettol (4.8% PCMX) removed 96.2% of the applied virus, comparable with povidone–iodine; 0.5% chlorhexidine gluconate–70% isopropanol; 1.5% chlorhexidine gluconate–15% cetrimide; and Ivory Soap. Surprisingly, tap water alone removed 83% of the applied virus particles. The concentration of PCMX in this preparation is consid-

TABLE 2 The Bacteriostatic Activity of PCMX in Distilled Water Against a Range of Skin Isolates

Organism	Minimal inhibitory concentration (MIC) ppm
Staphylococcus group I	167
Staphylococcus group I	167
Staphylococcus group II	167
Staphylococcus group IV	167
Staphylococcus group V	167
Micrococcus spp. (aerobic)	83
Streptococcus faecalis type C	83
S. faecalis type D	167
S. faecalis type D	167
S. faecalis type D	167
S. faecalis type A	167
Neisseria catarrhalis	167
Acinetobacter nitratum	83
Escherichia coli AA 9395	42
E. coli AA 9721	167
E. coli AA 9803	167
E. coli 10245	167
E. coli	167
Citrobacter spp.	167
Klebsiella aerogenes CA 22852	167
K. aerogenes CA 23535	167
K. aerogenes	167
K. aerogenes	167
Proteus mirabilis BA 23919	167
P. mirabilis BA 25702	167
P. mirabilis BA 25945	167
P. mirabilis (1)	167
P. mirabilis (2)	167
P. mirabilis	167
Providencia sp. (A)	167
Providencia sp. (B)	167
Pseudomonas aeruginosa 12211	167
P. aeruginosa	167
P. aeruginosa 65376	167
P. aeruginosa Thornham	167
Enterobacter species	167
Serratia marcescens	167
Mycobacterium smegmatis	167
Diphtheroid (lipolytic)	83
Diphtheroid (nonlipophilic)	83
Diphtheroid (anaerobic)	167

Source: This table was summarized from data published by Ottawa Chemical Division.

TABLE 3 Minimal Inhibitory Concentration (MIC) for Seamless Anti-Sept Bacterial Scrub Solution Using Gram-Positive Bacteria, Gram-Negative Bacteria, Acid-Fast Bacilli, and Selected Fungi

Test organism	MIC
Gram-positive bacteria	
Bacillus subtilis	1:4096
Corynebacterium xerosis	1:2048
Micrococcus luteus	1:516
Staphylococcus aureus (22)	1:1024
S. aureus (ATCC 25923)	1:2048
S. epidermidis	1:4096
Streptococcus faecalis	1:1024
Gram-negative bacteria	
Citrobacter freundii	1:128
Serratia marcescens	1:128
Enterobacter cloacae	1:128
Klebsiella pneumoniae	1:512
Escherichia coli	1:256
Salmonella typhimurium	1:256
Shigella sonnei	1:256
Enterobacter gergoviae	1:128
Pseudomonas aeruginosa	1:4
P. fluorescens	1:4
Acid-fast bacilli	
Mycobacterium smegmatis	1:1048
Fungi	
Candida albicans	1:1024
Aspergillus niger	1:2048

TABLE 4 Challenge Time–Kill Data for Ultradex Surgical Scrub (3:18% Chloroxylenol)

	Organism counts (CFU/ml or CFU/g)			
Test organisms	1.0 min	2.5 min	5.0 min	Control count
Staphylococcus aureus ATCC 6538	0	0	0	5.4×10^6
Pseudomonas aeruginosa ATCC 9027	0	0	0	9.1×10^6
Escherichia coli ATCC 8739	1.05	0	0	2.0×10^6
Streptococcus faecalis ATCC 85W1000	0	0	0	5.0×10^6
Candida albicans ATCC 10231	9.5×10^3	1.0×10^3	4.0×10^2	7.8×10^5

TABLE 5 The Fungicidal Activity of Dettol Against Hospital Isolates

Organism	Dilution of Dettol killing the inocula within 10 min at 20°C	ppm
Candida albicans	1:130	369
C. tropicalis	1:225	213
C. parapsilosis	1:200	240
C. krusei	1:180	266
C. guilliermondii	1:280	171
C. pseudotropicalis	1:180	266
Microsporum canis	1:150	320
Tricophyton mentagrophytes	1:150	320
T. rubrum	1:300	160
T. interdigitale	1:150	320

erably higher than that usually found in handscrub preparations. The PCMX has also been reported to effectively inactivate the virus causing foot-and-mouth disease [56].

C. Clinical Studies

Surgical preparation of the hands before surgery started with Lister's personal revolution in the operating room. Semmelweis had brilliantly and intuitively discovered the role of the hands in spreading disease when the doctor moved from patient to patient. As often as we try to improve or sophisticate Semmelweis' work, modern science, with its statistical approach, does not seem able to be as direct or convincing.

The approach to handwashing and scrubbing was straightforward until after World War II. Prudence had long dictated that materials in the environment that were known to be contaminated with potentially pathogenic organisms should be disinfected. The disinfectant was almost always harsh, smelly, and distinctly unpleasant. Thus, the origin of the idea or feeling that an antimicrobial must hurt and be really nasty to work is based in this history. The use of harsh products in a preparation was reluctantly accepted as a normal part of surgery, along with very high postinfection rates. Liquid green soap, a stiff brush, and an alcohol rinse with appropriate ablutions was the religious practice.

In 1937–1938, Price, at the University of Utah, began to look at the objective evaluation of the results of handwashing. He wanted to examine the number and kind of organisms constituting the microflora and, then, to find the effect that handwashing might have [57]. Lister made significant and unequaled progress in the implementation of sterile surgery. Even so, the postsurgical infection rate still remained high. The succeeding event that supplemented the sterile surgical field ideal was the innovation, by Halsted, of wearing rubber surgical gloves. The hands of the surgical personnel in the sterile field now did have the sterile surface so long idealized. The ideal of sterility soon was abandoned because gloves broke or were cut during the surgical procedure. This renting of the glove's sterile integrity is the basic rationale supporting surgical scrubbing. Once the glove is broken or punctured, the bacteria stream into the sterile field and into the surgical wound; the wound site, otherwise, may well be sterile or contain very few microorganisms. One may ask why the inadvertent addition of normal skin microflora to this site is a problem. It is a guess to assume that the surgeon's hands will always have an innocuous flora. Furthermore, some formerly safe nonpathogenic bacteria such as *Staphylococcus epidermidis* can be infectious in an open wound, in vascular donor sites, or in the brain.

Historically, the preparation of the surgical personnel began with Lister and phenol and was followed by liquid green soap and alcohol, and during World War I, Dakins Solution (hypochlorite). These chemicals, combined with a stiff-bristled brush, earned a deserved reputation of harshness. In the late 1940s, iodophors were discovered, and the perceived effectiveness of iodine became available in a less-irritating and more acceptable product than their long-used relative, iodine tincture. Still a few years later, hexachlorophene arrived, and a mild nonirritating scrub was available. During these stages of development, little or no testing was done, other than the demonstration of in vitro activity, particularly against staphylococci.

The Price work from the late 1930s [57] set a new standard for showing effectiveness as a handwashing product, in that he paid attention to the flora of the skin. He did another innovative procedure in that he tested his own hand flora repeatedly to establish fluctuations and a base value for testing.

A concept that is critically important in testing antimicrobial effectiveness on the hands involves the idea that, when the hands are washed and the wash water assayed for antimicrobial count, the number of bacteria or colony-forming units is a measure of the *fraction removed*. The total count of the microbial flora on any person's hands cannot be measured. So, in fact, a fraction removed from an unknown total is being measured.

One should remember that the total fluctuates from day to day. As this concept is examined, the conclusion that a precise estimate of the count before treatment of any kind, *at the time the testing is done*, becomes an increasingly essential test element. The importance of this idea will be apparent as we discuss the chronological development of effectiveness testing of surgical scrubs.

Price's method of sampling the microbial flora used successive basins of water, with sequential washing. For a minute, focus on the Price test, since its purpose is frequently misinterpreted. Price examined the composition of the normal flora and the nature and design of removal if the hands are washed consecutively.

The advent of the FDA's OTC Drug Review in 1972 provided an impetus for FDA to define specific testing procedures. The OTC panel developed definitions of product categories and specific-testing procedures. This author developed the glove juice test, based on suggestions from other investigators. The need to use this test was soon expressed when New Drug Applications (NDAs) were required by FDA for hexachlorophene-containing handwashing and surgical scrub products. The specific testing procedure can be found in detail in the *Federal Register*, CFR 121 [58] and in the American Society for Testing and Materials (ASTM) microbiology methods book [59].

An Ottawa Chemical Company study [60], as well as others, have shown that the substantively active antimicrobial is retained in the stratum corneum of the skin after use and acts to keep the count low during surgeries, up to 6 h, even when the conditions encourage bacterial growth. In the absence of a substantive antimicrobial for instance, with iodophors, the population reestablishes after 2–3 h.

This action is also implied in several studies with alcohol formulation and various concentrations of PCMX used for skin preoperative preparation. Studies by Byatt and Henderson [61], Frazer [62], Caplin and Chapman [63], and Davies et al. [64], all showed excellent action of chloroxylenol when formulated in an alcohol base, which seemed to be advantageous in defatting the skin for surgery by removing skin debris and acting with persistent or substantive results.

If one examines the results of glove juice tests (described in Chap. 2) with the available antimicrobial surgical scrub products, the measurement in surgical scrubbing is in large measure against the staphylococcal, diphtheroid, and micrococcal species on the skin.

In the actual scrub process, there are two factors working at the same time: the killing, or static, antibacterial activity of the chemical antimicrobial ingredient(s), and the physical removal of the bacteria. In the current glove juice test, the formulated product is tested, and no provision is made for testing the detergent formulation without the antimicrobial, so no real separation of these factors is possible. Alterations have been made in the test by ASTM, primarily randomizing hands into groups for testing and reducing the hourly samples.

The historical data concerning the formulation and activity of chloroxylenol has often been misinterpreted. The first major publication about a formulated product with chloroxylenol [22] is a classic, and is still pertinent. Chloroxylenol has been used in England and other British Commonwealth countries in Dettol (4.8% chloroxylenol with alcohol and terpineol) and its imitators. The uses have ranged from hard-surface disinfectants, through instrument disinfection and skin degerming. Some early applications were made in the United States for skin antimicrobials, but they were not extensive. A more common use then was as a preservative. There is still wide usage as a preservative in a wide variety of products.

It was only after the revelations about hexachlorophene toxicity that many companies began to look at PCMX for skin use—almost as if it were a new antimicrobial. The first of the newer handwashing products available with chloroxylenol were termed "lotion" soaps with fairly low (0.5–1%) concentrations of PCMX. Some special application handwashes with higher concentrations of PCMX for meat handlers have been formulated. A 3% formulation was marketed as a surgical scrub in 1978. One source of confusion about the effectiveness of PCMX is the wide range of concentrations being marketed, but for different uses. The vagaries experienced in formulating PCMX as an active antimicrobial have contributed to the surprising range of effectiveness deduced from studies.

The second study following the OTC–*Federal Register* protocol was performed with a 3% PCMX formulation. A summary of the results of this test showed that because of the increasing concern about the discrepancies in data from glove juice tests and the unnecessary complexity in the test, a new protocol was outlined that keeps many of the elements of the test, but also improves it. The central revisions were (a) the test is structured to use one hand of a subject for the test compound and one hand for the baseline estimation, so that the samples are taken simultaneously, reducing variation from the predetermined baseline or eliminating the use of an average baseline value; (b) there is a small variation from hand to hand; and (c) this moving of the baseline determination from pretest estimation into the test framework necessitates a change in the test sampling. In the changed protocol, the microbiological estimation of the immediate postscrub count, and that after a specific time period wearing gloves, have to be made on different days, to have a test that more realistically reflected the actual use of the scrub products in a hospital setting and, particularly, to demonstrate the substantive effect of PCMX in use. The test period was extended before the last sample was taken (2 weeks).

The results from several studies using the glove juice protocol or other clinical protocols will be reviewed.

A chloroxylenol-containing athlete's foot formulation was tested by a protocol conforming to the original glove juice procedure published as part of the OTC Antimicrobial I Panel. The results of the test with a formulation containing 3.75% chloroxylenol were a reduction from baseline immediately after scrubbing of approximately 1.5–2.0 $\log_{10}$. The reestablishment of the microbial flora after scrubbing and donning of gloves remained at the reduced levels the first 4 h, rising slightly during the last 2 h. On the fifth day of scrubbing, there was a greater reduction than on the first day, an indication of substantivity [65].

A second glove juice test, performed with a 3% chloroxylenol formulation, conformed to the original protocol in the *Federal Register* [66]. The initial reduction from the baseline was approximately 1 $\log_{10}$ and exceeded a 1 $\log_{10}$ reduction on day 5. Over the 6 h sampling times on each day, the count remained reduced for the 6 h, demonstrating that PCMX in an appropriate formulation is substantive to the skin.

A third study was published by Soulsby et al. [67] on a product containing 3% chloroxylenol and compared directly to Hibiclens with 4% chlorhexidine gluconate (CHG). The results showed essentially the same reduction with both products. There was an approximate a 1-$\log_{10}$ reduction from baseline after initial scrubbing and a demonstration of substantivity over the 6-h period. Larson [68] suggests that because of certain aspects of the study (different methods of application of the products) one can question its results.

A study by Paulson [69], using the glove juice protocol, compared the immediate, substantive, and residual effect of PCMX, 4% CHG, an iodophor, and an alcohol–detergent, all impregnated in scrub brushes, and 2% CHG solution. The PCMX-impregnated brush showed immediate activity greater than the alcohol-impregnated brush, equal to that of the 2% CHG solution, but less than that of the iodophor-4 and 4% CHG-impregnated brush. No significant persistent effects were noted on day 1; however, on days 2 and 5, persistent effects were noted at the 3-h sampling time, but not at 6 h. No residual effect was noted in this study.

Rickett and Colman performed a multiple use application of 2.5% PCMX to the feet, followed by evaluation of the balance of the flora of the feet [70]. Quantitative and qualitative evaluation of the flora was done before, during, and after use of the product. There was some reduction in the level of the flora, no increase in gram-negative organisms, and no change in the balance of the flora.

A Cade handwashing test was performed with a 2% PCMX bar soap, with a small group of subjects [71]. This type of test involves programmed washing over a 10-day period. Significant reduction was implied in the summary.

IV. SUMMARY

Huge numbers of clinical isolates have been tested against Dettol to determine the bactericidal concentration, including a wide range of fungi and *Trichomonas vaginalis* [54]. There is general agreement that the most resistant organism to chloroxylenol is *Pseudomonas aeruginosa*. Because of the necessity for formulation as an emulsion and for solubilization, formulations vary significantly in their MIC–MBC values [71–73]. It is also clear that PCMX is bactericidal.

The early history and development is intertwined with that of hexachlorophene. When first introduced in the United States in the 1940s, the effectiveness of chloroxylenol was tested in many commercial products, including shampoos, athlete's foot cream, acne cream, handwashing and surgical scrub products, preservative for a wide array of cosmetic and industrial products, and unusual applications such as film coatings. Chloroxylenol and a related product chlorocresol are widely used in Europe, and their wide use there is recognized as safe and effective.

Experts and investigators who have reviewed all of the toxicity data available on PCMX have hypothesized that PCMX is very rapidly metabolized to the glucuronide as it is absorbed from the skin and is then rapidly excreted [74]. Extremely high systemic and topical doses have been applied to humans without adverse effect. It is unclear with all the available evidence why the FDA continues to question the toxicity database.

REFERENCES

1. G. Sykes, Horizons in disinfection and antisepsis, Jubilee Memorial Lecture, *Chem. Ind.*, June 1:885–888 (1963).
2. G. Lockemann, and E. Heicken, Ueber die keimtotende Wirkung und ihrer Monochlor- und Monobromderivate. [On the bactericidal action of 6 isomers of xylenol and their monochlor- and monobrom-derivatives], First Division Originals from the Chemical Division of the Institut Robert Koch, Berlin, 1939.
3. G. Lockemann, and T. Kunzmann, Keimtotende Wirkung und chemische Konstitution der isomeren Xylenole und ihrer Monohalogenderivate. [Bactericidal action and chemical constitution of xylenol isomers and their monohalogen derivatives], *Ungewandie Chem. 46*(21):296–301 (1933).
4. G. Gladden, and W. W. Cocker, British Patent 529070, Nov. 13, 1940; U. S. Patent 2,350.677; June 6, 1944.
5. Van Auwers, Marbe, and Saurwein, *Fortschr. Chem. Physik. Physik. Chem. 18*(2): 1–45 (1925); *Chem. Abstr. 19*:2339 (1925).
6. J. Lister, On the antiseptic principle in the practice of surgery, *Br. Med. J. 2*:246 (1867).
7. T. D. Brock, *Milestones in Microbiology*, American Society for Microbiology, Washington, DC, 1961, pp. 80–118, 163–176.
8. S. S. Block, Historical review, *Disinfection Preservation and Sterilization* (C. A. Lawrence and S. S. Block, eds.), Lea & Febiger, Philadelphia, 1968.
9. G. Kuchenmeister, Ueber Desinfections—mittel in Allgemeinen, das Spirol und seine therapeutische Verwendung im Besondern. *Dtsch. Klin. Eingange Zwanzigot. Jahrhund. 12*:123–124 (1986).
10. R. Etinger-Tulczynska, and W. Ulrich, Uber die keimtotende Wirkung von Gemischen Verschiedenen Manochlorderivatives of Phenol-homologen. [On the bactericidal action of mixtures of various monochlor-derivatives of phenol homologs], Institute Robert Koch, 1932.
11. OTC topical antimicrobial products and drug and cosmetic products, *Fed. Reg. 38*(179):33115 (1974).
12. OTC topical antifungal drug products for over-the-counter human use: Establishment of a monograph, *Fed. Reg. 47*(56):12533–12356 (1982).
13. Cosmetic, Toiletry and Fragrance Association (CTFA), Final Report of the safety assessment for chloroxylenol, *Cosmet. Ingred. Rev.* (CIR). CTFA, Washington, DC.
14. W. F. Schorr, Cosmetic allergy. A comprehensive study of the many groups of chemical antimicrobial agents, *Arch. Dermatol. 104*:459–466 (1971).
15. M. D. Ray, K. E. Avis, and C. C. Flanigan, Microbiological evaluation of PCMX complexes, *J. Pharmacol. Sci. 57*:609–613 (1968).

16. R. Hare, E. Raik, and S. Gash, Efficiency of antiseptics when acting on dried organisms, *Br. Med. J. 1*:496–500 (1963).
17. J. N. Bruun, J. Boe, and C. O. Solberg, Disinfection of the hands of ward personnel. A comparison of six disinfectants, *Acta Med. Scand. 184*:417–423 (1968).
18. R. W. Ruhen, and W. G. R. M. de Boer, Bactericidal action of handwashing preparations, *Med. J. Aust. 1*:222–223 (1972).
19. Data submitted to FDA as part of the OTC Review, Antimicrobial I and Antimicrobial II Panels are available under Freedom of Information and can be obtained under Docket No. 75N-0183.
20. W. L. Guess, and M. K. Bruch, A review of available toxicity data on the topical antimicrobial chloroxylenol, *J. Toxicol. Cutan. Ocular Toxicol. 5*:233–262 (1986).
21. B. Zondek, The excretion of halogenated phenols and their use in the treatment of urogenital infections, *Percutan. Chemother. J. Urol. 48*:747–758.
22. L. Colebrook, and W. R. Maxted, Antisepsis in midwifery, *J. Obstet. Gynaecol. Br. Emp. 40*(6):3–32 (1933).
23. B. Zondek, Fate of halogenated phenols in the organism, *Biomed. J. 37*:592–595 (1943).
24. B. Zondek, and M. Finkelstein, Blood concentrations of *p*-chloro-xylenol in man following parenteral percutaneous and rectal application, *Proc. Soc. Exp. Biol. Med. 61*:200–202 (1946).
25. J. M. Joseph, Observations on the acute toxicity of two chlorinated phenols, *J. Am. Pharm. Assoc. 41*(11):595–596 (1952).
26. B. Hunter, J. L. Bridges, and A. J. Newman, Dettol RBA 666. Preliminary assessment of toxicity to rats, oral administration for four weeks, Huntingdon Research Centre, FDA Docket No. 75N-0183, 1973.
27. M. E. Havler, B. J. Jordan, S. Malam, and M. J. Rance, Metabolism studies of PCMX, Report No. 5369/2, Reckitt and Colman Co., FDA Docket No. 75N-0183, 1974.
28. H. Chesterman, P. Whitehead, and A. E. Street, RBA 666 oral toxicity studies in beagle dogs, Huntingdon Research Centre, FDA Docket No. 75N-0183, 1973.
29. H. Chesterman, R. Heywood, M. H. Barber, A. E. Street, and C. P. Cherry, RBA 666 oral toxicity studies in beagle dogs, Huntingdon Research Centre, FDA Docket No. 75N-0183, 1963.
30. J. R. Harr, Pennwalt Corp.—study director, Pennwalt submission to FDA Docket No. 75N-0183, 1978.
31. J. W. Leichliter, Subacute and chronic dermal application of Ottasept Extra to rabbits, Hill Top/Ottawa Chemical Co., FDA Docket No. 75N-0183, 1965.
32. S. J. Bradberry, and J. Hayden, Effect of Dettol on wound healing in rats, Bias Ltd. Report No. RC76132, Bias Ltd., FDA Docket No. 75N-0183, 1976.
33. M. E. Havler, and M. A. McLeavy, The metabolism of *p*-chloro-*m*-xylenol (PCMX) in Sprague–Dawley and Gunn Wistar rats, Reckitt and Colman Co., FDA Docket No. 75N-0183, 1977.
34. B. J. Jordan, J. D. Nichols, K. Robinson, and M. J. Rance, Human volunteer studies on Dettol bathing products, Reckitt and Colman Co., FDA Docket No. 75N-0183, 1973.
35. B. J. Jordan, J. D. Nichols, and M. J. Rance, Dettol bathing product—preliminary volunteer study, FDA Docket No. 75N-0183, 1973.
36. S. Stavchansky, Report to Dexide, Inc., and submission to FDA, June 19, 1985; FDA Docket No. 75N-0183, 1985.

37. Reckitt and Colman Pharmaceutical Division, Dettol and related antiseptics/disinfectants emergency calls to New Cross Hospital Poisons Unit, 1969–1976, Submission to FDA Docket No. 75N-0183, 1974.
38. C. D. Calnan, Contact dermatitis from drugs. Symposium on drug sensitization, *Proc. R. Soc. Med. 55*:39–42 (1962).
39. W. F. Schorr, Cosmetic allergy, a comprehensive study of the many groups of chemical antimicrobial agents, *Arch. Dermatol. 105*:459–466 (1971).
40. F. N. Marzulli, and H. I. Maibach, Antimicrobials: Experimental contact sensitization in man, *J. Soc. Cosmet. Chem. 24*:399–421 (1973).
41. T. J. Munton, and J. Prince, The bacteriostatic and bactericidal activity of Dettol against a range of recently isolated mesophilic strains including members of the normal flora and cutaneous pathogens of the skin, Submitted to FDA Docket No. 75N-0183, 1975.
42. W. B. Hugo, and A. D. Russell, *Disinfection, Preservation and Sterilization* (A. D. Russell, W. G. Hugo, and G. A. J. Ayliffe, eds.), Blackwell Scientific Publication, Oxford, 1982, Chap. 2.
43. H. S. Bean, and H. Berry, The bactericidal activity of phenols in aqueous solutions of soap. Part I. The solubility of water-insoluble phenol in aqueous solutions of soap, *J. Pharm. Pharmacol. 2*:484–490 (1950).
44. S. F. Bloomfield, M. Arthur, E. Looney, K. Begun, and H. Patel, Comparative testing of disinfectant and antiseptic products using proposed European suspension testing methods, *Lett. Appl. Microbiol. 13*:233–237 (1991).
45. H. S. Bean, and R. C. Farrel, The resistance of *Pseudomonas aeruginosa* in aqueous solutions of phenols, *J. Pharm. Pharmacol. 19*:183S–188S (1967).
46. A. D. Russell, and J. R. Furr, The antibacterial activity of a new chloroxylenol containing ethylenediamine tetra-acetic acid, *J. Appl. Bacteriol. 43*:253–260 (1977).
47. J. Prince, C. E. Deverill, and G. A. J. Ayliffe, A membrane filter technique for testing disinfectants, *J. Clin. Pathol. 28*:71–76 (1975).
48. J. Dankert, and I. K. Schut, The antibacterial activity of chloroxylenol in combination with ethylenediaminetetraacetic acid, *J. Hyg. (Cambr.) 76*:11–22 (1976).
49. Dexide, Inc., Glove juice test, Submission to FDA, FDA Docket No. 75N-0183, 1985.
50. Seamless Hospital Products, Company Brochure, Seamless Hospital Products Co., Wallingford, CT, 1988.
51. Dexide, Inc., Submission to FDA. Docket No. 75N-0183, 1986.
52. Pennwalt Co., MIC determinations of MIC of Ottasept Extra against fungi, Submission to FDA Docket N. 75*N*-0183, 1973.
53. T. J. Munton, and J. Prince, The fungicidal activity of Dettol against recently isolated fungal strains from hospital sources, Laboratory Report No. HL75/15, Submitted to FDA Docket No. 75N-0183, 1975.
54. Reckitt and Coleman, The trichomonacidal activity of Dettol in vitro experiments using culture strains of *Trichomonas vaginalis*, Submitted to FDA Docket No. 75N-0183, 1975.
55. S. A. Ansari, S. A. Sattar, V. S. Springthorpe, G. A. Wells, and W. Tostowaryk, In vivo protocol for testing efficacy of hand-washing agents against viruses and bacteria: Experiments with rotavirus and *Escherichia coli*, *Appl. Environ. Microbiol. 55*:3113–3118 (1989).

56. R. F. Sellers, The inactivation of foot-and-mouth disease virus by chemicals and disinfectants, *Vet. Rec. 83*:504–506 (1968).
57. P. B. Price, The bacteriology of normal skin; a new quantitative test applied to a study of the bacterial flora and disinfectant action of mechanical cleansing, *J. Infect. Dis. 63*:301–313 (1938).
58. FDA *Federal Register*, OTC topical antimicrobial products and drug and cosmetic products. Sept. 13, 39:33102–33141 (1974).
59. American Society for Testing and Materials, E1115-86, Evaluation of surgical hand scrub formulations. ASTM Standards on Materials and Environmental Microbiology, Ed. E35.15 Subcommittee, 1987.
60. Ottawa Chemical Co., Substantivity of Ottasept on skin, Submitted to FDA Docket No. 75N-0183, 1975.
61. N. Byatt, and A. Henderson, Preoperative sterilization of the perineum. A comparison of six antiseptics, *J. Clin. Pathol. 26*:921–924 (1973).
62. J. Frazer, The effect of two alcohol based antiseptics on artificially contaminated skin, *Microbios Lett. 119*:122 (1963).
63. H. Caplin, and D. C. Chapman, A comparison of three commercially available antiseptics against opportunist gram negative pathogens, *Microbios 16*(64):133–138 (1976).
64. J. Davies, J. R. Babb, G. A. J. Ayliffe, and M. D. Wilkins, Disinfection of the skin of the abdomen, *Br. J. Surg. 65*:855–858 (1978).
65. Pennwalt Co., Submission to FDA, glove juice test with Fresh Soap (Report 80 a,b), FDA Docket No. 75N-0183, 1973.
66. Dexide, Inc., Submission to FDA reviewing the effectiveness of chloroxylenol and the results of a glove juice test, FDA Docket No. 75N-0183, 1986.
67. M. E. Soulsby, J. B. Barnett, and S. Maddox, Brief report: The antiseptic efficacy of chloroxylenol-containing vs. chlorhexidine gluconate-containing surgical scrub preparations, *Infect. Control 7*:223–226 (1986).
68. E. Larson, and G. H. Talbot, An approach for selection of health care personnel handwashing agents, *Infect. Control 7*:419–424 (1986).
69. D. Paulson, Comparative evaluation of five surgical hand scrub preparations, *AORN J. 60*:246–256 (1994).
70. Reckitt and Colman, Balance of microbial flora of the feet after use 2.5 V/V dilution of Dettol, Report BL 75/75/13/RC 7567, Submitted to FDA Docket No. 75N-0183, 1975.
71. Ottawa Chemical Co., Cade handwashing test with 2 percent chloroxylenol, Submitted to FDA Docket No. 75N-0183, 1950.
72. H. S. Bean, and H. Berry, Bactericidal activity of chloroxylenol in potassium laurate solutions, *J. Pharm. Pharmacol. 5*:632–639 (1953).
73. M. Ganzler, Patent to Chesebrough, Inc., Germicidal petrolatum compositions, British Patent 885,959, Jan. 3, 1962.
74. W. Guess, and B. Jordan, Personal communication, 1988.

Index